T0202809

Communications
in Computer and Information Science

1927

Rationale

The CCIS series is devoted to the publication of proceedings of computer science conferences. Its aim is to efficiently disseminate original research results in informatics in printed and electronic form. While the focus is on publication of peer-reviewed full papers presenting mature work, inclusion of reviewed short papers reporting on work in progress is welcome, too. Besides globally relevant meetings with internationally representative program committees guaranteeing a strict peer-reviewing and paper selection process, conferences run by societies or of high regional or national relevance are also considered for publication.

Topics

The topical scope of CCIS spans the entire spectrum of informatics ranging from foundational topics in the theory of computing to information and communications science and technology and a broad variety of interdisciplinary application fields.

Information for Volume Editors and Authors

Publication in CCIS is free of charge. No royalties are paid, however, we offer registered conference participants temporary free access to the online version of the conference proceedings on SpringerLink (http://link.springer.com) by means of an http referrer from the conference website and/or a number of complimentary printed copies, as specified in the official acceptance email of the event.

CCIS proceedings can be published in time for distribution at conferences or as post-proceedings, and delivered in the form of printed books and/or electronically as USBs and/or e-content licenses for accessing proceedings at SpringerLink. Furthermore, CCIS proceedings are included in the CCIS electronic book series hosted in the SpringerLink digital library at http://link.springer.com/bookseries/7899. Conferences publishing in CCIS are allowed to use Online Conference Service (OCS) for managing the whole proceedings lifecycle (from submission and reviewing to preparing for publication) free of charge.

Publication process

The language of publication is exclusively English. Authors publishing in CCIS have to sign the Springer CCIS copyright transfer form, however, they are free to use their material published in CCIS for substantially changed, more elaborate subsequent publications elsewhere. For the preparation of the camera-ready papers/files, authors have to strictly adhere to the Springer CCIS Authors' Instructions and are strongly encouraged to use the CCIS LaTeX style files or templates.

Abstracting/Indexing

CCIS is abstracted/indexed in DBLP, Google Scholar, EI-Compendex, Mathematical Reviews, SCImago, Scopus. CCIS volumes are also submitted for the inclusion in ISI Proceedings.

How to start

To start the evaluation of your proposal for inclusion in the CCIS series, please send an e-mail to ccis@springer.com.

Jian Chen · Van-Nam Huynh · Xijin Tang ·
Jiangning Wu
Editors

Knowledge and Systems Sciences

22nd International Symposium, KSS 2023
Guangzhou, China, December 2–3, 2023
Proceedings

 Springer

Editors
Jian Chen
Tsinghua University
Beijing, China

Van-Nam Huynh 🄳
Japan Advanced Institute of Science
and Technology
Ishikawa, Japan

Xijin Tang 🄳
CAS Academy of Mathematics and Systems
Science
Beijing, China

Jiangning Wu
Dalian University of Technology
Dalian, China

ISSN 1865-0929   ISSN 1865-0937 (electronic)
Communications in Computer and Information Science
ISBN 978-981-99-8317-9   ISBN 978-981-99-8318-6 (eBook)
https://doi.org/10.1007/978-981-99-8318-6

This Springer imprint is published by the registered company Springer Nature Singapore Pte Ltd.
The registered company address is: 152 Beach Road, #21-01/04 Gateway East, Singapore 189721, Singapore

Paper in this product is recyclable.

Preface

The annual International Symposium on Knowledge and Systems Sciences aims to promote the exchange and interaction of knowledge across disciplines and borders to explore new territories and new frontiers. With more than 20 years of continuous endeavors, attempts to strictly define knowledge science may be still ambitious, but a very tolerant, broad-based and open-minded approach to the discipline can be taken. Knowledge science and systems science can complement and benefit each other methodologically.

The first International Symposium on Knowledge and Systems Sciences (KSS 2000) was initiated and organized by Japan Advanced Institute of Science and Technology (JAIST) in September of 2000. Since then with collective endeavours illustrated in KSS 2001 (Dalian), KSS 2002 (Shanghai), KSS 2003 (Guangzhou), KSS 2004 (JAIST), KSS 2005 (Vienna), KSS 2006 (Beijing), KSS 2007 (JAIST), KSS 2008 (Guangzhou), KSS 2009 (Hong Kong), KSS 2010 (Xi'an), KSS 2011 (Hull), KSS 2012 (JAIST), KSS 2013 (Ningbo), KSS 2014 (Sapporo), KSS 2015 (Xi'an), KSS 2016 (Kobe), KSS 2017 (Bangkok), KSS 2018 (Tokyo), KSS 2019 (Da Nang) and KSS 2022 (online and Beijing), KSS has developed into a successful platform for many scientists and researchers from different countries mainly in Asia.

Over the past 20 years, people interested in knowledge and systems sciences have aggregated into a community, and an international academic society (International Society for Knowledge and Systems Sciences) has existed for 20 years since its founding during KSS 2003 in Guangzhou. To celebrate the 20th anniversary of ISKSS at its birth place along with the breakthrough in AI technology as well as extensive applications across economic, industrial and social activities, KSS 2023 was organized by the International Society for Knowledge and Systems Sciences and co-organized by the Systems Engineering Society of China at South China University of Technology. It had the theme of "**Knowledge and Systems Science in the Age of Generative AI**" during December 2–3, 2023. To fit that theme, four distinguished scholars were invited to deliver the keynote speeches.

- Kaushik Dutta (University of South Florida, USA), "Responsible AI and Data Science for Social Good"
- Yongmiao Hong (AMSS, CAS; UCAS, China), "Forecasting Inflation Rates: A Large Panel of Micro-Level Data Approach"
- Yoshiteru Nakamori (JAIST, Japan), "Encouraging the Development of Knowledge Systems"
- Hiroki Sayama (Binghamton University, SUNY, USA), "Analysis, Visualization and Improvement of Human Collaboration Dynamics Using Computational Methods"

KSS 2023 received 56 submissions from authors studying and working in China, Japan, Spain and Thailand, and finally 20 submissions were selected for publication in the proceedings after a double-blind review process. The co-chairs of the International Program Committee made the final decision for each submission based on the review reports from the referees, who came from Australia, China, Japan, New Zealand, Thailand and UK. Each accepted submission received three reviews on average.

To make KSS 2023 happen, we received a lot of support and help from many people and organizations. We would like to express our sincere thanks to the authors for their remarkable contributions, all the technical program committee members for their time and expertise in paper review with a very very tight schedule, and the proceedings publisher Springer for a variety of professional help. It is the 6th time the KSS proceedings have been published as one CCIS volume after a successful collaboration with Springer during 2016–2019 and 2022. We greatly appreciate those four distinguished scholars for accepting our invitation to present keynote speeches at the symposium by hybrid modes. Last but not least, we are very indebted to the organizing group for their hard work.

We are happy with the thought-provoking and lively scientific exchanges in the essential fields of knowledge and systems sciences during the symposium.

December 2023

Jian Chen
Van-Nam Huynh
Xijin Tang
Jiangning Wu

Organization

Organizer

International Society for Knowledge and Systems Sciences

Host

South China University of Technology

Co-organizer

Systems Engineering Society of China

General Chair

Jian Chen — Tsinghua University, China

Program Committee Chairs

Van-Nam Huynh — Japan Advanced Institute of Science and Technology, Japan

Xijin Tang — CAS Academy of Mathematics and Systems Science, China

Jiangning Wu — Dalian University of Technology, China

Technical Program Committee

Lina Cao — North China Institute of Science and Technology, China

Jindong Chen — Beijing Information Science & Technology University, China

Yu-wang Chen — University of Manchester, UK

Xuefan Dong — Beijing University of Technology, China

Yucheng Dong	Sichuan University, China
Chonghui Guo	Dalian University of Technology, China
Xunhua Guo	Tsinghua University, China
Rafik Hadfi	Kyoto University, Japan
Xiaohui Huang	CAS Academy of Mathematics and Systems Science, China
Van-Nam Huynh	Japan Advanced Institute of Science and Technology, Japan
Vincent Lee	Monash University, Australia
Jie Leng	The People's Bank of China, China
Weihua Li	Auckland University of Technology, New Zealand
Yongjian Li	Nankai University, China
Zhenpeng Li	Taizhou University, China
Zhihong Li	South China University of Technology, China
Yan Lin	Dalian Maritime University, China
Bo Liu	CAS Academy of Mathematics and Systems Science, China
Dehai Liu	Dongbei University of Finance & Ecomonics, China
Yijun Liu	CAS Institute of Science and Development, China
Jinzhi Lu	Beihang University, China
Mina Ryoke	University of Tsukuba, Japan
Bingzhen Sun	Xidian University, China
Xijin Tang	CAS Academy of Mathematics and Systems Science, China
Jing Tian	Wuhan University of Technology, China
Yuwei Wang	CAS Insitute of Automation, China
Cuiping Wei	Yangzhou University, China
Jiangning Wu	Dalian University of Technology, China
Haoxiang Xia	Dalian University of Technology, China
Xiong Xiong	Tianjin University, China
Nuo Xu	Communication University of China, China
Zhihua Yan	Shanxi University of Finance and Economics, China
Thaweesak Yingthawornsuk	King Mongkut's University of Technology Thonburi, Thailand
Wen Zhang	Beijing University of Technology, China
Yaru Zhang	China Mobile Information Technology Center, China
Xiaoji Zhou	China Aerospace Academy of Systems Science and Engineering, China

Abstracts Keynotes

Responsible AI and Data Science for Social Good

Kaushik Dutta

School of Information Systems and Management, Muma College of Business,
University of South Florida, USA
duttak@usf.edu

Abstract. Artificial intelligence (AI)-based solutions have the potential
to address many of the world's most challenging technological and societal problem). The responsible development, use, and governance of AI
have become an increasingly important topic in AI research and practice in recent years. As AI systems are becoming more integrated into
our daily lives and decision-making processes, it is essential to ensure
that they operate responsibly and ethically. To address social problems
with an emphasis on human values and ethics, AI and data science solutions should be built based on five major principles; unbiased results,
transparency, accountability, social benefit, and privacy. By incorporating these principles into the AI algorithm development process, we can
design a responsible AI that is fair and accountable, and that benefits all
users without causing harm or bias. A responsible algorithm should also
ensure that the AI models are transparent and explainable so that users
can understand how the algorithm works and how it makes decisions and
it should serve as a tool for positive change in society. Moreover, it is
important to continually monitor and update the AI algorithm to ensure
its ongoing responsible use. We want to solve the UN's sustainable development challenges and have the potential to impact society from the lens
of responsible AI and data science using design and develop new ML
algorithms, cutting-edge software engineering & data science methods,
computation tools & techniques, AI models, and real-world case studies that can help operationalize responsible AI and data science to solve
challenging societal problems.

Forecasting Inflation Rates: A Large Panel of Micro-Level Data Approach

Yongmiao Hong[1,2]

[1] Academy of Mathematics and Systems Science, Chinese Academy of Sciences
[2] University of Chinese Academy of Sciences, China
ymhong@amss.ac.cn

Abstract. Inflation forecast is of great importance to economic agents, including households, businesses, and policy makers. To predict inflation, economists commonly rely on aggregate macro and financial time series data. However, the process of aggregation often results in the loss of valuable micro information, which diminishes important characteristics like heterogeneity, interaction and structural breaks. This paper proposes a novel micro approach to forecast inflation based on a large panel of individual stock prices. The results demonstrate that this micro forecasting method significantly enhances the accuracy of inflation forecasting when compared to conventional econometric models like benchmark univariate time series and factor models. By employing machine learning algorithms, we can effectively aggregate the information contained in individual stock prices, leading to more precise inflation forecasts. Additionally, a comparison between the predictive capacities of individual stock prices, representative aggregate asset prices, and the commonly used macroeconomic database FRED-MD shows that individual stock prices provide additional valuable information for predicting future inflation rates. Moreover, the improvement achieved through the micro forecasting method becomes increasingly prominent as the forecast horizon increases. Our paper introduces a novel micro-level methodology for predicting macro time series, highlighting the untapped potential of large models that warrants further exploration.

Encouraging the Development of Knowledge Systems

Yoshitery Nakamori

Japan Advanced Institute of Science and Technology, Japan
nakamori0212@gmail.com

Abstract. This presentation will begin by considering the origin of the name of this academic society. Our initial intention was to support the development of a new discipline called knowledge science, using the ideas and methods of systems science, which already boasted many achievements. We soon realized that systems science should also incorporate ideas from knowledge science, and the co-evolution of both sciences became the goal of this society. As time passed, the idea that a new academic system should be created by merging parts of both sciences emerged.

Young members of this society probably do not know what knowledge science is in the first place. So, in this presentation, I will explain the definition of knowledge science again. I've thought about this a lot over the past 20 years, and this is my definition of the final version. Correspondingly, I define knowledge technology. It is defined as meeting a wide range of technologies that support human intellectual activities, including some systems technologies.

I then define a knowledge system that can be constructed through the collaboration of knowledge science and systems science. The knowledge system was proposed 20 years ago by Professor Zhongtuo Wang of the Dalian University of Technology. Recently, I revisited it and redefined it to be more specific. In this presentation, I will give its definition with some examples. In particular, I would like to emphasize that the knowledge system is a problem-solving system that deeply involves humans, has a knowledge creation process, and incorporates artificial intelligence as necessary.

I believe that the research theme of the members of this society is the development of knowledge technology or problem-solving using knowledge technology. In this presentation, I would like to conclude by appealing to all members to turn their attention to the development of knowledge systems. In particular, I recommend thinking about knowledge systems for those aiming to launch large-scale research projects.

Analysis, Visualization and Improvement of Human Collaboration Dynamics Using Computational Methods

Hiroki Sayama

Binghamton University State University of New York, USA
sayama@binghamton.edu

Abstract. Artificial Intelligence (AI), Machine Learning (ML), and other related modern computational technologies are often thought of as ways to replace human labor with automated processes. However, they have significant potential to augment and amplify human capabilities, especially in collaborative work settings. In this talk, we aim to highlight some promising areas of social application of AI/ML in this space. Specifically, we provide an overview of our recent work on the applications of AI/ML for (1) analyzing group dynamics in virtual teams, (2) analyzing and visualizing human collective ideation and innovation, and (3) improving social connectivities for better collaborative outcomes.

Contents

Opinion Mining and Knowledge Technologies

Data Mining, Machine Learning and Deep Learning

A Hybrid Supervised Learning Approach for Intrusion Detection Systems

Tianhao Liu, Wuyue Fan, Gui Wang, Weiye Tang, Daren Li, Man Chen, and Omar Dib[✉]

Department of Computer Science, Wenzhou-Kean University, Wenzhou, China
{tiliu,fanwu,wangui,tangwei,lidar,cheman,odib}@kean.edu

Abstract. The Internet's rapid development has raised significant concerns regarding network attacks' increasing frequency and evolving nature. Consequently, there is an urgent demand for an effective intrusion detection system (IDS) to safeguard data. Artificial intelligence subfields, including machine learning and deep learning, have emerged as valuable tools in addressing this issue, yielding substantial achievements. Nonetheless, the escalating utilization of the Internet and technological advancements have led to a notable rise in the frequency and complexity of network attacks. Consequently, traditional IDSs encounter challenges detecting cyber attacks and countering advanced threats. Developing robust and efficient intrusion detection systems becomes crucial to overcome these limitations. This study explores current network vulnerabilities and proposes an IDS based on supervised learning algorithms. This framework integrates decision trees (DT), random forests (RF), extra trees (ET), and extreme gradient boosting (XGBoost) methods to identify network attacks. Through experimental evaluation on the Edge-IIoTset dataset, the system demonstrates solid performance by effectively and accurately detecting network attacks.

Keywords: cybersecurity · intrusion detection system · machine learning · ensemble learning · hyperparameter optimization

1 Introduction

The rapid development and extensive Internet utilization have brought transformative changes to our daily lives and work environments. However, this connectivity has also exposed us to increasing cybersecurity threats [11]. Malicious actors employ diverse methods to breach systems, compromise sensitive information, manipulate data, and disrupt operations, resulting in significant risks and losses for individuals, organizations, and society [18].

Intrusion Detection Systems (IDS) have emerged as a crucial security measure to counter these evolving threats. IDS proactively aims to detect and mitigate potential security vulnerabilities and attacks. IDS identifies patterns, abnormal behaviors, or indicators associated with known attack patterns by analyzing

J. Chen et al. (Eds.): KSS 2023, CCIS 1927, pp. 3–17, 2023.
https://doi.org/10.1007/978-981-99-8318-6_1

network traffic, system logs, and other pertinent data sources. IDS generates alerts or takes proactive measures to prevent further damage or unauthorized access upon detecting suspicious activity. Implementing IDS enhances organizational security and enables swift responses to potential threats.

Despite its benefits, IDS faces several challenges. The sheer volume of network traffic data poses a significant processing and analysis burden, requiring efficient capabilities to swiftly and accurately detect potential attacks [10]. Additionally, network attacks continually evolve, with attackers employing new forms and patterns to bypass traditional defense mechanisms. This necessitates regular updates and adaptations in IDS to effectively identify new and previously known attacks. Striking a balance between detecting genuine threats and minimizing false positives is crucial to avoid unnecessary interference and alert fatigue.

To address these challenges, machine learning (ML) algorithms have demonstrated their effectiveness in designing IDS. ML techniques leverage labeled and unlabeled data to identify hidden patterns, detect anomalies, and accurately predict potential attacks. By automating the analysis process, ML-based IDS enables efficient and effective detection of intrusion attempts, contributing to proactive network and system protection in cybersecurity.

In this article, we introduce an IDS framework that leverages the strengths of multiple ML algorithms, such as decision trees (DT), random forests (RF), extra trees (ET), and extreme gradient boosting (XGBoost), along with a stacking ensemble model, serving as both binary and multi-class classifiers. Our models have been fine-tuned using state-of-the-art hyperparameter optimization techniques, BO-TPE, and Optuna. The impact of such HPO techniques on the model's performance has been also investigated. Our framework includes a comprehensive set of data pre-processing techniques for the recent Edge-IIoTset dataset [12], introducing an innovative approach to fortify network security.

The paper is organized as follows: Sect. 2 provides a comprehensive review of critical IDS studies conducted from 2016 to 2023. Section 3 presents the adopted dataset. We elaborate on the proposed intrusion detection framework in Sect. 4. Section 5 presents the experimental study and results. Finally, Sect. 6 concludes the paper and outlines future directions.

2   Related Work

Several studies have explored the application of machine learning (ML) models in the field of Intrusion Detection (ID) for various network environments, including vehicular ad hoc networks (VANETs). Alshammari et al. [4] employed K-Nearest Neighbors (KNN) and Support Vector Machine (SVM) techniques to detect DoS and Fuzzy attacks in VANETs. While their approach demonstrated effectiveness, it focused on specific attack types and relied on datasets provided by the Hacking and Countermeasure Research Lab (HCRL) [15], which may limit its generalizability. Similarly, Aswal et al. [5] investigated the suitability of six traditional ML algorithms, such as Naive Bayes (NB), KNN, Logistic Regression (LR), Linear Discriminant Analysis (LDA), Classification And Regression

Trees (CART), and SVM, for detecting bot attacks in the Internet of Vehicles (IoV). Their evaluation on the CICIDS2017 dataset showcased the performance of the CART model, but they acknowledged the potential variability of algorithm performance across different configurations and datasets.

In the study by Gao et al. [13], an adaptive ensemble learning model was proposed for intrusion detection using the NSL-KDD dataset. The MultiTree algorithm, constructed by adjusting training data proportion and using multiple decision trees, achieved an accuracy of 84.2%. Additionally, an adaptive voting algorithm combining several base classifiers achieved a final accuracy of 85.2%. While the ensemble model improved detection accuracy, the study did not extensively discuss the limitations and potential drawbacks of the approach.

Reviewing studies utilizing various datasets, including CSE-CIC IDS-2018, UNSW-NB15, ISCX-2012, NSL-KDD, and CIDDS-001, Kilincer et al. [17] applied classical ML algorithms like SVM, KNN, and DT after performing max-min normalization on the datasets. While some studies reported successful results, there is a need for more critical analysis and comparison of these methods to identify their limitations and potential areas of improvement.

Deep learning (DL) models have also gained attention in IDS research. Yang and Shami [22] proposed a transfer learning and ensemble learning-based IDS for IoV systems, achieving high detection rates and F1 scores. However, their evaluation only considered two well-known public IDS datasets, Car-Hacking and CICIDS2017, limiting the generalizability of their results. Aloqaily et al. [3] introduced the D2H-IDS method based on deep belief and decision tree-based approaches for ID in connected vehicle cloud environments. The method achieved promising results but focused only on the NSL-KDD dataset and a limited number of attack types. The study could benefit from a broader evaluation and analysis of its performance across different datasets and attack scenarios.

The tree-based stacking ensemble technique (SET) proposed by Rashid et al. [20] demonstrated effectiveness on NSL-KDD and UNSW-NB15 datasets. However, the impact of feature selection techniques on the model's performance and the additional computational overhead induced by the stacking method should be further investigated and discussed.

DL algorithms, including CNN, LSTM, and hybrid CNN-LSTM models, were applied by Alkahtani and Aldhyani [2] in a robust ID framework. The models achieved high accuracy rates, but it is crucial to analyze the trade-offs between accuracy and computational complexity and investigate the generalizability of the results to different datasets.

Bertoli et al. [7] described the AB-TRAP framework for intrusion detection, which achieved strong performance with minimal CPU and RAM usage. However, the evaluation focused on specific datasets and addressed concerns related to model realization and deployment, leaving room for further investigation into the framework's performance across diverse network traffic scenarios.

To tackle availability issues and mitigate DDoS attacks in VANETs, Gao et al. [14] proposed a distributed NIDS that relied on a real-time network traffic collection module, Spark, for data processing and an RF classifier for classification.

Although their system demonstrated high accuracy rates, a more comprehensive analysis of its scalability and robustness is needed.

Addressing the class imbalance problem in intrusion detection, Liu et al. [19] proposed the DSSTE algorithm. While their approach outperformed other methods on NSL-KDD and CSE-CIC-IDS2018 datasets, further analysis is required to assess its performance on different attacks and datasets.

Injadat et al. [16] proposed a multi-stage optimized ML-based NIDS framework, emphasizing the impact of oversampling and feature selection techniques on performance and time complexity. The framework demonstrated potential for reducing training sample and feature set sizes, but a more comprehensive evaluation is necessary to understand its limitations and generalizability.

Basati et al. [6] introduced a lightweight and efficient neural network for NIDS, addressing computational complexity concerns in IoT devices. Although their model achieved efficiency, its evaluation focused on limited datasets, warranting further investigation into its performance across diverse scenarios.

In terms of hyper-parameter optimization (HPO), Yang and Shami [21] reviewed available frameworks and recommended Bayesian Optimization Hyper-Band (BOHB) for ML models when randomly selected subsets are highly representative of the dataset. Despite their computational time requirements, heuristic approaches such as Genetic Algorithm (GA) and Particle Swarm Optimization (PSO) are typically more suitable for larger configuration spaces.

While these studies have indeed added value to the field of IDS using ML and DL models, there is still a crucial requirement for comprehensive evaluations and comparisons using recent datasets such as the Edge-IIoTset dataset [12]. It's imperative to examine the impact of data preprocessing techniques, innovative hyperparameter optimization methods, acknowledge limitations and potential drawbacks, and conduct a more in-depth analysis of the trade-offs between accuracy, computational complexity, and scalability. These analyses are instrumental in propelling the development of more effective IDS frameworks.

3   Cyber Security Dataset

Including high-quality and diverse legitimate entries in a dataset is crucial when constructing the normal behavioral profile of a system. This allows security solutions to recognize known attack patterns and identify emerging ones accurately. Moreover, the presence of malicious entries is vital for security solutions to detect and respond to potential threats effectively. By incorporating legitimate and malicious data, security systems can comprehensively understand system behavior and effectively adapt to evolving attack patterns.

The data set used in this study, Edge-IIoTset [12], was created using a dedicated IoT/IIoT test platform containing various devices, sensors, protocols, and configurations associated with IoT/IIoT environments. The IoT data is generated from over 10 IoT devices, such as low-cost digital sensors for temperature and humidity, ultrasonic sensors, water level detection, pH Sensor Meters, soil moisture sensors, heart rate sensors, and flame sensors. Specifically, the Edge-IIoTset dataset includes normal scenarios and fourteen attacks targeting IoT and IIoT connectivity protocols. Figure 1 presents the various categories in the dataset along with the number of instances in every category.

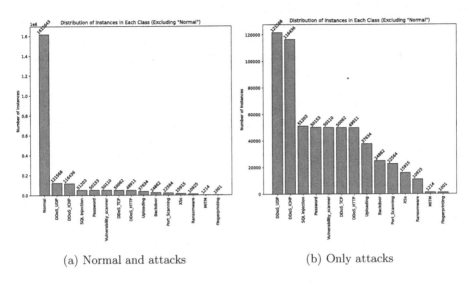

(a) Normal and attacks   (b) Only attacks

Fig. 1. Number of Instances in Different Classes

The attack scenarios in the Edge-IIoTset dataset are grouped as follows:

1. **DoS/DDoS:** Attackers aim to deny services to legitimate users, individually or in a distributed manner. This category includes four commonly used techniques: TCP SYN Flood, UDP flood, HTTP flood, and ICMP flood.
2. **Information gathering:** This category involves obtaining intelligence about the targeted victim. The dataset includes three steps for information-gathering: port scanning, OS fingerprinting, and vulnerability scanning.
3. **Man in the middle:** This attack intends to compromise and alter communication between two parties who assume they are in direct communication. The dataset focuses on two commonly used protocols, DNS and ARP, for executing this attack.
4. **Injection:** These attacks aim to compromise the integrity and confidentiality of the targeted system. The dataset includes three approaches: XSS, SQL injection, and uploading attacks.
5. **Malware:** This category includes attacks that have gained public attention due to extensive damage and reported losses. The dataset includes three types of malware attacks: backdoors, password crackers, and ransomware.

By examining these attacks, we aim to gain insights into the vulnerabilities and risks associated with IoT and IIoT systems and further enhance our understanding of potential security challenges in this domain.

4   Proposed Framework

In this article, we harness the power of supervised ML algorithms to efficiently detect cyber attacks. Our work incorporates the latest advancements in ML-based intrusion detection, providing an innovative approach to enhance network security. Figure 2 demonstrates the architecture of the proposed ML architecture, comprising four main stages: data pre-processing, feature engineering, tree-based models, and Stacking ensemble model.

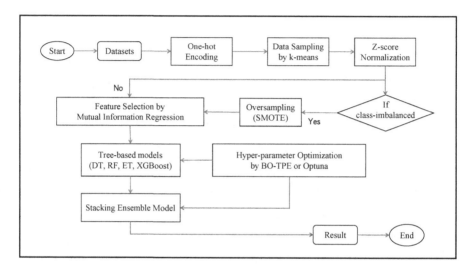

Fig. 2. An Intrusion Detection Architecture Using Machine Learning

4.1   Data Pre-processing

In data pre-processing, we process the dataset through multiple techniques, which provides convenience for subsequent steps. Missing values can occur due to errors, omissions, or other reasons during the data collection. To handle missing values, we can either delete records that contain missing values or use imputation methods to fill in the missing values. Given our dataset's large size, deleting records containing missing values will not significantly impact the overall dataset. Hence, we will adopt the approach of deleting records with missing values.

When encountering categorical features, we need to transform them into a format that ML algorithms can handle. One commonly used method for this transformation is called one-hot encoding. One-hot encoding is a technique that converts categorical variables into binary vector representations. We create a new binary feature column for each distinct category in the original feature.

Since the dataset is still too large after data cleaning and one-hot encoding, we used the k-means algorithm to form a representative reduced dataset. The k-means algorithm is a commonly used unsupervised learning clustering algorithm.

Its objective is to divide a dataset into k distinct clusters, where each data point is assigned to the cluster whose centroid is most similar to it. The algorithm randomly chooses k initial cluster centroids as starting points. Then, assign each data point to the cluster represented by the nearest centroid based on distance. Next, update the positions of each cluster centroid based on the assigned data points by computing the average of all data points in each cluster. Repeat this assignment and update the process until the positions of the cluster centroids no longer change or reach a predefined number of iterations. Finally, we randomly select a subset of data points from each cluster to form a reduced, but representative dataset. This undersampling technique greatly reduces the running time of the subsequent steps without compromising the model's accuracy.

Z-score normalization is a commonly used data normalization technique. This method aims to eliminate the dimensional differences between data, enabling the comparison and processing of different features on the same scale, thereby improving the effectiveness of model training. For each feature, we first calculate the mean and standard deviation of that feature across the entire dataset. Then, for each data point, we subtract the mean of the feature and divide it by the standard deviation, resulting in the corresponding Z-score.

The data set after the above processing may be subject to data imbalance, and the SMOTE algorithm (Synthetic Minority oversampling Technique) can be used to oversample the minority classes [9]. The SMOTE algorithm selects k nearest neighbors from the minority class and computes their feature differences. Then, the eigenvalues are interpolated along these differences in the feature space to generate new synthetic samples. This approach helps improve the model's ability to learn from the minority class, thus improving the overall classification accuracy, robustness, and performance.

4.2   Feature Engineering

Feature engineering is a crucial step in machine learning and data mining, involving transforming, selecting, and creating new features from raw data to extract useful information and improve model performance. Mutual information regression is a feature selection method that measures the dependence and non-linear relationship between two variables.

During the feature engineering stage, we use mutual information regression to assess the correlation between each feature and the target variable, selecting highly correlated features. Through mutual information regression, we can quantify the predictive power of each feature for the target variable and identify features that exhibit a strong correlation with the target variable.

Choosing highly correlated features brings several benefits. Firstly, it reduces the dimensionality of the feature space, leading to lower model complexity and computational costs. Secondly, highly correlated features often contain important information relevant to the target variable, improving predictive capability. Additionally, by reducing unrelated features, we can mitigate the impact of noise and redundant information on the model, enhancing its robustness and generalization ability.

4.3 Supervised Learning Algorithms for IDS

After performing data pre-processing and feature engineering, the labeled datasets obtained are utilized for training an ensemble learning model for developing a signature-based Intrusion Detection System (IDS).

We utilize multiple tree models to train the preprocessed data in the model training phase. These tree models include decision trees, random forests, extreme random trees, and XGBoost, all ensemble learning methods based on tree structures and exhibit strong modeling capability and predictive performance.

Decision trees recursively select features and split nodes to construct a tree-like structure for classification or regression tasks. Decision trees have the advantage of interpretability and can handle categorical and continuous features.

Random forests consist of multiple decision trees. In random forests, each decision tree is constructed by randomly sampling the original data with replacement, and during feature selection at each node, only a subset of features is considered. By ensembling multiple decision trees, random forests can reduce the risk of overfitting and improve the model's generalization ability.

Like random forests, extremely randomized trees introduce even more randomness in feature selection. Specifically, extremely randomized trees do not choose the optimal split feature at each node, but rather randomly select features for splitting. This additional randomness can reduce the model's variance and increase its diversity.

XGBoost is a powerful model based on gradient-boosting trees. It iteratively trains multiple trees and optimizes the model's predictive capability using the gradient information of the loss function. XGBoost offers high predictive performance and computational efficiency and can handle large-scale datasets and high-dimensional features.

This research paper also focuses on optimizing the important hyperparameters of the four tree-based machine-learning algorithms. To achieve this, we employ two prominent Hyperparameter Optimization (HPO) methods: BO-TPE (Bayesian Optimization with Tree-Parzen Estimator) [21] and Optuna [1]. By leveraging these HPO techniques, we aim to fine-tune the hyperparameters of the machine learning algorithms, thereby enhancing their performance and effectiveness in the context of our study. Using BO-TPE and Optuna allows us to systematically explore the hyperparameter space, leading to more optimal configurations for the tree-based algorithms.

BO-TPE utilizes two density functions, $l(x)$ and $g(x)$, as generative models for the variables. By specifying a predefined threshold $y*$ to differentiate relatively good and poor results, the objective function of TPE is modeled by Parzen windows as follows [21]:

$$p(x|y, D) = \begin{cases} l(x), & if \ y < y* \\ g(x), & if \ y > y* \end{cases} \tag{1}$$

The functions $l(x)$ and $g(x)$ represent the probability of detecting the next hyperparameter value in the well-performing and poor-performing regions. BO-TPE identifies the optimal hyperparameter values by maximizing the $l(x)/g(x)$

ratio. The Parzen estimators are organized in a tree structure, allowing the retention of specific conditional dependencies among hyperparameters. Moreover, BO-TPE optimizes various types of hyperparameters [21]. Therefore, BO-TPE is utilized to optimize the hyperparameters of tree-based machine-learning models with many hyperparameters.

Optuna [1] is a generic Grid search library that uses sequence model-based optimization (SMBO) techniques. Optuna combines Bayesian optimization, random sampling, and pruning strategies to search hyperparametric spaces effectively. Optuna provides a simple and flexible way to automatically define hyperparametric search spaces, evaluation functions, and optimization goals and select new hyperparametric combinations for evaluation. Optuna also supports parallel optimization, speeding up the optimization process by simultaneously performing hyperparametric searches on multiple computing resources.

After obtaining the four tree-based machine learning models, we utilize stacking, an ensemble learning method, to enhance the model performance. Combining multiple base learners generally exhibits better generalizability than a single model. Stacking is a commonly used ensemble learning technique that employs the output labels estimated by the four base learners (Decision Tree, Random Forest, Extra Trees, and XGBoost) as input features to train a robust meta-learner, which makes the final predictions. By leveraging stacking, we can utilize the information from all four base learners to reduce individual learner errors and obtain a more reliable and robust meta-classifier. Our proposed system selects the best-performing model among the four base models as the algorithm to construct the meta-learner, likely achieving the best performance.

4.4   Validation Metrics

To ensure the generalizability of the proposed framework and mitigate overfitting, a combination of cross-validation and hold-out methods is employed in the attack detection experiments. The following steps outline the train-test-validation split and model evaluation procedures for each dataset:

1. Train-Test Split: An 80%-20% train-test split is performed, where 80% of the samples are used to create a training set, and the remaining 20% form the test set. The test set remains untouched until the final hold-out validation.
2. Cross-Validation: A 5-fold cross-validation is implemented on the training set. This approach allows for evaluating the proposed model on different dataset regions. In each iteration or fold of the cross-validation process, 90% of the original training set is used for model training. At the same time, the remaining 10% serves as the validation set for model testing.
3. Hold-Out Validation: The model trained in Step 2 is tested on the untouched test set to evaluate its performance on a new dataset. This step provides an additional assessment of the model's generalizability and effectiveness.

The performance evaluation of the proposed IDS involves several metrics, namely accuracy, precision, recall, and F1-score [16]. These metrics collectively provide a comprehensive assessment of the system's effectiveness.

Accuracy (Acc) is calculated using the formula:

$$Acc = \frac{TP_{Attack} + TN_{Attack}}{TP_{Attack} + TN_{Attack} + FP_{Attack} + FN_{Attack}} \quad (2)$$

Precision (Pr) is determined as follows:

$$Pr = \frac{TP_{Attack}}{TP_{Attack} + FP_{Attack}} \quad (3)$$

Recall (Re) is computed as follows:

$$Re = \frac{TP_{Attack}}{TP_{Attack} + FN_{Attack}} \quad (4)$$

F1-score (F1) is a harmonic mean of precision and recall:

$$F1 = \frac{2 \times Precision \times Recall}{Precision + Recall} \quad (5)$$

Using these metrics, the performance of the IDS can be thoroughly evaluated, providing insights into its ability to detect and classify attacks accurately.

5   Experimental Study

5.1   Experimental Setup

The development of the proposed Intrusion Detection System (IDS) involved utilizing Python libraries renowned for their effectiveness in feature engineering, machine learning algorithms, and hyperparameter optimization. The specific libraries employed were Pandas for comprehensive data manipulation and preprocessing, Scikit-learn for implementing machine learning algorithms and evaluating their performance using appropriate metrics, XGBoost [8] for harnessing the capabilities of gradient boosting models, and Optuna [1] for automating hyperparameter optimization. We employed a robust server computer, the Dell Precision 3660 Tower, to conduct experiments and train the IDS models. This machine had a high-performance i9-12900 central processing unit (CPU) featuring 16 cores and a base clock speed of 2.40 GHz. Furthermore, the server boasted ample memory capacity, specifically 64 GB, ensuring the system's capability to handle the computationally intensive demands of the training process.

5.2   Performance Analysis

The subsection presents the folded cross-validation results of the proposed model on the Edge-IIoTset dataset, summarized in Table 1 and 2. Table 1 provides an overview of the results obtained without implementing techniques such as SMOTE, feature selection, and scaling. In contrast, Table 2 encompasses the results by including these techniques. Both tables showcase the performance of the original models, denoted as "Original", representing models without any hyperparameter tuning. Additionally, the performance of the models using the BO-TPE and OPTUNA hyperparameter optimization methods is individually presented. This allows for a comparative analysis of the impact of hyperparameter tuning on the model's performance. In addition to classification metrics, we report each model's training and prediction time. These time measurements provide valuable insights into the computational efficiency of the models.

Table 1. Results without using SMOTE, feature selection, and scaling techniques

	Method	Acc (%)	Pr (%)	Rc (%)	F1 (%)	Train Time (S)	Pre Time (S)
Original	RF	98.12	98.12	98.12	98.11	11.6	0.84
	DT	98.00	98.00	98.00	98.00	0.58	0.01
	ET	98.05	98.05	98.05	98.05	11.65	0.91
	XGBoost	98.39	98.57	98.39	98.39	6.27	0.03
	Stacking	98.25	98.26	98.35	98.36	5.90	0.02
BO-TPE	RF	98.39	98.46	98.39	98.39	14.10	0.45
	DT	98.37	98.51	98.37	98.37	0.44	0.02
	ET	98.40	98.42	98.40	98.40	6.60	0.30
	XGBoost	**98.48**	**98.64**	**98.48**	**98.47**	6.35	0.02
	Stacking	98.43	98.45	98.43	98.43	3.90	0.02
OPTUNA	RF	98.33	98.35	98.33	98.32	4.51	0.25
	DT	98.04	98.05	98.04	98.04	0.166	0.01
	ET	98.33	98.34	98.33	98.33	4.46	0.31
	XGBoost	98.46	98.49	98.46	98.46	5.50	0.02
	Stacking	98.15	98.15	98.15	98.15	5.81	0.02

The performance of the algorithms without hyperparameter tuning is shown in the "Original" section of Table 1. Moreover, we plot the accuracy of the various models in Fig. 3. Random Forest (RF), Decision Tree (DT), Extra Tree (ET), XGBoost, and Stacking models are evaluated. The XGBoost model achieves the highest accuracy of 98.39% and the highest F1 score of 98.39% among all the original models. It also exhibits the shortest training time of 6.27 s and a prediction time of 0.03 s. This underscores the algorithm's feasibility in real-time scenarios, taking both quality and running time into account. In the "BO-TPE" section, the models are optimized using the BO-TPE hyperparameter optimization method. The RF, DT, ET, XGBoost, and Stacking models show improved performance compared to their original counterparts. Once again, the XGBoost model achieves the highest accuracy of 98.48% and the highest F1 score of 98.47% among the BO-TPE optimized models. However, the training and prediction times are slightly longer than the original XGBoost model. This section also demonstrates that BO-TPE can enhance the classification accuracy of tree-based models without significantly increasing their training time.

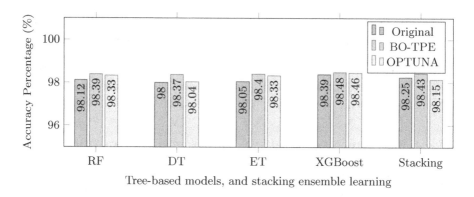

Fig. 3. Performance Comparison of Machine Learning Models

The models are optimized in the "OPTUNA" section using the Optuna hyperparameter optimization method. Like the BO-TPE results, the RF, DT, ET, XGBoost, and Stacking models demonstrate improved performance compared to the original models. The XGBoost model achieves the highest accuracy of 98.46% and the highest F1 score of 98.46% among the OPT-optimized models. The optimized models' training and prediction times are also within reasonable ranges. It's worth noting that the initial weights of the models can influence the training time across the three sections. Additionally, it's important to emphasize that BO-TPE and OPTUNA are not directly comparable methods; neither method dominates the other in all aspects. To summarize, results in Table 1 indicate that both hyperparameter optimization techniques, BO-TPE and Optuna, contribute to improved model performance. The XGBoost model consistently outperforms the other algorithms, achieving the highest accuracy and F1 score in both the original and optimized scenarios. The stacking model shows comparable performance to XGBoost in some cases but falls slightly behind. Both XGBoost and stacking exhibit fast prediction times, rendering them well-suited for real-time intrusion detection applications.

The updated findings in Table 2 showcase the results after incorporating additional processing steps such as SMOTE, feature selection, and scaling. The performance metrics demonstrate a reduction to varying degrees compared to the previous table. Furthermore, the training time experiences a significant increase. Despite these changes, it is important to note that the XGBoost method maintains its superiority over the random forest, decision tree, and extra tree algorithms. The XGBoost model achieves remarkable performance with an accuracy of 98.05%, precision of 98.38%, recall of 98.05%, and F1 score of 98.09%. Although the Stacking model also shows promising results, it falls behind the XGBoost model in terms of overall performance.

Comparing the two tables, it is evident that the impact of feature engineering and processing steps on model performance can be mixed. While these steps aim to enhance performance, they can also introduce some challenges. In Table 2, the accuracy, precision, recall, and F1 score exhibit varying degrees of decrease, indi-

Table 2. Results using SMOTE, feature selection, and scaling techniques

	Method	Acc (%)	Pr (%)	Rc (%)	F1 (%)	Train Time (S)	Pre Time (S)
Processing	RF	94.00	95.13	94.00	94.16	14.74	0.10
	DT	94.42	94.72	94.42	94.52	7.93	0.01
	ET	95.87	96.12	95.87	95.96	11.25	0.13
	XGBoost	98.05	98.38	98.05	98.09	32.63	0.01
	Stacking	96.86	96.98	96.86	96.90	90.64	0.03
BO-TPE	RF	94.59	94.98	94.59	94.73	57.80	0.14
	DT	94.30	94.57	94.30	94.38	8.93	0.01
	ET	96.13	96.35	96.13	96.20	12.94	0.09
	XGBoost	**98.04**	**98.18**	**98.04**	**98.09**	425.30	0.05
	Stacking	96.87	96.98	96.86	96.90	133.68	0.04
OPTUNA	RF	94.44	94.82	94.44	94.57	695.99	0.92
	DT	94.30	94.64	94.30	94.41	6.58	0.01
	ET	96.27	96.51	96.28	96.36	152.10	0.96
	XGBoost	97.86	98.00	97.86	97.91	517.29	0.04
	Stacking	96.84	96.98	96.83	96.86	16.90	0.01

cating that these additional steps may have removed essential information from the dataset. The introduction of SMOTE for data balancing and the mutual information regression method for feature engineering can potentially introduce noise and discard relevant features, respectively. This can result in a reduction in performance metrics. In cases where the dataset already contains perfect information, applying these techniques may not optimize the results as expected. Moreover, the processing steps increase training times, as indicated by the significantly higher training time values in Table 2. This is attributed to the additional calculations and transformations in the feature engineering and processing stages. Despite the trade-offs associated with feature engineering and processing, the XGBoost algorithm consistently outperforms the other algorithms in both tables. This highlights the robustness and effectiveness of XGBoost in handling the complexities of the given dataset. The XGBoost model demonstrates its ability to adapt and leverage available features to achieve superior results. Therefore, when considering the inclusion of feature engineering and processing steps, it is crucial to carefully evaluate their impact on performance metrics and training times, considering the specific dataset characteristics. In such scenarios, the XGBoost algorithm offers a reliable and effective solution, particularly when accuracy and computational efficiency are important considerations.

6 Conclusions

We proposed a multi-classification intrusion detection system (IDS) to enhance system security by effectively detecting various known cyber attacks. The proposed IDS includes data preprocessing, feature engineering, and four major machine learning algorithms. Hyperparameter optimization techniques, such as BO-TPE and Optuna, are employed to improve the model's performance. The proposed IDS demonstrates its effectiveness in detecting known attacks, achieving high accuracy and precision rates of 98.48% and 98.64% respectively, with an average F1 score and recall rate of 98.48% and 98.47%, respectively, on the Edge-IIoT dataset. Among the evaluated models, including Random Forest, Extra Trees, Decision Tree, XGBoost, and stacking, XGBoost consistently demonstrated the highest level of performance. Notably, both XGBoost and stacking exhibited fast prediction times, making them well-suited for real-time ID applications. Our results also highlight that hyperparameter optimization techniques like BO-TPE and Optuna significantly improve the classification accuracy of tree-based models without drastically increasing their training time. It's important to note that these two methods excel in different aspects and are not directly comparable. In future work, we recommend addressing the challenge of detecting unknown attacks. To achieve this, we propose exploring unsupervised and online learning methods to refine our detection methods further. This will ensure continuous adaptation to evolving attack patterns and enhance our real-time detection capabilities.

References

1. Akiba, T., Sano, S., Yanase, T., Ohta, D., Koyama, M.: Optuna: a next-generation hyperparameter optimization framework. Software (2019). https://optuna.org/
2. Alkahtani, H., Aldhyani, T.H.: Intrusion detection system to advance internet of things infrastructure-based deep learning algorithms. Complexity **2021**, 1–18 (2021)
3. Aloqaily, M., Otoum, S., Al Ridhawi, I., Jararweh, Y.: An intrusion detection system for connected vehicles in smart cities. Ad Hoc Netw. **90**, 101842 (2019)
4. Alshammari, A., Zohdy, M.A., Debnath, D., Corser, G.: Classification approach for intrusion detection in vehicle systems. Wirel. Eng. Technol. **9**(4), 79–94 (2018)
5. Aswal, K., Dobhal, D.C., Pathak, H.: Comparative analysis of machine learning algorithms for identification of bot attack on the internet of vehicles (IoV). In: 2020 International Conference on Inventive Computation Technologies (ICICT), pp. 312–317. IEEE (2020)
6. Basati, A., Faghih, M.M.: DFE: efficient IoT network intrusion detection using deep feature extraction. Neural Comput. Appl. **34**, 1–21 (2022)
7. Bertoli, G.D.C., et al.: An end-to-end framework for machine learning-based network intrusion detection system. IEEE Access **9**, 106790–106805 (2021)
8. Chen, T., He, T.: XGBoost: eXtreme gradient boosting (2015). https://CRAN.R-project.org/package=xgboost. R package version 0.4-2
9. Chen, Z., et al.: Machine learning based mobile malware detection using highly imbalanced network traffic. Inf. Sci. **433–434**, 346–364 (2018). https://doi.org/10.1016/j.ins.2017.04.044
10. Diallo, E.H., Dib, O., Agha, K.A.: The journey of blockchain inclusion in vehicular networks: a taxonomy. In: 2021 Third International Conference on Blockchain Computing and Applications (BCCA), pp. 135–142 (2021). https://doi.org/10.1109/BCCA53669.2021.9657050

11. Diallo, E.H., Dib, O., Al Agha, K.: A scalable blockchain-based scheme for traffic-related data sharing in VANETs. Blockchain: Res. Appl. **3**(3), 100087 (2022)

12. Ferrag, M.A., Friha, O., Hamouda, D., Maglaras, L., Janicke, H.: Edge-IIoTset: a new comprehensive realistic cyber security dataset of IoT and IIoT applications for centralized and federated learning. IEEE Access **10**, 40281–40306 (2022)

13. Gao, X., Shan, C., Hu, C., Niu, Z., Liu, Z.: An adaptive ensemble machine learning model for intrusion detection. IEEE Access **7**, 82512–82521 (2019)

14. Gao, Y., Wu, H., Song, B., Jin, Y., Luo, X., Zeng, X.: A distributed network intrusion detection system for distributed denial of service attacks in vehicular ad hoc network. IEEE Access **7**, 154560–154571 (2019)

15. Hacking, Lab, C.R.: Can intrusion dataset (2017). http://ocslab.hksecurity.net/Dataset/CAN-intrusion-dataset

16. Injadat, M., Moubayed, A., Nassif, A.B., Shami, A.: Multi-stage optimized machine learning framework for network intrusion detection. IEEE Trans. Netw. Serv. Manage. **18**(2), 1803–1816 (2020)

17. Kilincer, I.F., Ertam, F., Sengur, A.: Machine learning methods for cyber security intrusion detection: datasets and comparative study. Comput. Netw. **188**, 107840 (2021)

18. Liu, J., Xue, H., Wang, J., Hong, S., Fu, H., Dib, O.: A systematic comparison on prevailing intrusion detection models. In: Takizawa, H., Shen, H., Hanawa, T., Hyuk Park, J., Tian, H., Egawa, R. (eds.) PDCAT 2022. LNCS, vol. 13798, pp. 213–224. Springer, Cham (2022). https://doi.org/10.1007/978-3-031-29927-8_17

19. Liu, L., Wang, P., Lin, J., Liu, L.: Intrusion detection of imbalanced network traffic based on machine learning and deep learning. IEEE Access **9**, 7550–7563 (2020)

20. Rashid, M., Kamruzzaman, J., Imam, T., Wibowo, S., Gordon, S.: A tree-based stacking ensemble technique with feature selection for network intrusion detection. Appl. Intell. **52**, 1–14 (2022)

21. Yang, L., Shami, A.: On hyperparameter optimization of machine learning algorithms: theory and practice. Neurocomputing **415**, 295–316 (2020)

22. Yang, L., Shami, A.: A transfer learning and optimized CNN based intrusion detection system for internet of vehicles. arXiv preprint arXiv:2201.11812 (2022)

A Novel Approach for Fake Review Detection Based on Reviewing Behavior and BERT Fused with Cosine Similarity

Junren Wang[1], Jindong Chen[1,2]([✉]), and Wen Zhang[3]

[1] Beijing Information Science and Technology University, Beijing 100192, China
{2022020710,j.chen}@bistu.edu.cn
[2] Intelligent Decision Making and Big Data Application Beijing International Science and Technology Cooperation Base, Beijing, China
[3] Beijing University of Technology, Beijing, China
zhangwen@bjut.edu.cn

Abstract. The pre-trained models such as BERT for fake review detection have received more attention. Most of research has overlooked the role of behavioral features. Additionally, the improvements of the pre-trained models have increased the computational overhead. To enhance BERT's ability to extract contextual features while minimizing computational overhead, we propose an end-to-end model for fake review detection that combines textual and behavioral feature information. Firstly, the review average cosine similarity and the review corpus are jointly fed into BERT to obtain textual feature vectors. Then, the underlying patterns in the behavioral information are extracted by CNN to construct behavioral feature vectors. Finally, the two feature vectors are concatenated for fake review detection. The entire model and each component were evaluated on YELP dataset. Compared with the original BERT model, the F1 score and AUC score of the BERT model fused cosine similarity are improved by 6.30% and 6.37%, respectively. Our model achieves a further improvement of 12.47% and 12.48% in F1 score and AUC score, and shows good performance in precision and recall. The experimental results show that the proposed model improves the effectiveness of BERT for detecting fake reviews, and is particularly suitable for scenarios where can capture behavioral features.

Keywords: Fake review detection · BERT · Behavioral features · Feature fusion

1 Introduction

Internet's popularity and e-commerce have made online product reviews increasingly common. Online product reviews play an important role in consumers' purchasing decisions [1]. A survey conducted in 2023 revealed that 87% of consumers would not consider businesses with an average rating below 3 stars, while

J. Chen et al. (Eds.): KSS 2023, CCIS 1927, pp. 18–32, 2023.
https://doi.org/10.1007/978-981-99-8318-6_2

only 6% of consumers stated that average star ratings would not impact their choices [2]. Hence, consumers often rely more on the opinions and experiences of other buyers rather than advertisements.

Due to the significant impact of online product reviews on consumer purchasing decisions, the issue of fake reviews in online product reviews is becoming increasingly severe [3]. Online merchandising platforms have made significant investments to fight against fake reviews. However, the issue of fake reviews remains serious. A 2023 survey reveals a decline in consumer trust on online reviews [2]. Therefore, developing effective methods for detecting fake reviews has become increasingly important.

Early research on fake review detection focused on mining new features (textual or behavioral), which was time-consuming and labor-intensive [4,5]. With the adoption of word embeddings, researchers began utilizing convolutional neural networks and other neural network variations to learn textual features from reviews [6–9]. With the rise of pre-trained models, certain studies have commenced utilizing pre-trained models such as BERT [10] for fake review detection. However, there are two limitations in the current research on using pre-trained models for fake review detection. Firstly, the pre-trained models only utilize textual information and overlook the role of behavioral features, leading to sub-optimal performance on certain datasets [11]. Secondly, the improvements of the pre-trained models have increased the computational overhead.

To address the first issue, the pre-trained model can be combined with behavioral features to achieve better detection of fake reviews. The latent patterns of behavioral features can be extracted by utilizing CNN. The advanced representations obtained from CNN can be combined with the contextual representations obtained from pre-trained models for detecting fake review. To address the second issue, it is necessary to improve methods without modifying the architecture of the pre-trained models.

Consequently, this paper proposes an end-to-end model for fake review detection. The model has two novelties: 1) It combines BERT and CNNs, integrating textual information and reviewer's behavioral information. 2) It fuses the average cosine similarity of review categories into BERT, aiming to enhance the learning of contextual representations without significantly increasing computational costs.

The remainder of the paper is organized as follows: Sect. 2 presents related work, Sect. 3 describes our model and methodology, Sect. 4 is the experimental section containing the dataset, data pre-processing, indicator design and experimental results, and Sect. 5 concludes the paper.

2   Related Works

Machine learning methods are a common method to detect fake reviews, which is based on the science of information system design and utilizes features to detect fake reviews. Existing research on fake review detection can be broadly classified into two categories. The first category focuses solely on utilizing review text for fake review detection. The second category involves combining textual features and behavioral features for detecting fake reviews. For the first category of methods, ranging from word embedding techniques to attention mechanisms [6], generative adversarial networks (GANs) [8], long short-term memory networks (LSTM) [9], and further to bidirectional long short-term memory networks (BILSTM) [12], along with a series of pre-trained models [10,13–16], the emergence of these new technologies and models has enabled researchers to explore their application in the field of fake review detection. However, these researches have mainly focused on utilizing the review text while neglecting the behavioral features. Therefore, we choose to combine pre-trained models with behavioral features for fake review detection.

For the second category of methods, previous research primarily employed traditional supervised and unsupervised ML techniques, combined with raw behavioral and textual features for fake review detection [5]. Noekhah et al. [7] proposed a graph-based model (MGSD) for multi-iterative graph-based opinion spam detection. Manaskasemsak [17] proposed a graph partitioning approach (BeGP) and its extension (BeGPX) for fake review detection, which jointly review feature, reviewer features, emotion representation and word embedding representation. Javed et al. [18] proposed a technique that utilizes three different models trained based on the concept of multi-view learning. Duma et al. [19] proposed a novel Deep Hybrid Model for detecting fake reviews. This model incorporates latent text feature vectors, aspect ratings, and overall ratings in a joint learning framework. Hajek et al. [20] introduced a review representation model that leverages behavioral and sentiment-dependent linguistic features to effectively capture the domain context. However, previous research has seldom combined pre-trained models with behavioral features for the specific task of fake review detection.

Existing research has several gaps. Firstly, some research combines behavioral and textual features for fake review detection, but they overlook the dependencies between features. Secondly, some research utilizes pre-trained models for fake review detection, but they neglect behavioral features. Lastly, fine-tuning pre-trained models itself requires significant computational resources, and some approaches aimed at improving the model's performance further exacerbate the computational costs. In light of these gaps, it is important to explore methods that integrate pre-trained models with behavioral features. Indeed, it is necessary to explore improvement methods that incur lower computational costs while obtaining better textual representations.

3   Fake Review Detection Design

The proposed model can be divided into three main components: a textual feature extractor that fuses review average cosine similarity to BERT, a reviewer's behavioral feature extractor, and fake review detection, as shown in Fig. 1. The basic process of each component is described in detail in subsequent subsections.

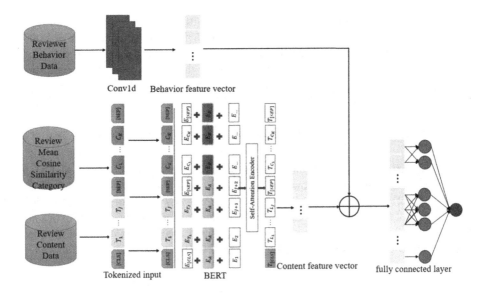

Fig. 1. Proposed fake review detection design.

3.1 Fusing Cosine Similarity into BERT

The emergence of BERT model set off a boom in pre-trained models, benefiting from its design and training on vast corpora, which enables its strong performance on diverse downstream tasks. BERT can be utilized in either pre-trained or fine-tuned modes. To utilize the contextual knowledge gained by BERT during its training, we employ the fine-tuning approach and incorporate cosine similarity category number to enhance feature learning by BERT.

Cosine Similarity Category. In an analysis conducted by Zhang [21], several features were examined for their impact on fake review detection rates, with the average cosine similarity of a reviewer's reviews being identified as one of the more important linguistic features. Therefore, it is fused with BERT to enhance its effectiveness.

The formula for cosine similarity is shown in Eq. 1, where A and B denote the vectors of two reviews. The average cosine similarity is obtained by averaging the cosine similarity of each pair of review of a reviewer. Due to the average cosine similarity of a review is a continuous numerical variable, it cannot be directly incorporated into BERT along with review. Therefore, we employ a decision tree classification algorithm [22] to discretize the average cosine similarity. The discretization process is as follows:1) The continuous average cosine similarity of the review is transformed into a discrete target variable. Since the average cosine similarity of reviews is a decimal number between 0 and 1, the value after multiplying the average cosine similarity of all reviews by 1000 and rounding it up was used as the target variable. 2) A decision tree classifier is constructed using the Gini coefficient [23] as the splitting criterion. The minimum number of samples required to form a leaf node is set to 5% of the total number of samples. 3) The decision tree model is fitted and trained on the data, resulting in Boolean dummy variables as the output. These dummy variables are concatenated together. 4) The concatenated dummy variables are then converted back into decimal category numbers.

$$COS_{AB} = \frac{A \bullet B}{||A|| * ||B||} \tag{1}$$

The distribution of the obtained cosine similarity categories among real and fake reviews is counted, as shown in Fig. 2. Among them, label0 represents fake reviews, label1 represents true reviews. The graph shows that in some categories the number of fake reviews far exceeds the number of real reviews, while in others the opposite is true. Therefore, it is hypothesized that fusing cosine similarity category into BERT can help BERT to obtain better text representations.

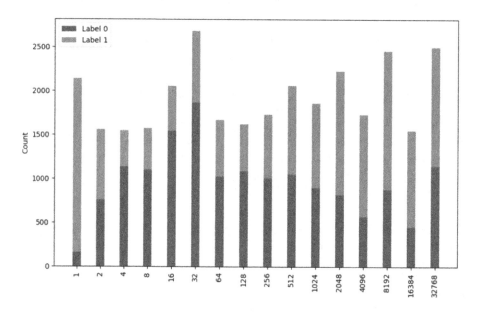

Fig. 2. The distribution of average cosine similarity categories.

Review Text Feature Vector. The structural framework of the text feature extractor based on BERT fused with cosine similarity category is shown in Fig. 3. The text content of the review and the review average cosine similarity category number are put together using [SEP] and then feed into the BERT model together, with the review text as the first sentence and the review average cosine similarity category as the second sentence. Tokenize the review text with

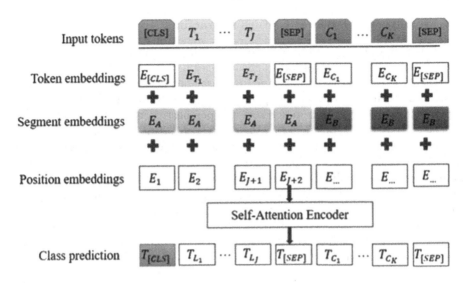

Fig. 3. Fusing cosine similarity category into BERT.

T_i, and T_1 and T_J represent the first token and the last token, respectively. Also, mark the Token of review average cosine similarity number as C_l and C_1 and C_K represent the first and last token respectively. [CLS] and [SEP] are special tokens in BERT, with [CLS] marking the beginning of the sentence and [SEP] used to separate the Tokens of the two sentences. E_{T_1} to E_{T_J} denotes the vector of tokens of the review text after token embedding, the E_{C_1} to E_{C_K} denotes the vector of tokens of the average cosine similarity category of the review after token embedding, the $E_{[CLS]}$ and $E_{[SEP]}$ are the vectors of [CLS] and [SEP] after token embedding. E_A and E_B denote the vectors of the first and second sentences of the input after segment embedding respectively. E_1 to $E_{...}$ indicates the vector of all tokens aligned together after position embedding. T_{L_1} to T_{L_J} represents the vector of the three embedding vectors of tokens that represent the review text after the self-attentive mechanism. T_{C_1} to T_{C_K} represents the vector of the three embedding vectors of tokens that represent the average cosine similarity category of the review after the self-attentiveness mechanism.

It is noteworthy that despite the apparent brevity of the category number, it may actually be 1024 or a larger multiple of 2, so that the word corresponding to this number may not be available in BERT's dictionary. In this case, BERT's tokenizer cuts the word into smaller words, e.g., 1024 into 10 and 24. Therefore, we conducted other experiments to avoid cutting the word by converting the number of digits of a decimal category to the number of digits of the highest bit of its corresponding binary number, e.g., converting 1024 to 10. The former method will be referred to as CC/N and the latter method will be referred to as CC/L. After several rounds of training the final pooled representation of the last layer of hidden states of the BERT output is obtained as the obtained review text feature vector.

3.2 Reviewer Behavioral Feature Extractor

The behavioral features of the reviewer play an important role in detecting fake reviews, and the presence of local dependencies in these features can affect the algorithm's effectiveness. In this paper, it is assumed that many of the reviewer's behavioral features exhibit local dependencies due to the fact that fake reviewers attempt to imitate real reviewers. From this perspective, the behavioral features of fake reviewers are considered to be locally dependent. Previous research [2] has shown that incorporating feature interactions can improve the accuracy of fake review detection. Based on these considerations, it is assumed that behavioral features are locally dependent. Given the assumption of local dependencies in reviewer behavioral features, this paper utilizes CNN (Convolutional Neural Network) for feature extraction. CNN operations can explicitly encode these relationships into the extracted features while preserving local dependencies. The convolutional kernel slides over the feature columns, capturing local patterns as it moves, which reflect the local dependencies present in the data.

For the selection of behavioral features, we follow Zhang.'s study and choose Rating score (RS), Rating deviation (RD), Rating entropy (RE), Review length (RL), Review gap (RG), Review count (RC), Review time difference (RTD), and User Tenure (UT) as the behavioral features [24]. These features are arranged sequentially. Table 1 provides specific information about the selected behavioral features, along with previous researches that have used them. The reviewer's behavioral feature extractor consists of N one-dimensional convolutional filters, each containing multiple convolutional kernels of different sizes. This design aims to capture as much feature information as possible. Figure 4 illustrates the process of learning behavioral features. The behavioral features of the reviewer, associated with each review, are represented as a one-dimensional vector of length e. The filters are then applied to generate feature maps, with each convolutional kernel producing a corresponding feature map. Denote the weight matrix of the filter by $Wrb \in R(m \times 1)$ where m denotes the size of the convolution kernel $m \in [1,6]$. Denotes the reviewer's behavior data vector by $X_{rb} \in R^{e \times 1}$, where e denotes the length of the reviewer's behavior data vector. For the j-th convolution kernel of the i-th filter, the feature map of this convolution kernel is computed using the following convolution operator:

$$m_{i,j}^{rb} = f\left(W_{rb}^{i,j} \cdot X_{rb} + b_{rb}\right), i \in 1, 2, \ldots, N, j \in [1, 6] \tag{2}$$

where f denotes the RELU function, which is chosen as the activation function because it works better, and because the size of the convolution kernel ranges from 1 to 6, the value of j is also in the range 1 to 6.

$$M_i^{rb} = \left[\max\left(m_{i,1}^{rb}\right), \max\left(m_{i,2}^{rb}\right), \ldots \max\left(m_{i,6}^{rb}\right)\right] \tag{3}$$

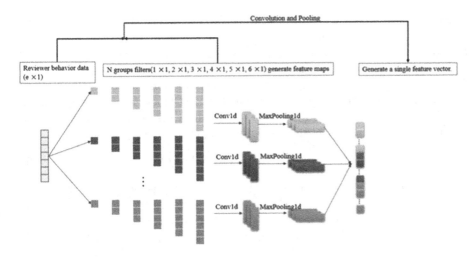

Fig. 4. Reviewer behavior feature extractor.

For the ith filter, in order to reduce the size of the features and prevent overfitting, the feature maps generated by each convolution kernel are subjected to one-dimensional maximum pooling, and finally the pooled feature maps are stitched together as one filter to obtain the feature maps.

$$M^{rb} = \left[M_1^{rb}, M_2^{rb}, \ldots M_N^{rb}\right] \tag{4}$$

Finally, the feature maps obtained from the N filters are put together to obtain a vector of behavioral features of the reviewer.

Table 1. Reviewer behavior features.

Feature name	Data type	Range	Used in the Prior Studies
Rating score (RS)	Continuous	[1, 5]	Zhang W et al. [24]
Rating deviation (RD)	Continuous	[0, 2.94]	Jindal and Liu [28]
Rating entropy (RE)	Continuous	[0, 1.92]	Ye and Akoglu et al. [27]
Review length (RL)	Continuous	[6, 3687]	Zhang D et al. [21]
Review gap (RG)	Continuous	[0, 2968]	Fei et al. [25]
Review count (RC)	Continuous	[2, 142]	Zhang L et al. [26]
Review time difference (RTD)	Continuous	[0, 3231]	Zhang W et al. [24]
User Tenure (UT)	Continuous	[0,3278]	Ma M et al. [29]

3.3   Fake Review Detection

The final review feature vector is composed of two components: the review text feature vector produced by the BERT model and the behavioral feature vector generated by the reviewer's behavioral feature extractor. The classifier used for fake review detection comprises a fully-connected layer that takes the review feature vector as input and classifies the review as either true or fake. In the training process, the cross-entropy function is commonly employed as the loss function for multi-classification problems. This function quantifies the discrepancy between the predicted outcome and the true label, serving as a training signal for the model. Its objective is to minimize the difference between the predicted and actual labels, thereby improving the model's performance.

$$H(P,Q) = -\sum (X)P(X) \times \log(Q(X)) \tag{5}$$

Cross-entropy is defined as shown in Eq. 5, where H denotes cross-entropy, P and Q represent the probability distribution representing the true label and the probability distribution of the model output, respectively, and X denotes the category $P(X)$ and $Q(X)$ are the probabilities of category X in distributions P and Q, respectively. The cross-entropy function was used as the loss function to optimize our model throughout the training and testing process.

4   Experiments

4.1   Dataset and Pre-processing

We utilize Yelp dataset obtained from http://Yelp.com, which comprises a total of 608,458 reviews. Among these reviews, 528,019 are categorized as genuine reviews, while 80,439 are categorized as fake reviews. Each review entry includes the ID of the reviewer, the ID of the restaurant, the review date, the textual content of the review, the star rating assigned, and a label indicating whether the review is genuine or fake.

Due to the imbalanced nature of the samples within Yelp dataset, where the number of genuine reviews significantly outweighs the number of fake reviews, we employ under sampling techniques to create a more balanced dataset. This involves selecting a subset of genuine reviews and an equivalent number of fake reviews. Additionally, reviews from inactive reviewers are excluded by us to enhance the effectiveness of the reviewer's behavioral information. The construction process of the dataset is as follows: 1) Select the fake review data from the dataset and remove reviews posted by reviewers who have fewer than 3 reviews associated with them. 2) Choose genuine review data from the dataset and randomly select reviews posted by reviewers who have at least 2 reviews until the difference between the number of selected genuine reviews and selected fake reviews is less than 4. 3) We calculate the behavioral and textual features for each reviewer and incorporate them into the dataset. The final dataset consists of 30,942 reviews, with 15,469 labeled as genuine and 15,473 labeled as fake.

4.2   Evaluation Metrics

To evaluate our approach, the performance of models using the CC/N method and the CC/L method on the dataset was tested by us, comparing them with other pre-trained baseline models. In this paper, five metrics are used by us to measure the performance of the models, namely precision (P), recall (R), F1 score, area under the curve (AUC) and accuracy (ACC). Where the metrics P, R, $F1$, ACC are defined in the following equations.

$$P = \frac{TP}{TP + FP} \tag{6}$$

$$R = \frac{TP}{TP + FN} \tag{7}$$

$$F1 = \frac{2P \bullet R}{P + R} \tag{8}$$

$$ACC = \frac{TP + TN}{TP + FP + TN + FP} \tag{9}$$

where TP denotes fake review that were correctly predicted, TN denotes true review that were correctly predicted, FP denotes fake review that were incorrectly predicted, and FN denotes true review that were incorrectly predicted.

4.3   Experimental Settings

For the baseline model of BERT, we use the pre-trained uncased BERT-base [30], which contains 12 Transformer blocks, 12 attention heads, and a hidden layer of size 768. The BertAdam [10] optimizer was proposed by a team of researchers at Google in the original BERT paper proposed, the BertAdam optimizer combines the specific needs of the Adam optimizer and the BERT model, using techniques such as adaptive learning rate adjustment and weight decay to improve the training of the model. BertAdam was therefore chosen as the optimizer for the model and the optimizer's warm-up ratio is set to 0.1 as recommended by the Google research team.

The selection of the learning rate has a significant impact on the model training process. Therefore, this paper conducted additional experiments to determine the most suitable learning rate. Several commonly used learning rates, namely 1e−4, 1e−5, 5e−5, 6e−5, and 5e−6, were compared as the learning rate for the optimizer. When the learning rate is too large, the learning curve becomes steep, making it difficult to reach the optimal solution. On the other hand, a learning rate that is too small slows down convergence and may result in getting trapped in local optimal solutions. It is validated with test that a learning rate of 5e−5 is selected as the optimal choice for the optimizer, striking a balance between convergence speed and avoiding local optimal solutions.

In the training of the model, the model with the highest AUC value on the validation set was selected by us to save, and training was stopped early if the AUC value did not improve 50 times in a row. The training was done on a 3090 graphics card and the length of the Token input to BERT was set to a maximum of 512.

4.4   Experimental Results

Table 2 presents the performance of our model and the baseline models on the dataset. The selected baseline models include well-known pre-trained models, aiming to showcase the performances of models that solely utilize textual reviews for fake review detection. Five baseline pre-trained models were chosen, namely BERT [10], ALBERT [13], ROBERTA [14], DISTILBERT [15], and ERNIE [16]. The BERT model achieved the best result, which can be attributed to the appropriate choice of learning rate and optimizer. Our model, using BERT as its base, also adopted the recommended learning rate and optimizer. Similarly, the same learning rate was applied to the other baseline models tested.

Table 2. Comparing the performance of our model and baseline model on the YELP dataset.

Algorithm for BF	Algorithm for Text	Acc	Precision	Recall	F1-score	AUC
None	BERT	68.44%	0.6850	0.6843	0.6841	0.6843
None	ALBERT	67.03%	0.6703	0.6703	0.6703	0.6703
None	ROBERTA	67.44%	0.6746	0.6743	0.6742	0.6743
None	DISTILBERT	68.02%	0.6808	0.6801	0.6798	0.6801
None	ERNIE	67.44%	0.6743	0.6743	0.6743	0.6743
Conv1d	BERT+CC/N	80.50%	0.8061	0.8049	0.8047	0.8049
Conv1d	BERT +CC/L	**80.92%**	**0.8111**	**0.8091**	**0.8088**	**0.8091**

Notes: BERT, Bidirectional Encoder Representations from Transformers; DISTIL-BERT, A Distilled Version of BERT; ALBERT, A Light Version of BERT; ROBERTA, A robustly optimized Bert pre-trained approach; ERNIE, A Continual Pre-training Framework for Language Understanding.

Our model which employing the CC/L method demonstrated substantial improvements in various metrics compared to the baseline BERT model. Specifically, our model exhibited a 12.48% increase in accuracy (from 68.44% to 80.92%), 12.61% increase in precision (from 0.6850 to 0.8111), 12.48% increase in recall (from 0.6843 to 0.8091), 12.48% increase in F1 score (from 0.6843 to 0.8091), and 12.48% increase in AUC (from 0.6843 to 0.8091). These improvements highlight the enhanced classification performance of our model, with each component contributing to its effectiveness. Further details regarding the individual contributions of each component will be discussed in the subsequent section on ablation experiments.

The model which employing CC/N method, on the other hand, achieved an accuracy of 80.50%, precision of 0.8061, recall of 0.8049, F1 score of 0.8047, and AUC score of 0.8049. Comparatively, the CC/L method achieved superior performance with an accuracy of 80.92%, precision of 0.8111, recall of 0.8091, F1 score of 0.8088, and AUC score of 0.8091. These results indicate that the CC/L method is superior to the CC/N method, potentially due to the behavior of BERT's Tokenizer. Since the Tokenizer divides Tokens not present in the word list into smaller parts and looks them up in the word list, the CC/N method may generate multiple Tokens for the conversion of cosine similarity category numbers, potentially affecting the classification performance.

In this paper, multiple metrics were employed to evaluate the model's performance, including accuracy, precision, recall, F1 score, and AUC score. The marginal differences observed in these metrics across the models can be attributed to the balanced nature of our dataset, which almost consists of an equal number of positive and negative examples.

4.5   Dimensionality Reduction Analysis of Text Feature Vectors

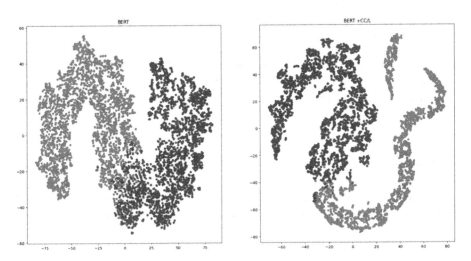

Fig. 5. Visualizations of the vectors of both BERT and the fused cosine similarity to BERT on the test set. (Color figure online)

Figure 5 shows the results of dimensionality reduction analysis on text feature vectors. The figure on the left shows the visualization of text feature vectors output by BERT using t-SNE in two-dimensional space. The image on the right is the result of using t-SNE to visualize the text feature vectors output by BERT using the CC/L method to fuse cosine similarity in two-dimensional space. Each data point represents an example, where blue color represents negative examples and red color represents positive examples. Compared to BERT, the vectors obtained from the CC/L method demonstrate better separation in the vector space. This validates our hypothesis that by incorporating cosine similarity into BERT, the model learns better representations.

4.6   Ablation Study

In order to assess the impact of each component on the overall effectiveness of our model, we conducted experiments on different variants of the model and compared their performance in terms of accuracy, precision, recall, F1 score, and AUC. Table 3 summarizes the key findings from these experiments. Firstly, our model utilizing the CC/L method achieved the best results, indicating the superiority of our approach. The inclusion of behavioral features proves to be particularly valuable in enhancing the model's effectiveness. Secondly, model 2 outperformed model 1, suggesting that the contribution of behavioral features of the reviewer is more significant than the average cosine similarity category. This finding reinforces the notion that behavioral features play a crucial role in fake review detection, supporting our previous hypothesis regarding the local dependency of these features.

Table 3. Performance of our model vs. its variations on fake review detection.

No	Algorithm for BF	Algorithm for Text	Acc	Precision	Recall	F1-score	AUC
1	None	BERT+CC/L	74.81%	0.7517	0.7480	0.7471	0.7480
2	Conv1d	BERT	76.71%	0.7671	0.7692	0.7671	0.7666
3	Conv1d	BERT+CC/L	**80.92%**	**0.8111**	**0.8091**	**0.8088**	**0.8091**

In summary, the results from Table 3 affirm the effectiveness of our model, with the model employing the CC/L method yielding the best performance. The inclusion of behavioral features, which exhibit local dependencies, proves to be a valuable contribution in improving the model's performance in fake review detection.

5   Conclusions

In this paper, we propose a novel end-to-end model for fake review detection, which consists of a text feature extractor fusing review cosine similarity category to BERT and a reviewer's behavioral feature extractor. The experiments on the YELP dataset demonstrate the superiority of our model. The main contribution of this paper is to propose a method that can better exploit the potential of BERT to enhance the learning ability of BERT in fake review detection for review text features with almost no increase in computational overhead, and to propose a deep learning model that combines pre-trained models with behavioral features. In the future, we will explore the effectiveness of combining other features with BERT and other pre-trained models, and hope to better exploit the potential of pre-trained models such as BERT.

Acknowledgement. The work is supported by National Key Research and Development Program Project "Research on data-driven comprehensive quality accurate service technology for small medium and micro enterprises" (Grant No.

2019YFB1405303), The Project of Cultivation for Young Top-motch Talents of Beijing Municipal Institutions "Research on the comprehensive quality intelligent service and optimized technology for small medium and micro enterprises" (Grant No. BPHR202203233), National Natural Science Foundation of China "Research on the influence and governance strategy of online review manipulation with the perspective of E-commerce ecosystem" (Grant No. 72174018)

References

1. Zhu, F., Zhang, X.: Impact of online consumer reviews on sales: the moderating role of product and consumer characteristics. J. Mark. **74**(2), 133–148 (2010)
2. Murphy, R.: Local Consumer Review Survey 2023 (2023). https://www.brightlocal.com/research/local-consumer-review-survey/
3. Lozano, M.G., Brynielsson, J., Franke, U., et al.: Veracity assessment of online data. Decis. Support Syst. **129**, 113132 (2020)
4. Jindal, N., Liu, B.: Analyzing and detecting review spam. In: Seventh IEEE International Conference on Data Mining (ICDM 2007), pp. 547–552. IEEE (2007)
5. Li, F.H., Huang, M., Yang, Y., et al.: Learning to identify review spam. In: Twenty-Second International Joint Conference on Artificial Intelligence (2011)
6. Ren, Y., Zhang, Y.: Deceptive opinion spam detection using neural network. In: Proceedings of COLING 2016, The 26th International Conference on Computational Linguistics: Technical Papers, pp. 140–150 (2016)
7. Noekhah, S., Binti Salim, N., Zakaria, N.H.: Opinion spam detection: using multi-iterative graph-based model. Inf. Process. Manag. **57**(1), 102140 (2020)
8. Aghakhani, H., Machiry, A., Nilizadeh, S., et al.: Detecting deceptive reviews using generative adversarial networks. In: 2018 IEEE Security and Privacy Workshops (SPW), pp. 89–95. IEEE (2018)
9. Wang, C.C., Day, M.Y., Chen, C.C., et al.: Detecting spamming reviews using long short-term memory recurrent neural network framework. In: Proceedings of the 2nd International Conference on E-Commerce, E-Business and E-Government, pp. 16–20 (2018)
10. Devlin, J., Chang, M.W., Lee, K., et al.: BERT: pre-training of deep bidirectional transformers for language understanding. arXiv preprint arXiv:1810.04805 (2018)
11. Mohawesh, R., Xu, S., Tran, S.N., et al.: Fake reviews detection: a survey. IEEE Access **9**, 65771–65802 (2021)
12. Liu, W., Jing, W., Li, Y.: Incorporating feature representation into BiLSTM for deceptive review detection. Computing **102**, 701–715 (2020). https://doi.org/10.1007/s00607-019-00763-y
13. Lan, Z., Chen, M., Goodman, S., et al.: ALBERT: a lite BERT for self-supervised learning of language representations. arXiv preprint arXiv:1909.11942 (2019)
14. Liu, Y., Ott, M., Goyal, N., et al.: RoBERTa: a robustly optimized BERT pre-training approach. arXiv preprint arXiv:1907.11692 (2019)
15. Sanh, V., Debut, L., Chaumond, J., et al.: DistilBERT, a distilled version of BERT: smaller, faster, cheaper and lighter. arXiv preprint arXiv:1910.01108 (2019)
16. Zhang, Z., Han, X., Liu, Z., et al.: ERNIE: enhanced language representation with informative entities. arXiv preprint arXiv:1905.07129 (2019)
17. Manaskasemsak, B., Tantisuwankul, J., Rungsawang, A.: Fake review and reviewer detection through behavioral graph partitioning integrating deep neural network. Neural Comput. Appl. **35**, 1169–1182 (2023). https://doi.org/10.1007/s00521-021-05948-1

18. Javed, M.S., Majeed, H., Mujtaba, H., et al.: Fake reviews classification using deep learning ensemble of shallow convolutions. J. Comput. Soc. Sci. **4**, 883–902 (2021). https://doi.org/10.1007/s42001-021-00114-y
19. Duma, R.A., Niu, Z., Nyamawe, A.S., et al.: A Deep Hybrid Model for fake review detection by jointly leveraging review text, overall ratings, and aspect ratings. Soft. Comput. **27**(10), 6281–6296 (2023). https://doi.org/10.1007/s00500-023-07897-4
20. Sahut, J.M., Hajek, P.: Mining behavioural and sentiment-dependent linguistic patterns from restaurant reviews for fake review detection (2022)
21. Zhang, D., Zhou, L., Kehoe, J.L., et al.: What online reviewer behaviors really matter? Effects of verbal and nonverbal behaviors on detection of fake online reviews. J. Manag. Inf. Syst. **33**(2), 456–481 (2016)
22. De Caigny, A., Coussement, K., De Bock, K.W.: A new hybrid classification algorithm for customer churn prediction based on logistic regression and decision trees. Eur. J. Oper. Res. **269**(2), 760–772 (2018)
23. Quinlan, J.R., Rivest, R.L.: Inferring decision trees using the minimum description length principle. Inf. Comput. **80**(3), 227–248 (1989)
24. Zhang, W., Xie, R., Wang, Q., et al.: A novel approach for fraudulent reviewer detection based on weighted topic modelling and nearest neighbors with asymmetric Kullback-Leibler divergence. Decis. Support Syst. **157**, 113765 (2022)
25. Fei, G., Mukherjee, A., Liu, B., et al.: Exploiting burstiness in reviews for review spammer detection. In: Proceedings of the International AAAI Conference on Web and Social Media, vol. 7, no. 1, pp. 175–184 (2013)
26. Mukherjee, A., Kumar, A., Liu, B., et al.: Spotting opinion spammers using behavioral footprints. In: Proceedings of the 19th ACM SIGKDD International Conference on Knowledge Discovery and Data Mining, pp. 632–640 (2013)
27. Ye, J., Akoglu, L.: Discovering opinion spammer groups by network footprints. In: Appice, A., Rodrigues, P.P., Santos Costa, V., Soares, C., Gama, J., Jorge, A. (eds.) ECML PKDD 2015. LNCS (LNAI), vol. 9284, pp. 267–282. Springer, Cham (2015). https://doi.org/10.1007/978-3-319-23528-8_17
28. Jindal, N., Liu, B.: Opinion spam and analysis. In: Proceedings of the 2008 International Conference on Web Search and Data Mining, pp. 219–230 (2008)
29. Ma, M., Agarwal, R.: Through a glass darkly: information technology design, identity verification, and knowledge contribution in online communities. Inf. Syst. Res. **18**(1), 42–67 (2007)
30. Wolf, T., Debut, L., Sanh, V., et al.: Transformers: state-of-the-art natural language processing. In: Proceedings of the 2020 Conference on Empirical Methods in Natural Language Processing: System Demonstrations, pp. 38–45 (2020)

LinkEE: Linked List Based Event Extraction with Attention Mechanism for Overlapping Events

Xingyu Chen and Weiwen Zhang[✉]

School of Computer Science and Technology, Guangdong University of Technology, Guangzhou, China
chenxingyu@mail2.gdut.edu.cn, zhangww@gdut.edu.cn

Abstract. Event extraction (EE) is an essential task in natural language processing (NLP), which aims to extract information about the events contained in a sentence. Most of the existing works are not appropriate for Chinese overlapping event extraction, where a Chinese overlapping event sentence contains multiple event types, trigger words and arguments. In this paper, we propose LinkEE, an attention-based linked list model for event extraction, which organizes event types, trigger words and arguments as the nodes in the linked list, and leverages conditional fusion functions as edges to pass specific information. Bi-directional feature information of the input is learned through Bi-LSTM, while contextual information is captured by Hydra attention to enhance the text vector representation. Experimental results on the public dataset FewFC illustrate that the proposed model LinkEE is superior to previous methods.

Keywords: Chinese overlapping event · Bi-LSTM · Hydra attention · Linked list

1 Introduction

Event extraction (EE) is the main task of information extraction (IE) and also the most challenging problem in natural language processing (NLP). Event extraction is widely used in areas such as automated question and answer [24], abstract generation [9]), information retrieval [10] and building knowledge bases for trend analysis [6,13,19]. Event extraction focuses on extracting valuable event information from various textual information and presenting it in a structured format. Specifically, given a sentence, the task is to extract the event types, trigger words and arguments in that sentence.

One of the challenges of event extraction is that sentences can have overlapping events. Overlapping event sentences [20] contain information about multiple events. As illustrated in Fig. 1, the overlapping event sentences include two events: (1) "Investment" event with the trigger word "buying" and the arguments "Sub", "Obj", "Money" and "Date"; (2) "Transfer of shares" event with the trigger word "selling" and the arguments "Sub-Org", "Target-Company", "Money" and "Date". However, most of the existing methods of event extraction (EE) do not consider such overlapping events [7,15,17]. Traditional joint methods result in missing some information on overlapping event extraction due to the fact that they can only extract a single event; although pipeline methods have the potential to extract overlapping events, the accuracy is degraded by error propagation. Further investigation should be performed in addressing the issue of Chinese overlapping events in event extraction.

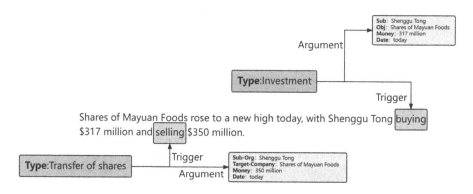

Fig. 1. An example of event extraction that includes two event types, two event trigger words and multi-event arguments.

In this paper, we introduce LinkEE, a linked list event extraction model based on an attention mechanism for overlapping events. Specifically, the LinkEE model consists of four modules: a BERT encoder, an event type decoder, a trigger word decoder and an event argument decoder. We analogise the extraction of overlapping events to the completing of linked lists. For each event, we firstly detect its event type by the event type decoder and feed it as the starting node of the linked list. The textual information encoded by BERT is fused with the event type information through conditional layer normalization (CLN) [23] and passed to the trigger word decoder. It is used to predict the trigger word node under a specific event type. Furthermore, the input text information by the trigger word decoder is fused with the trigger word information through CLN and passed to the event arguments parameter decoder. It is used to predict the event arguments node under a specific event type and a specific trigger word. Bi-directional feature information of the input is learned through Bi-LSTM, while contextual information is captured using Hydra attention to enhance the text vector representation. We evaluate the effectiveness of the proposed model on the Chinese financial event extraction dataset FewFC [28], with approximately 35.2% of overlapping events within the entire dataset.

The contributions of this paper are summarized as follows:

1. We propose LinkEE, which addresses the issue of overlapping events using the structure of linked list. The event type, event trigger word, and event argument parameters are used as the nodes of the linked list, and CLN are used as the edges of the linked list to pass the information of the nodes.
2. We adopt Bi-LSTM to process the sequence information of Chinese overlapping event sentences and utilize Hydra attention to capture more valuable contextual information to enhance the representation of text vectors.
3. Experimental results on the Chinese financial event dataset FewFC show that LinkEE outperforms existing methods in Chinese overlapping event scenarios.

2   Related Work

Current studies of event extraction can be roughly divided into two categories: pipelined methods [3,4,8,11,14,25,26] and traditional joint methods [12,15,17,18,21]. For one thing, the pipelined methods usually perform extraction of trigger words and arguments as subtasks at different stages separately. Although those methods have some advantages in resolving overlapping events, errors can be generated by the upstream subtasks, which directly affects the output of the downstream subtasks. This is known as error propagation, which decreases the accuracy of event extraction. For another, traditional joint methods jointly extracts trigger words and arguments. They tag sentences to solve the event extraction task by means of sequence tagging. Although the error propagation problem can be avoided, the traditional joint methods can only extract one of the tags from overlapping events. As there may be multiple tags in the sentence, it will lead to incomplete extraction of event information. Therefore, both the pipeline methods and traditional joint methods cannot solve overlapping event scenarios well.

A few works consider overlapping events, but the results are not satisfactory. For example, Yang et al. [26] uses the BERT encoder to solve the problem of overlapping event arguments by extracting the arguments based on trigger word. Chen et al. [3] proposed a dynamic multi-pool convolutional neural network (DMCNN) that captures key information using dynamic multi-pool layers based on event trigger words and arguments to achieve multiple event extraction tasks. Sha et al. [21] proposed a Dependency Bridge Recurrent Neural Network (DBRNN) to extract events by using tree structure and sequence structure and enhancing the model by using dependency bridges to carry semantic information. Zeng et al. [27] proposed a CNN+Bi-LSTM model for Chinese event extraction, using trigger word recognition as a sequence annotation problem. Nevertheless, all of those studies cannot extract overlapping events effectively. Hence, the task of event extraction for overlapping events is particularly important. Du [7] introduced a conditional random field (CRF) feature function to learn sequence features and achieve event extraction through constraints. All these methods try to solve the Chinese overlapping event problem in their own way, but the result is still not very satisfactory. Therefore, the Chinese overlapping event extraction problem needs to be further investigated.

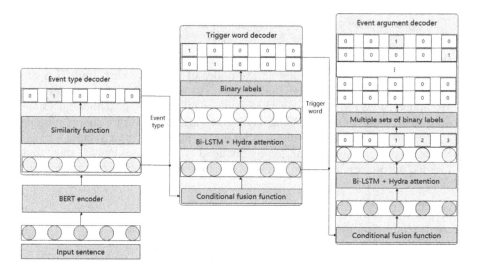

Fig. 2. An overview of our proposed method LinkEE, which consists of a BERT encoder, an event type decoder, a trigger word decoder and an event arguments decoder.

3 Our Approach

The proposed extraction model of overlapping events consists of four stages, including a BERT encoder, event type decoder, trigger word decoder and event arguments decoder. Figure 2 illustrates the structure of the LinkEE model. It adopts the structure of a linked list, with the three nodes of the linked list as event type, trigger word and event arguments, and CLN as an edge to pass specific information. The textual representation is also enhanced by capturing contextual information through Bi-LSTM and Hydra attention.

3.1 BERT Encoder

We leverage Devlin's BERT pre-training model [5], where BERT is a bidirectional Transformer architecture language representation model. It is based on a Mask mechanism to allow the BERT encoder to learn and train itself, retaining semantic information as much as possible in the sentence to output a vector representation of the sentence. We represent N sentences as $x = \{x_1, x_2, \ldots, x_N\}$. We feed the token into BERT and obtain the hidden state $H = \{h_1, h_2, \ldots, h_N\}$ for the input representation of the subtasks.

3.2 Event Type Decoder

An event type detection decoder is designed to identify event types. Motivated by non-trigger event detection [16], we likewise introduce an attention mechanism to capture contextual information for forecasting event types. Specifically, we use a random initialization matrix $t_{type} \in T$ as the event type representation. The similarity between the candidate event type T and the hidden state $h \in H$ of BERT's output is learned by defining a similarity function δ. The representation $s_{t_{type}}$ of a sentence S based on the type of event can be obtained as follows:

$$\delta\left(t_{type}, h_i\right) = V^T \tanh\left(W\left[t_{type}; h_i; \left|t_{type} - h_i\right|; t_{type} \odot h_i\right]\right) \tag{1}$$

$$s_{t_{type}} = \sum_{i=1}^{N} \frac{e^{\delta(t_{type}, h_i)}}{\sum_{j=1}^{N} e^{\delta(t_{type}, h_j)}} h_i \tag{2}$$

where W and V are learnable parameters, $|.|$ is the absolute value operator, $\odot$ is the corresponding element product operator, and $[.;.]$ denotes the concatenation of vectors.

The prediction of event types is based on the similarity between the text sentence $s_{t_{type}}$ and the event type information t_{type}, which is realized by the similarity function δ. We can then obtain the probability of each event type t_{type} appearing in the sentence as follows

$$\hat{t}_{type} = p(t_{type}|s) = \sigma(\delta(t_{type}, s_{t_{type}})) \tag{3}$$

where σ denotes the sigmoid function.

We choose the event type with $\hat{t}_{type} > \varepsilon_1$, where $\varepsilon_1 \in [0, 1]$ is a threshold value. Each event type forecasted in the sentence s is used as the starting node of the linked list.

3.3   Trigger Word Decoder

As a sentence may have multiple event types, we use the event type passed by the previous subtask as a condition to extract the trigger words under a specific event type in order to identify trigger words for different event types. The trigger word decoder contains a conditional fusion function module, a Bi-LSTM and Hydra attention module and a binary tagging module for trigger word. We refer to CLN, with the event type t_{type} passed as a condition, integrated into the hidden state H, to obtain the conditional fusion token $C^{t_{type}}$ as follows:

$$c_i = CLN(t_{type}, h_i) \tag{4}$$

$$CLN(t_{type}, h_i) = \gamma_t \odot \left(\frac{h_i - \mu}{\sigma}\right) + \beta_t \tag{5}$$

$$\gamma_t = W_\gamma t_{type} + b_\gamma \tag{6}$$

$$\beta_t = W_\beta t_{type} + b_\beta \tag{7}$$

where CLN is based on layer normalization [1], which dynamically generates gain γ_t and bias β_t based on the conditional information passed from the previous subtask. μ and σ are the mean and variance of h_i, and $\odot$ is the corresponding element product operator.

Most previous studies have used trigger word identification as a separate classification problem, however this is not sufficient for overlapping trigger word extraction. To address the extraction of overlapping trigger words, therefore, we design a Bi-LSTM and a Hydra attention module to enhance the conditional token representation. The Bi-LSTM is formulated as follows:

$$H_{LSTM} = [\overrightarrow{(LSTM)}(C^{t_{type}}) + \overleftarrow{(LSTM)}(C^{t_{type}})] \tag{8}$$

where $C^{t_{type}}$ is the representation vector consisting of c_i. H_{LSTM} is the hidden state of the Bi-LSTM output.

In particular, we use multi-head attention to improve the accuracy of the output vector representation of the Bi-LSTM module. We introduce Hydra attention [2], which maintains the same amount of computation by using the same number of attention heads as features. By changing the order of computation in the linear calculation, the computational cost is reduced while the accuracy is improved. To change the order of computation without changing the result, we use the multiplicative attention mechanism:

$$A(Q_k, K_k, V_k) = softmax(\frac{Q_k K_k^T}{\sqrt{D}})V_k \quad \forall k \in \{1, \dots, N\} \tag{9}$$

$$A(Q_k, K_k, V_k; \phi) = \phi(Q_k)(\phi(K_k)^T V_k) \quad \forall k \in \{1, \dots, N\} \tag{10}$$

where k denotes the kth attention head. ϕ denotes the linearly decomposable function by which the order of computation is changed so as to accomplish the computation of the multi-headed attention mechanism. We then can obtain $A^{t_{type}}$:

$$A^{t_{type}} = HydraAtt(H_{LSTM}) \tag{11}$$

where H_{LSTM} is the output of the Bi-LSTM.

To forecast trigger word, we use a pair of binary tags. For each token $A^{t_{type}}$, we forecast if it relates to the beginning or ending position of the trigger word t, which is given as follows:

$$\hat{t}_i^{st_{type}} = p(t_s \mid w_i, t_{type}) = \sigma(W_{t_s}^T A_i^{t_{type}} + b_{t_s}) \tag{12}$$

$$\hat{t}_i^{et_{type}} = p(t_e \mid w_i, t_{type}) = \sigma(W_{t_e}^T A_i^{t_{type}} + b_{t_e}) \tag{13}$$

where σ denotes the sigmoid function and $A_i^{t_{type}}$ denotes the ith token representation of the $A^{t_{type}}$ vector. We pick the token $\hat{t}_i^{st_{type}} > \varepsilon_2$ as the start position and $\hat{t}_i^{et_{type}} > \varepsilon_3$ as the end position, where ε_2 , $\varepsilon_3 \in [0, 1]$ are scalar thresholds.

To extract the event trigger word t, we consider the beginning position and the nearest ending position, forming a complete trigger word token between them. If two start position tokens appear before the end token appears, we choose the one with higher probability as the start position. Thus, the model extracts its associated trigger word node from the event type node information passed by the CLN.

3.4   Event Arguments Decoder

Since a sentence might contain various event types, as well as trigger word under a particular event type, we extract the arguments for a particular event type conditional on the event type and the event trigger words under the particular event type. We design the event arguments decoder with a conditional fusion function module, a Bi-LSTM and Hydra attention module, and a multi-group binary tagging module for trigger words.

In addition, unlike trigger word decoders, arguments depend on specific event types, i.e., different event types require different arguments to be extracted. Hence, we use the parameter supplement template to extract arguments by enumerating the list of arguments under a particular event type, as shown in Table 1.

Table 1. Argument completion templates, depending on the type of event passed by the subtask, i.e., completing its arguments by a specific event type. The table shows which thesis parameters are to be completed for "investment", "prosecution", "winning", etc.

Type	Arg1	Arg2	Arg3	Arg4	Arg5
Investment	Money	Date	Sub	Obj	
Prosecution	Obj-per	Date	Sub-per	Obj-org	Sub-org
...	...				
Bid winner	Amount	Date	Sub	Obj	

In order to make the template deductive, we parameterise the template so that it can be learned together with the model parameters. Specifically, given the event type t_{type}, we establish the dependencies $I(r, t_{type})$ between the event type and the event argument parameters:

$$I(r, t_{type}) = \sigma(w_r^T t_{type} + b_r) \tag{14}$$

where σ is the sigmoid function and w_r, b_r are the variables associated with the event arguments.

We apply conditional fusion via CLN and incorporate the trigger word information into the conditional text c_i in Eq. (4). The trigger word vector representation t is taken as the mean of the starting and ending positions representation. The same Bi-LSTM and Hydra attention are used to enhance the conditional text representation $C^{t_{type}t}$. To mark the trigger word position, we use the relative distance representation P, which is used to represent the relative distance between the present tag and the trigger word. Finally, the tag representation $A^{t_{type}t}$ used for the extraction of the event arguments is obtained as:

$$A^{t_{type}t} = \left[HydraAtt\left(C^{t_{type}t}\right); P\right] \tag{15}$$

In order to forecast the arguments, we use multiple sets of binary labels. For each token $A_i^{t_{type}t}$, we forecast if it relates to the beginning or ending position of the arguments r as follows:

$$\hat{r}_i^{st_{type}t} = p(a_r^s \mid w_i, t_{type}, t) = I(r, t_{type})\sigma(w_{r_s}^T A_i^{t_{type}t} + b_{r_s}) \quad (16)$$

$$\hat{r}_i^{et_{type}t} = p(a_r^e \mid w_i, t_{type}, t) = I(r, t_{type})\sigma(w_{r_e}^T A_i^{t_{type}t} + b_{r_e}) \quad (17)$$

where σ denotes the sigmoid function, $A_i^{t_{type}t}$ denotes the ith token representation of the $A^{t_{type}}$ vector, a_r^s denotes the starting position of the argument and a_r^e denotes the ending position of the argument. We pick the token $\hat{r}_i^{st_{type}t} > \varepsilon_4$ as the start position and $\hat{r}_i^{et_{type}t} > \varepsilon_5$ as the end position, where ε_4 , $\varepsilon_5 \in [0,1]$ are scalar thresholds.

To extract all event arguments r, as in the previous subtask, we consider the beginning position and the nearest ending position, forming a complete event arguments between them. Thus, the model extracts its associated event arguments nodes from the event type nodes and trigger word node information passed by the CLN.

4   Experiments

4.1   Dataset and Evaluation Metric

Similar to the work [22], we conduct experiments on the Chinese financial event extraction dataset FewFC [28]. We train/validat/test the data in a ratio of 8:1:1. Details of the dataset are shown in Table 2.

Table 2. Statistics of the dataset. Each column indicates the number of sentences with overlapping elements (Overlap), sentences without overlapping elements (Normal), all sentences (Sentence) and all events (Event).

	Overlap	Normal	Sentence	Event
Training	1560	5625	7185	10277
Validation	205	694	899	1281
Testing	210	688	898	1332
All	1975	7007	8982	12890

To evaluate the performance of the LinkEE model, we consider four tasks by following previous work [3,8] as below:

1. **Trigger Identification (TI):** if the forecasted trigger word matches the anticipated trigger word, the trigger word is successfully identified.
2. **Trigger Classification (TC):** if the trigger word is successfully identified and allocated to the appropriate type, then the trigger word is successfully classified.

3. **Arguments Identification (AI):** if the event type is successfully identified and the forecast event arguments match the anticipated arguments, the arguments are successfully identified.
4. **Arguments Classification (AC):** if the forecasted event arguments are successfully identified and the forecast event arguments roles match with golden roles, the event arguments are successfully classified.

In this paper, we use Precision (P), Recall (R) and F-measure (F1) as evaluation metrics.

4.2   Implementation Details

We employ the Chinese BERT model as a text encoder, which has 12 layers, 768 hidden units and 12 attentions. Table 3 shows the hyperparameters of the LinkEE model, where the values are set based on the work [22].

Table 3. Hyperparameters of the model

Hyperparameters	Value
Type embedding dimension	768
Position embedding dimension	64
Dropout rate of decoders	0.3
Batch size	8
Training epoch	20
Initial learning rate of BERT	$2e^{-5}$
Learning rate of decoders	$1e^{-4}$
ε_1	0.5
ε_2	0.5
ε_3	0.5
ε_4	0.5
ε_5	0.5

4.3   Comparison Methods

We compare LinkEE with the previous work, where BERT-softmax and BERT-CRF are traditional joint methods and the rest are pipeline methods as follows:

1. **BERT-softmax** [5] uses a BERT pre-training model to encode the text, extracting event trigger word and event arguments.
2. **BERT-CRF** [8] constrains dependencies between tags via CRF based on the use of the BERT text encoder.
3. **PLMEE** [26] uses the BERT encoder to solve the problem of overlapping event arguments by extracting the arguments based on trigger word in a pipeline method.

4. **MQAEE** [11] sufficiently captures the dependencies between event types, trigger words and event arguments. We restructure **MQAEE** into three models to solve the extraction problem for overlapping events: 1) **MQAEE-1** first forecasts the types and then forecasts the overlapping trigger word and overlapping arguments based on the types. 2) **MQAEE-2** first forecasts overlapping trigger word with types and then forecasts overlapping arguments based on the types. 3) **MQAEE-3** sequentially forecasts types, overlapping trigger word based on the types, and overlapping arguments based on the types and trigger word.
5. **CasEE** [22] uses BERT encoding to obtain dependencies among event type, trigger word and arguments via CLN.

4.4   Comparative Results

The comparative results for all methods on FewFC are shown in Table 4. In the trigger word extraction task, compared to the traditional joint method BERT-CRF, LinkEE can achieve improvements of 3.4% and 6.5% in terms of the F1 scores of TI and TC, respectively. In addition, the recall metric of LinkEE is improved by a relative margin of around 10%-20%. As the traditional joint method uses a sequence annotation strategy for overlapping event extraction, it will suffer from label conflicts, i.e., multiple events in overlapping sentences generate multiple labels. However, that method can only forecast one of these labels, resulting in incomplete extraction information. Compared to the pipelined method CasEE, LinkEE is 0.3% higher in F1 scores for TI and 0.2% lower in F1 scores for TC, slightly outperforming CasEE overall.

Table 4. Event extraction results for the FewFC dataset.

	TI(%)			TC(%)			AI(%)			AC(%)		
	P	R	F1	P	R	F1	P	R	F1	P	R	F1
BERT-softmax	89.0	79.0	84.0	80.2	61.8	69.8	74.6	62.8	68.2	72.5	60.2	65.8
BERT-CRF	**90.8**	80.8	85.5	**81.7**	63.6	71.5	**75.1**	64.3	69.3	**72.9**	61.8	66.9
PLMEE	83.7	85.8	84.7	75.6	74.5	75.1	74.3	67.3	70.6	72.5	65.5	68.8
MQAEE-1	90.1	85.5	87.7	77.3	76.0	76.6	62.9	71.5	66.9	51.7	70.4	59.6
MQAEE-2	89.1	85.5	87.4	79.7	76.1	77.8	70.3	68.3	69.3	68.2	66.5	67.3
MQAEE-3	88.3	86.1	87.2	75.8	76.5	76.2	69.0	67.9	68.5	67.2	65.9	66.5
CasEE	89.4	87.7	88.6	77.9	78.5	**78.2**	72.8	73.1	72.9	71.3	71.5	71.4
LinkEE	87.5	**90.3**	**88.9**	75.1	**81.2**	78.0	71.4	**75.8**	**73.5**	70.0	**74.2**	**72.1**

In the arguments extraction task, compared to the traditional joint method BERT-CRF, LinkEE can achieve improvements of 4.2% and 5.2% in terms of the F1 scores of AI and AC, respectively. For the recall metric, LinkEE also performs better. Compared to the pipeline method PLMEE, LinkEE can have improvements of 2.9% and 3.3% in terms of the F1 scores of AI and AC, respectively. Compared to MQAEE-2, LinkEE has the improvement of 4.2% and 4.8%. In addition, compared to CasEE, the improvement of LinkEE is 0.6% and 0.7%.

Besides, LinkEE obtains better results in the evaluation metric of recall. Since most of the sequence labeling methods can only predict a single label on overlapping event extraction, which leads to poor recall performance. CasEE alleviates label conflicts to a certain extent and leads to higher recall. However, since we add a variant of the attention mechanism that better captures contextual information, we can achieve better prediction of multiple labels, and thus the recall of LinkEE is significantly higher than that of CasEE.

This demonstrates that LinkEE can effectively capture textual feature information by Bi-LSTM and hydra attention. Therefore, LinkEE can enhance the efficiency of capturing important information, and improve extraction accuracy for overlapping event extraction.

4.5   Analysis on Overlap/Normal Data

To further measure the effectiveness of LinkEE in solving the overlap problem, we divided the test data into two groups: events with overlap and events without overlap.

Table 5. Results of events with overlap in testing. F1 scores are reported for each evaluation metric.

	TI(%)	TC(%)	AI(%)	AC(%)
BERT-softma	76.5	49.0	56.1	53.5
BERT-CRF	77.9	52.4	61.0	58.4
PLMEE	80.7	66.6	63.2	61.4
MQAEE-1	87.0	73.4	69.4	62.3
MQAEE-2	83.6	70.4	62.1	60.1
MQAEE-3	87.5	73.7	64.3	62.2
CasEE	89.0	74.9	71.5	70.3
LinkEE	**89.0**	**75.5**	**71.7**	**78.8**

Events With Overlap. As shown in Table 5, our method outperforms previous methods on overlapping events. Compared with sequence labeling methods, our method is able to predict multiple labels. Compared with CasEE, we add Bi-LSTM and Hydra attention to the subtask to capture the information of the context more effectively, which achieves the better results.

Events Without Overlap. As shown in Table 6, our model still performs acceptably without overlapping events. BERT-softmax gives better results on the arguments because it avoids the problem of error propagation. MQAEE-2 gives better results on the trigger words, probably because it is jointly predicting the trigger words and the types, resulting in a reduction of error propagation.

Table 6. Results of events without overlap in testing. F1 scores are reported for each evaluation metric.

	TI(%)	TC(%)	AI(%)	AC(%)
BERT-softma	86.9	79.9	**76.2**	**74.1**
BERT-CRF	88.4	80.8	74.9	72.8
PLMEE	86.4	79.7	75.7	74.0
MQAEE-1	88.0	78.5	65.1	57.7
MQAEE-2	**89.0**	**82.0**	74.2	72.3
MQAEE-3	87.1	77.6	71.3	69.6
CasEE	88.4	80.2	74.0	72.3
LinkEE	88.5	79.8	75.0	73.3

4.6   Discussion for Model Variants

In order to demonstrate the superiority of the model modules, different experiments are carried out on LinkEE.

Table 7 illustrates the capabilities of the decoder variants on the trigger word extraction. (1) Bi-LSTM replaced with LSTM: Due to the complexity of Chinese overlapping events, Bi-LSTM is better than LSTM in processing the sentence sequence information. (2) Hydra attention replaced with multi attention: From Table 7, it can be seen that the multi-head attention mechanism is able to capture overlapping trigger word semantic information, but hydra attention performs better. This may be due to the fact that hydra attention increases the number of attention heads and changes the order of computation. It improves the accuracy of the experiment. (3) Comparison with CasEE for this extraction module: Compared with CasEE, the trigger word decoder in this paper is obviously more effective. Thus, the results of the experiments verify the effectiveness of the LinkEE modules.

Table 8 illustrates the capabilities of the decoder variants on the event arguments extraction. (1) Bi-LSTM replaced with LSTM: Similarly, Bi-LSTM is better than LSTM in processing the sentence sequence information. (2) Hydra attention replaced with multi attention: hydra attention performs better than multi attention. Hydra attention improves the accuracy of the model. (3) Com-

Table 7. Results for the trigger word extraction decoder variant as well as CasEE. The evaluation metrics are precision (P), recall (R) and F1 score in the TC metrics.

Variants	P(%)	R(%)	F1(%)
repl. lstm	88.9	**90.9**	89.9
repl. attention	**91.3**	89.2	90.2
CasEE	90.1	90.2	90.1
LinkEE	89.9	90.8	**90.4**

parison with CasEE for this extraction module: Compared with CasEE, the event arguments decoder in this paper is obviously more effective. Thus, the results of the experiments verify the effectiveness of the LinkEE modules.

Table 8. Results for the event arguments extraction decoder variant as well as CasEE. The evaluation metrics are precision (P), recall (R) and F1 score in the TC metrics.

Variants	P(%)	R(%)	F1(%)
repl. lstm	84.7	82.9	83.8
repl. attention	84.6	83.1	83.9
CasEE	84.1	**83.7**	83.9
LinkEE	**84.8**	83.4	**84.1**

5 Conclusion

We propose LinkEE, a linked list model for attention-based overlapping event extraction. It organises event types, trigger word and arguments as nodes in a linked list and uses conditional fusion functions as edges to convey specific information. Bi-directional feature information of the input is learned through Bi-LSTM, while contextual information is captured using Hydra attention to enhance the text vector representation. The model predicts the node of the current module by passing specific event information through the CLN to implement the extraction of Chinese overlapping events. Experiments on the public dataset FewFC show that the F1 score of our model LinkEE is 0.6% and 0.7% higher than that of CasEE in under the AI and AC, respectively.

In the future, we will further optimize the message passing of linked lists and simplify the model to improve the accuracy and efficiency of Chinese overlapping event extraction.

Acknowledgments. This research is supported by National Natural Science Foundation of China under Grant 62002071, Guangzhou Science and Technology Planning Project under Grant 202201011835, Guangdong Basic and Applied Basic Research Foundation under Grant 2023A1515011577, Top Youth Talent Project of Zhujiang Talent Program under Grant 2019QN01X516, and Guangdong Provincial Key Laboratory of Cyber-Physical System under Grant 2020B1212060069.

References

1. Ba, J., Kiros, J.R., Hinton, G.E.: Layer normalization. arXiv abs/1607.06450 (2016)
2. Bolya, D., Fu, C.Y., Dai, X., Zhang, P., Hoffman, J.: Hydra attention: efficient attention with many heads. In: Karlinsky, L., Michaeli, T., Nishino, K. (eds.) ECCV 2022. LNCS, vol. 13807, pp. 35–49. Springer, Cham (2023). https://doi.org/10.1007/978-3-031-25082-8_3
3. Chen, Y., Xu, L., Liu, K., Zeng, D., Zhao, J.: Event extraction via dynamic multi-pooling convolutional neural networks. In: Proceedings of the 53rd Annual Meeting of the Association for Computational Linguistics and the 7th International Joint Conference on Natural Language Processing (Volume 1: Long Papers), pp. 167–176 (2015)
4. Chen, Y., Chen, T., Ebner, S., White, A.S., Van Durme, B.: Reading the manual: event extraction as definition comprehension. In: Proceedings of the Fourth Workshop on Structured Prediction for NLP. Association for Computational Linguistics, Online (2020)
5. Devlin, J., Chang, M.W., Lee, K., Toutanova, K.: BERT: pre-training of deep bidirectional transformers for language understanding. In: Proceedings of the 2019 Conference of the North American Chapter of the Association for Computational Linguistics: Human Language Technologies (Volume 1: Long and Short Papers), Minneapolis, Minnesota, pp. 4171–4186. Association for Computational Linguistics (2019). https://doi.org/10.18653/v1/N19-1423. https://aclanthology.org/N19-1423
6. Ding, X., Zhang, Y., Liu, T., Duan, J.: Using structured events to predict stock price movement: an empirical investigation. In: Proceedings of the 2014 Conference on Empirical Methods in Natural Language Processing (EMNLP), pp. 1415–1425 (2014)
7. Du, J., Luo, L., Sun, Z.: Research on event extraction method based on a lite BERT and conditional random field model. In: 2021 IEEE 11th International Conference on Electronics Information and Emergency Communication (ICEIEC), pp. 112–117. IEEE (2021)
8. Du, X., Cardie, C.: Event extraction by answering (almost) natural questions. In: Proceedings of the 2020 Conference on Empirical Methods in Natural Language Processing (EMNLP), pp. 671–683 (2020)
9. Filatova, E., Hatzivassiloglou, V.: Event-based extractive summarization. In: Proceedings of ACL Workshop on Summarization (2004)
10. Jungermann, F., Morik, K.: Enhanced services for targeted information retrieval by event extraction and data mining. In: Kapetanios, E., Sugumaran, V., Spiliopoulou, M. (eds.) NLDB 2008. LNCS, vol. 5039, pp. 335–336. Springer, Heidelberg (2008). https://doi.org/10.1007/978-3-540-69858-6_36
11. Li, F., et al.: Event extraction as multi-turn question answering. In: Findings of the Association for Computational Linguistics, EMNLP 2020, pp. 829–838 (2020)
12. Li, Q., Ji, H., Huang, L.: Joint event extraction via structured prediction with global features. In: Proceedings of the 51st Annual Meeting of the Association for Computational Linguistics (Volume 1: Long Papers), pp. 73–82 (2013)
13. Li, Z., Ding, X., Liu, T.: Constructing narrative event evolutionary graph for script event prediction. In: International Joint Conference on Artificial Intelligence (2018)
14. Liu, J., Chen, Y., Liu, K., Bi, W., Liu, X.: Event extraction as machine reading comprehension. In: Proceedings of the 2020 Conference on Empirical Methods in Natural Language Processing (EMNLP), pp. 1641–1651 (2020)

15. Liu, J., Chen, Y., Liu, K., Zhao, J.: Event detection via gated multilingual attention mechanism. In: Proceedings of the AAAI Conference on Artificial Intelligence, vol. 32 (2018)
16. Liu, S., Li, Y., Zhang, F., Yang, T., Zhou, X.: Event detection without triggers. In: Proceedings of the 2019 Conference of the North American Chapter of the Association for Computational Linguistics: Human Language Technologies (Volume 1: Long and Short Papers), pp. 735–744 (2019)
17. Nguyen, T.H., Cho, K., Grishman, R.: Joint event extraction via recurrent neural networks. In: Proceedings of the 2016 Conference of the North American Chapter of the Association for Computational Linguistics: Human Language Technologies, pp. 300–309 (2016)
18. Nguyen, T.M., Nguyen, T.H.: One for all: neural joint modeling of entities and events. In: Proceedings of the AAAI Conference on Artificial Intelligence, vol. 33, pp. 6851–6858 (2019)
19. Rospocher, M., et al.: Building event-centric knowledge graphs from news. J. Web Semant. **37**, 132–151 (2016)
20. Rui-Fang, H., Shao-Yang, D.: Joint Chinese event extraction based multi-task learning. J. Softw. **30**(4), 1015–1030 (2019)
21. Sha, L., Qian, F., Chang, B., Sui, Z.: Jointly extracting event triggers and arguments by dependency-bridge RNN and tensor-based argument interaction. In: Proceedings of the AAAI Conference on Artificial Intelligence, vol. 32 (2018)
22. Sheng, J., et al.: CasEE: a joint learning framework with cascade decoding for overlapping event extraction. In: Findings of the Association for Computational Linguistics: ACL-IJCNLP 2021, pp. 164–174. Association for Computational Linguistics, Online (2021). https://doi.org/10.18653/v1/2021.findings-acl.14. https://aclanthology.org/2021.findings-acl.14
23. Su, J.: Conditional text generation based on conditional layer normalization (2019)
24. Thenmozhi, D., Kumar, G.: An open information extraction for question answering system. In: 2018 International Conference on Computer, Communication, and Signal Processing (ICCCSP), pp. 1–5 (2018). https://doi.org/10.1109/ICCCSP.2018.8452854
25. Wadden, D., Wennberg, U., Luan, Y., Hajishirzi, H.: Entity, relation, and event extraction with contextualized span representations. In: Proceedings of the 2019 Conference on Empirical Methods in Natural Language Processing and the 9th International Joint Conference on Natural Language Processing (EMNLP-IJCNLP), pp. 5784–5789 (2019)
26. Yang, S., Feng, D., Qiao, L., Kan, Z., Li, D.: Exploring pre-trained language models for event extraction and generation. In: Proceedings of the 57th Annual Meeting of the Association for Computational Linguistics, pp. 5284–5294 (2019)
27. Zeng, Y., Yang, H., Feng, Y., Wang, Z., Zhao, D.: A convolution BiLSTM neural network model for Chinese event extraction. In: Lin, C.-Y., Xue, N., Zhao, D., Huang, X., Feng, Y. (eds.) ICCPOL/NLPCC 2016. LNCS (LNAI), vol. 10102, pp. 275–287. Springer, Cham (2016). https://doi.org/10.1007/978-3-319-50496-4_23
28. Zhou, Y., Chen, Y., Zhao, J., Wu, Y., Xu, J., Li, J.: What the role is vs. what plays the role: semi-supervised event argument extraction via dual question answering. In: Proceedings of the AAAI Conference on Artificial Intelligence, vol. 35, pp. 14638–14646 (2021)

End-to-End Aspect-Based Sentiment Analysis Based on IDCNN-BLSA Feature Fusion

Xinyuan Liu[1], Jindong Chen[1,2(✉)], and Wen Zhang[3]

[1] Beijing Information Science and Technology University, Beijing 100192, China
{2021020677,j.chen}@bistu.edu.cn
[2] Intelligent Decision Making and Big Data Application Beijing International Science and Technology Cooperation Base, Beijing, China
[3] Beijing University of Technology, Beijing, China
zhangwen@bjut.edu.cn

Abstract. The existing end-to-end Aspect-Based Sentiment Analysis (ABSA) algorithms focus on feature extraction by a single model, which leads to the loss of the important local or global information. In order to capture both local and global information of sentences, an end-to-end ABSA method based on features fusion is proposed. Firstly, the pre-trained model BERT is applied to obtain word vectors; secondly, Iterated Dilated Convolutions Neural Networks (IDCNN) and Bi-directional Long Short-Term Memory (BiLSTM) with Self-Attention mechanism (BLSA) are adopted to capture local and global features of sentences, and the generated local and context dependency vectors are fused to yield feature vectors. Finally, Conditional Random Fields (CRF) is applied to predict aspect words and sentiment polarity simultaneously. On Laptop14 and Restaurant datasets, our model's F1 scores increased by 0.51%, 3.11% respectively compared with the best model in the comparison experiment, and 0.74%, 0.78% respectively compared with the single model with the best effect in the ablation experiment. We removed each important module in turn in subsequent experiments and compared it with our model. The experimental results demonstrate the effectiveness of this method in aspect word recognition and its better generalization ability.

Keywords: Aspect-based sentiment analysis · feature fusion · IDCNN · BLSA · BERT

1 Introduction

Sentiment analysis [1] refers to the processing and analysis of texts with subjective sentiment tendencies to determine whether their sentiment towards something is positive, neutral, or negative [2]. Sentiment analysis can be divided into document-level [3], sentence-level [4] and aspect-based sentiment analysis (ABSA) according to the granularity of the processed text. Document-level or

J. Chen et al. (Eds.): KSS 2023, CCIS 1927, pp. 48–62, 2023.
https://doi.org/10.1007/978-981-99-8318-6_4

sentence-level sentiment analysis is to judge the overall sentiment tendency of the whole text or the whole sentence, but cannot analyze the specific aspect of the sentence separately, and it cannot meet more accurate and detailed analysis requirements. For example, in the restaurant review "Great food but the service was dreadful", the reviewer has positive sentiment towards the "food" aspect and negative sentiment towards the "service" aspect. Hence, for sentences containing multiple aspects, ABSA is required to analyze each aspect of a sentence.

ABSA can be divided into two sub-tasks: aspect extraction and sentiment analysis. Previous researchers merged two subtasks into one task, which can be seen as a sequence labeling problem. The task can be described as follows: given a sentence containing aspect words and their corresponding sentiments, the aspect words in the sentence are extracted and sentiment polarity is determined through an overall model. Mitchell et al. [5] proposed a model for jointly training two sub-tasks, using a set of target boundary labels, such as "BIOES", to label the aspects of sentences, utilizing a set of sentimental labels (such as POS, NEG, NEU) to label the sentimental polarity of the aspect. In order to better complete these two subtasks, Tang et al. [6] proposed Targe-Dependent-LSTM (TD-LSTM) for handling aspectual sentiment classification tasks based on LSTM. The applies two LSTMs to separately model the aspect word and its context, effectively exploiting the role of LSTM in aspectual sentiment classification. Meanwhile, in order to solve the problem of target word-context relationship, a Target-Connection-LSTM (TC-LSTM) model was proposed. TC-LSTM integrated the information of target word and context on the basis of TD-LSTM and achieved better results. Based on the LSTM, Wang et al. [7] proposed a LSTM model based on attention mechanism to improve the performance of aspect level sentiment analysis. Xing et al. [8] proposed to introduce the input layer into the attention based Convolutional Neural Network (CNN), which has been improved compared with LSTM. These models provide ideas for feature extraction in end-to-end ABSA tasks. However, in the case of a large amount of data annotation, if any independent task encounters problems, it will lead to a decrease in overall effectiveness.

To solve the above issues, Zhang et al. [9] proposed a unified tagging scheme y = B-POS, I-POS, E-POS, S-POS, B-NEG, I-NEG, E-NEG, S-NEG, B-NEU, I-NEU, E-NEU, S-NEU, O. Except O, each tag contains two parts of tagging information: the boundary of target mention, and the target sentiment. For example, B-POS denotes the beginning of a positive target mention, and S-NEG denotes a single word negative opinion target. Li et al. [10] investigated the complete task of ABSA and designed an end-to-end framework consisting of a two-layer LSTM, the upper layer network was used to generate the final annotation results based on the labeling scheme, while the lower layer network was used to assist in predicting the target boundary. As this method uses traditional word vector models such as Word2vec or GloVe in the training process, the generated text vectors are static text vectors and cannot dynamically obtain contextually relevant semantic information. Therefore, Li et al. [11] proposed a BERT-based neural network model that addressed the issue of context independence in the E2E-ABSA layer. They achieved this by fine-tuned BERT and designing downstream models to generate the final results. This work provides valuable insights for our study.

For the unified tagging scheme, sentiment classification and aspect extraction may involve different semantic relationships, and unifying them into one task may lead to conflicts and semantic ambiguity, and may not accurately capture the subtle relationships between them. Therefore, we need to further enhance the feature extraction and classification capabilities of the model. Researchers have attempted to improve models such as CNN and BiLSTM to capture sentence features more fully and achieved significant results. In the same type of sequence labeling task, Strubell et al. [12] proposed IDCNN, thus greatly alleviated the problem of excessive parameter size and overfitting in CNN. IDCNN has been applied to entity recognition and achieved significant results. However, it has not yet been widely applied in ABSA task. Due to its excellent performance in entity recognition, we can also apply IDCNN to end to end ABSA tasks with the same annotation method as entity recognition.

Each of these models has its own advantages and disadvantages in extracting sentence features. LSTM, especially after adding the self-attentive mechanism (hereinafter referred to as BLSA), it can focus more on the important information and effectively use the contextual information for sentiment analysis, but BLSA, may ignore some local important information [13]. IDCNN can handle various positions of aspects in the text, so as to effectively discover sentiment information. However, IDCNN is relatively poor in capturing long-distance dependencies in text. In order to better capture sentence features, Shang et al. [14] proposed a method for CNN-BiLSTM fusion and achieved results in entity recognition tasks. This method to some extent solves the problem of traditional methods being unable to capture the dependency relationship before and after text sequences. However, the two models used for feature extraction in this method can be further improved to achieve better feature extraction results.

More comprehensive feature extraction methods can improve the effectiveness of ABSA tasks. In order to better extract features from text, we propose a feature fusion method using IDCNN and BLSA as feature extraction models, respectively. Firstly, pre-trained model BERT is used to generate vectors with contextual information. Secondly, BLSA is applied to capture contextual dependencies in sentences. At the same time, IDCNN is used to capture local sentimental information as a supplement to BLSA feature extraction, and then the two are fused. Finally, the sentence features captured by the two models are fused to obtain sentence features that have both contextual dependencies and local sentimental information, and the fused sentence features are ultimately fed into the CRF to obtain the final result.

2 Model Overview

2.1 Model Architecture

The general architecture of the model is shown in Fig. 1. The input to the model is a collection of words in a sentence $X = \{x_1, x_2, ..., x_t\}$, which is transformed into the corresponding text embedding. The representation of each sentence containing contextual information is computed by a BERT component with 12 transformer layers $H^T = \{h_1^T, h_2^T, ..., h_n^T\}$, h_n^T is the corresponding feature vector for a word. The representations generated by BERT are fed into two downstream models, obtaining the feature representation $H^I = \{h_1^I, h_2^I, ..., h_n^I\}$ for IDCNN and $H^B = \{h_1^B, h_2^B, ..., h_n^B\}$ for BLSA. After the two downstream models extracted features separately, the features extracted from the two are fused to obtain sentence features with both local information and contextual relationships. The final text features are fed into the CRF layer to obtain the final labels.

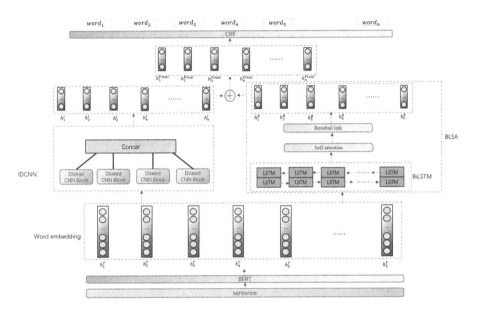

Fig. 1. The structure of our model

2.2   BERT as Embedding Layer

In this paper, the fine-tuned BERT as a pre-trained model is applied to generate word vectors, whose structure is shown in Fig. 2. BERT splices three features of a sequence of words of input length t as $E = \{e_1, ..., e_i, ..., e_t\}$, which contains partial embedding, positional embedding and word embedding. The partial embedding is used to identify the first and second half of the sentence; the positional embedding represents the relative position of each word in the sentence; and the word embedding is the corresponding vector representation of each word. The input features are then refined feature by feature through multiple Transformer layers to obtain a representation of the utterance containing contextual information. Although BERT is based on transformer, it does not use the entire transformer structure, only the Encoder part of the Transformer structure, and builds multiple layers of encoders together to form the basic network structure. Due to its multi-headed attention mechanism, each word in a sentence contains contextual information.

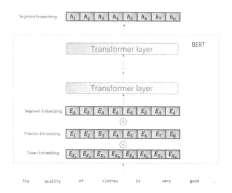

Fig. 2. The structure of the fine-tuned BERT

2.3   Downstream Model Design

IDCNN. Yu et al. [15] proposed the Dilated Convolutions Neural Networks (Dilated CNN). Dilated CNN adds a dilation width d to the convolution kernel. Kernel has a weight matrix $k = [k_{-l}, k_{-l+1}, \ldots, k_l]$, the convolution operation can be written as:

$$c_t = \sum_{i=-l}^{l} k_i \cdot h_{t+i \cdot d}^T + b_c \tag{1}$$

Strubell et al. applied inflated convolutional neural networks to the named entity recognition task and proposed IDCNN. IDCNN consists of multiple structurally identical expanded convolutional modules stitched together, with multiple layers of expanded convolution inside each module. Let $H^T = \{h_1^T, h_2^T, \ldots, h_n^T\}$, h_n^T be the contextual representation obtained from the BERT, which is used as input to the IDCNN, and the IDCNN structure is implemented as follows:

$$c_t^{(1)} = D_1^{(0)} h_1^T \tag{2}$$

$$c_t^{(j)} = r\left(D_{2^t}^{(j-1)} c_t^{(j-1)}\right) \tag{3}$$

$$c_t^{(n+1)} = r\left(D_1^{(n)} c_t^{(n)}\right) \tag{4}$$

where: $D_t^{(j)}$ is the expanded convolutional network with an expanded width of d at the jth layer; $c_t^{(j)}$ is the feature obtained by convolving the jth layer of the network; $r()$ denotes the ReLu activation function.

Four identical expansion convolution modules with the same structure as Strubell to splice together are used in our paper, each module is composed of three layers of expansion convolutions with expansion widths of 1, 1, and 2. The representation after splicing is as $H_I = \{h_1^I, h_2^I, ..., h_n^I\}$.

The feature extraction process for each block is described as follows: the first layer Dilated CNN uses a convolution check with an expansion rate of 1 to perform convolution operations on the input sequence. This layer calculates the local features containing each word and its adjacent words. The output is a feature sequence of the same length as the input sequence. The second layer Dilated CNN operates the same as the first layer, performing convolution operations on the output of the first layer to further extract features. The output is still a feature sequence of the same length as the input sequence. The third layer Dilated CNN uses a convolution check with an expansion rate of 2 to perform convolution operations on the output of the second layer. This layer extracts features on a larger scale, including each word, adjacent words every other word, and adjacent words every two words. The final output is a feature sequence of the same length as the input sequence.

BiLSTM. Recurrent Neural Network (RNN) was proposed by Elman [16]. The model can process sequential data by feeding the data into the model line by line according to the time series. Hochreiter et al. [17] proposed Long Short-Term Memory (LSTM), which used gate structure features to simulate the forgetting and memory mechanisms of the human brain to overcome the problem of gradient disappearance during the training of long sequences. Let $H^T = \{h_1^T, h_2^T, ..., h_n^T\}$ be the contextual representation obtained from BERT, which is used as the input to the LSTM, and the LSTM structure is implemented as follows:

$$f_t = \sigma\left(W_f\left[h_{t-1}, h_t^T\right] + b_f\right) \tag{5}$$

$$u_t = \sigma\left(W_u\left[h_{t-1}, h_t^T\right] + b_u\right) \tag{6}$$

$$o_t = \sigma\left(W_o\left[h_{t-1}, h_t^T\right] + b_u\right) \tag{7}$$

$$\tilde{c}_t = tanh\left(\left(W_c\left[h_{t-1}, h_t^T\right] + b_c\right)\right) \tag{8}$$

$$c_t = f_t \cdot c_{t-1} + u_t \cdot \tilde{c}_t \tag{9}$$

where c_{t-1} is the cell state input at moment $t-1$; h_{t-1} is the output at moment $t-1$; h_1^T is the input at the current moment; h_t is the output; c_t is the cell state. $\tilde{c}_t$ is the candidate neuron; σ is the Sigmoid function; f_t, f_u, f_o, f_c and b_t, b_u, b_o, b_c are the weight coefficients and offset matrices of each corresponding component, respectively.

Bidirectional LSTM (Bi-LSTM) learns historical information and future information about a word or phrase by using two layers of LSTM neurons to learn the sequence data from left to right and from right to left respectively. Combining the two types of information allows for a better description of the contextual content.

Self-attention Networks. Attention mechanisms were first proposed and used in the field of image vision, Bahdanau et al. [18] first combined attention mechanisms with natural language processing tasks, and Vaswani et al. [19] first proposed a Self-Attention Mechanism. In the Self-Attention Mechanism, each element obtains attention weights for all other elements, and the final contextual representation is derived by weighting and summing in the encoder.

In this paper, we use two self-attentive mechanisms consisting of self-attentive layers and residual links [20] to help BiLSTM capture sentence features and normalize them, denoted as L_N. This self-attentiveness is referred to as "SAN" in this paper. Let $H = \{h_1, h_2, ..., h_n\}$ be the feature representation obtained from the BiLSTM, which is used as the input to the SAN. The SAN is represented is below:

$$H^B = L_N \left(H^T + SLF - ATT\left(Q, K, V\right) \right) \tag{10}$$

$$Q, K, V = H^T W^Q, H^T W^K, H^T W^V \tag{11}$$

Feature Fusion. Feature fusion is the stitching or merging of features extracted from the hidden layers of two or more models. Feature layer fusion is useful for fusing models with different structures, such as feature layer fusion between convolutional and recurrent neural networks, to better exploit the feature representation capabilities of different layers of the network and thus improve model effectiveness. The feature fusion of different models can improve their predictive ability, alleviate overfitting, and improve model stability.

We use IDCNN and BLSA models with different structures of CNN structure as downstream models to extract features in sentences. Since the two models have complementary advantages, we fuse their features to obtain feature vectors with local information and context dependency. Let $H^B = \{h_1^B, h_2^B, ..., h_n^B\}$ and $H^I = \{h_1^I, h_2^I, ..., h_n^I\}$ be the features representation obtained from BiLSTM and IDCNN. The fusion method is as follows:

$$H^{Final} = W_I \cdot H^I + W_B \cdot H^B \tag{12}$$

$$W_I + W_B = 1 \tag{13}$$

where H^{Final} is the representation of the two models after fusion; W_I, and W_B are the weights of IDCNN and BLSA. Finally, the fused features are input to CRF to obtain the final predicted labels.

Conditional Random Fields. Conditional Random Fields (CRF) [21] has good results in sequence modelling. CRF has been widely used to solve sequence annotation problems together with neural network models [22–24]. CRF establishes the relationship between annotations for each position of the annotation. Specifically, the CRF model considers the relationship between each annotation and its adjacent annotations simultaneously. In the sequential annotation problem, CRF considers the interdependencies between labels and is able to capture the dependencies between the global labels and the current observation. During training, the CRF model optimizes the parameters by maximizing the log-likelihood function so that the probability of the labeled sequence output by the model is maximized with respect to the true labeled sequence. Overall, the CRF model is able to perform better annotation prediction based on the contextual information of each annotation in the annotation sequence. Therefore, this paper uses CRF to complete the prediction of the final labels. The labels $Y = \{y_1, ..., y_t\}$ the sequence level scores s (x, y) and the likelihood values $p(y, x)$ are calculated as follows:

$$s\left(x, y\right) = \sum_{t=0}^{T} M_{y_t, y_{t+1}}^{A} + \sum_{t=0}^{T} M_{t,y_t}^{P} \tag{14}$$

$$p\left(y, x\right) = softmax(x, y) \tag{15}$$

where M^A is the randomly initialised transition matrix used to model the dependencies between adjacent predictions; M^P is the emission matrix for the linear transformation of the sentence features extracted by the downstream model described above. All possible tag sequences are normalized using the softmax function. When the final decoding is done, the highest scoring tag sequence is found as the output by the Viterbi algorithm and is represented as follows:

$$y^* = argmax \; s\left(x, y\right) \tag{16}$$

3 Experiment Results and Discussions

3.1 Dataset and Parameters Settings

The experimental data comes from SemEval's Laptop14 and Restaurant datasets, where Restaurant is a dataset filtered and merged by Restaurant14, Restaurant15, and Restaurant16. In this paper, experiments were conducted on the above two benchmark datasets for aspect-based sentiment analysis, where each aspect was labelled as one of three sentimental polarities: positive, neutral and negative. The statistics are summarized in Table 1.

The pre-trained model used in this paper is "bert-base-uncase", where the number of transformer layers is 12. The random seeds were fixed to ensure the interpretability of the experimental results. We tuned the parameters of the model, the main parameters of the experiments are shown in Table 2.

Table 1. Summary of data sets.

Dataset		Train	Dev	Test	Total
Laptop	sentences	2741	304	800	3845
	aspect	2041	256	634	2931
Restaurant	sentences	3490	387	2158	6035
	aspect	3893	413	2287	6593

Table 2. Parameter settings.

Parameters	Laptop	Restaurant
W_I	0.3	0.3
W_B	0.7	0.7
Epoch	30	30
Batch Size	25	16
Hidden Size	768	768
Sequence Length	128	128
Learning Rate	2e−5	2e−5
Dropout Rate	0.1	0.1
Optimizer	Adam	Adam
Random Seed	43	43

3.2 Performances Comparison

In order to verify the performances of the models, this paper has conducted comparison experiments with existing models and has performed ablation experiments based on its own model. The comparison experimental models include 2-layer-LSTM and the representative LSTM-CRF sequence tagging model. The experimental results are shown in Table 3.

Table 3. Comparison of performances among different models.

Models	Laptop			Restaurant		
	P	R	F1	P	R	F1
2-Layer-LSTM	61.27	**54.89**	57.90	68.46	71.01	69.80
LSTM-CRF-1	58.61	50.47	57.24	66.10	66.30	66.20
LSTM-CRF-2	58.66	51.26	54.71	61.56	67.26	64.29
LM-LSTM-CRF	53.31	59.40	56.19	68.46	64.43	66.38
BERT-(IDCN+BLSA)-CRF	**63.64**	54.73	**58.41**	**72.11**	**73.74**	**72.91**

As can be seen from Table 3, the three evaluations metrics of the model in this paper have better performance in both datasets. With the introduction of the pre-trained model BERT, the model performance has a large improvement on both datasets compared to the representative LSTM-CRF sequence tagging model in the comparison experiments. Compared with the best 2-layer LSTM in the comparative model, F1 score increased by 0.51% on Laptop14, and 3.11% on Restaurant. This can prove the effectiveness of our model.

3.3 Ablation Experiments

The ablation experimental models include BERT-Linear, BERT-CRF, BERT-BiLSTM-CRF, BERT-BLSA-CRF, BERT-IDCNN-CRF, BERT-CNN-CRF, BERT-(IDCNN+BiLSTM)-CRF and BERT-(CNN+BLSA)-CRF. In this paper, three evaluation metrics, namely precision, recall and Micro- F1 three evaluation metrics as the experimental results, and the experimental results are shown in Table 4.

Table 4. Experimental results of different models.

Models	Laptop			Restaurant		
	P	R	F1	P	R	F1
BERT-Linear	60.11	51.58	55.51	71.27	72.61	71.93
BERT-CRF	60.68	52.94	56.55	69.80	72.53	71.13
BERT-BiLSTM-CRF	61.19	53.47	57.03	70.62	72.04	71.32
BERT-BLSA-CRF	62.16	53.63	57.57	69.36	**74.79**	71.96
BERT-IDCNN-CRF	59.93	50.95	55.07	70.67	70.51	70.58
BERT-CNN-CRF	58.79	50.86	54.54	70.61	69.90	70.25
BERT-(CNN+BLSA)-CRF	62.83	53.12	57.57	70.95	73.32	72.11
BERT-(IDCNN+BiLSTM)-CRF	62.39	53.63	57.67	71.08	73.22	72.13
BERT-(IDCN+BLSA)-CRF	**63.64**	**54.73**	**58.41**	**72.11**	73.74	**72.91**

In the ablation experiments, it can be found that the removal of BLSA for capturing contextual relationships leads to the most severe degradation in model performance. After removing BLSA, the F1 score of the Laptop dataset decreased by 3.34% and the Restaurant score decreased by 2.33%, indicating that features containing contextual information play a key role for aspect-based sentiment analysis. Secondly, by removing the IDCNN for local sentiment information, the F1 score of the model decreased by 0.84% on the Laptop14 and 0.95% on the Restaurant dataset, indicating that capturing local sentiment information is also a key factor in improving the performance of the model. Furthermore, after replacing the IDCNN module with CNN, the F1 score of the model decreased by 0.84% in Laptop14 and 0.8% in Restaurant, indicating that IDCNN performs slightly better in ABSA tasks than CNN. Finally, removing the SAN from the BLSA, which were used to enhance the model's focus on important information and contextual relationships in the text, also reduced the model performance, with the model's F1 score dropping by 0.74% on the Laptop14 and 0.78% on the Restaurant dataset, suggesting that the SAN can help BiLSTM further capture contextual information. Thus, all parts of the model have their important roles in the aspect-based sentiment analysis task.

3.4   Generalization Issue

In order to verify the generalization ability of the model on different datasets, we selected the best performing single model BERT-BLSA-CRF model and baseline model BERT Linear, and compared them with the model in this paper on Restaurant14, Restaurant15 and Restaurant16, respectively, with Micro-F1 as the evaluation metric. At the same time, we summarized the different sentimental aspects of three datasets.

Table 5. Different sentimental aspects of three datasets.

Dataset	Restaurant14				Restaurant15				Restaurant16			
	S	+	0	−	S	+	0	−	S	+	0	−
Train	2736	2750	459	723	1183	1263	33	405	1799	2015	107	612
Dev	304	380	50	249	130	148	11	53	200	230	11	76
Test	2204	2373	306	355	685	617	25	223	676	707	63	158

Table 6. F1 scores for the three models in Restaurant14, Restaurant15 and Restaurant16.

Models	Restaurant14(F1)	Restaurant15(F1)	Restaurant16 (F1)
BERT-Linear	70.79	55.30	64.36
BERT-BLSA-CRF	70.73	**57.95**	67.97
BERT-(IDCN+BLSA)-CRF	**71.90**	57.91	**68.32**

As can be seen from Table 5, On restaurant14 and restaurant16, there is a significant difference in the proportion of the three sentiments between the test and validation sets, while on restaurant15, the proportion of the three sentiments is similar. Our model performs better on restaurant14 and restaurant16, which can be explained by the significant distribution differences among the three sentiments in the training and testing sets of the two datasets.

As can be seen from Table 6, on the Restaurant14, Restaurant15 and Restaurant16, the F1 scores of this paper's model improved by 1.11%, 2.51% and 3.96% respectively compared to the baseline model BERT-Linear; compared to BERT-BLSA-CRF, on the Restaurant14 and Restaurant16 by 1.17% and 0.35%, respectively, and by 0.04% on the Restaurant15. It can be seen that on all three datasets, the model in this paper has a greater improvement compared to the baseline model BERT-Linear; the overall performance is better compared to the BERT-BLSA-CRF model; while BERT-BLSA-CRF performs poorly on the Restaurant14, with no improvement compared to the baseline model BERT-Linear, but instead The performance of BERT-BLSA-CRF on the Restaurant14 was poor, with no improvement over the baseline model BERT-Linear, but rather a 0.06% decrease. It can be concluded that the model in this paper has stronger generalizability compared to a single model.

3.5   Case Studies

In order to further investigate the impact of various parts of the model on its performance, we removed the IDCNN module and BLSA module separately and compared them with our model for case analysis. We selected the model trained on Restaurant with the best performance for aspect sentiment prediction, and selected two classic cases for analysis. The cases and results are shown in the Table 7.

Table 7. Typical case studies.

Sentences	Models	Predicted results
The pizza has small portions and the taste is terrible	BERT-BLSA-CRF	pizza, NEG
		portions, NEG
		taste, NEG
	BERT-(IDCNN+BLSA)-CRF	portions, NEG
		taste, NEG
	BERT-IDCNN-CRF	portions, NEG
		taste, NEG
I tend to judge a sushi restaurant by its sea urchin, which was heavenly at sushi rose	BERT-BLSA-CRF	sea, POS
	BERT-(IDCNN+BLSA)-CRF	sea urchin, POS
	BERT-IDCNN-CRF	urchin, POS

From the first case, both models are correct in the sentiment polarity judgement, but for the aspect extraction, BERT-BLSA-CRF extracts the redundant aspect "pizza", which means that BLSA extracts all the entities related to the sentiment words according to the context dependency, but the entities such as "pizza" is not what we want to extract and determine the sentiment polarity, we only need to extract the "portions" and "taste" aspects of "pizza". We only need to extract the "portions" and "taste" of "pizza" and make a sentiment polarity judgement. BERT-IDCNN-CRF and BERT-(IDCNN+BLSA)-CRF can correctly predict the two aspects of this sentence and sentimental polarity. This indicates that, with the addition of IDCNN, the local features can be extracted to find the aspect words more precisely.

From the second sample, BERT-BLSA-CRF only found the aspect word sea and while BERT-IDCNN-CRF only found the aspect word urchin. It can be seen that the two models may lose some information when facing difficult sentences, leading to problems in phrase recognition. BERT - (IDCNN+BLSA) - CRF can correctly recognize the phrases appearing in this sentence, indicating that the fusion of IDCNN and BLSA can enable the model to capture information in the sentence more fully, thereby improving the accuracy of the model's prediction.

4 Conclusions

In ABSA task, it is crucial to extract the features of the corresponding aspect in the text. In this paper, in order to extract the features in sentences more fully, we use IDCNN and BLSA as downstream models to extract the features of sentences respectively, and fuse the features of the two downstream models, so as to take advantage of the strengths of both models and make the captured sentence features more adequate, thus improving the performance of the model.

For modelling text sequences, IDCNN and BLSA each have their own advantages and disadvantages. IDCNN uses a CNN structure, which has the advantage of parallel computation and can easily speed up the training process, and is also suitable for shorter text inputs, and is usually less prone to problems such as gradient disappearance and gradient explosion than recurrent neural network structures such as LSTM. The BLSA, on the other hand, uses a BiLSTM and Self-Attention structure, which focuses more on sequence order compared to IDCNN, has a stronger modelling capability, and is able to capture the textual information at each time step and the relationship between them. Therefore, IDCNN and BLSA also have complementary advantages in modelling text sequences.

IDCNN and BLSA are both models designed for text classification tasks and are effective in discovering sentiment information in text in sentiment analysis. IDCNN is able to extract effective text features from the output of the convolutional layer for classification by a downstream classifier, while BLSA is able to better capture contextual information and relationships in text, improving the classification capability of the model, especially at the aspect level. In sentiment analysis, the feature extraction and classification capabilities of the model need to be further enhanced because of the requirement to target different aspects of the text for sentiment analysis.

In summary, IDCNN and BLSA have many complementary advantages for aspect-based sentiment analysis tasks, with unique strengths in text feature extraction, sequence modelling and text classification. The combined use of both IDCNN and BLSA models allows for better sentiment analysis.

Acknowledgement. The work is supported by National Key Research and Development Program Project "Research on data-driven comprehensive quality accurate service technology for small medium and micro enterprises" (Grant No. 2019YFB1405303).

The Project of Cultivation for Young Top-motch Talents of Beijing Municipal Institutions "Research on the comprehensive quality intelligent service and optimized technology for small medium and micro enterprises" (Grant No. BPHR202203233).

National Natural Science Foundation of China "Research on the influence and governance strategy of online review manipulation with the perspective of E-commerce ecosystem" (Grant No. 72174018).

References

1. Yuan, J.H., Wu, Y., Lu, X.: Recent advances in deep learning based sentiment analysis. China Technol. **63**(10), 1947–1970 (2020)
2. Chen, L., Guan, J.L., He, J.H., Peng, J.Y.: Advances in emotion classification research. Comput. Res. Dev. **54**(06), 1150–1170 (2017)
3. Yadollahi, A., Shahraki, A.G., Zaiane, O.R.: Current state of text sentiment analysis is from opinion to emotion mining. ACM Comput. Surv. **50**(2), 1–33 (2018)
4. Meena, A., Prabhakar, T.V.: Sentence level sentiment analysis in the presence of conjuncts using linguistic analysis. In: Amati, G., Carpineto, C., Romano, G. (eds.) ECIR 2007. LNCS, vol. 4425, pp. 573–580. Springer, Heidelberg (2007). https://doi.org/10.1007/978-3-540-71496-5_53
5. Mitchell, M., Aguilar, J., Wilson, T., Durme, B.V: Open domain targeted sentiment. In: Proceedings of the 2013 Conference on Empirical Methods in Natural Language Processing, USA, pp. 1643–1654. Association for Computational Linguistics (2013)
6. Tang, D., Qin, B., Feng, X.C., Liu, T.: Effective LSTMs for target-dependent sentiment classification. In: Matsumoto, Y., Prasead, R. (eds.) Proceedings of COLING 2016, the 26th International Conference on Computational Linguistics: Technical Papers, COLING, Japan, vol. 1311, pp. 3298–3307. The COLING 2016 Organizing Committee (2016). https://doi.org/10.48550/arXiv.1512.01100
7. Wang, Y.Q., Huang, M., Zhu, X.Y., Zhao, L.: Attention-based LSTM for aspect-level sentiment classification. In: Su, J., Duh, K. (eds.) Proceedings of the 2016 Conference on Empirical Methods in Natural Language Processing, EMNLP, vol. 1058, pp. 606–615. Association for Computational Linguistics, Texas (2016). https://doi.org/10.18653/v1/D16-1058
8. Xing, Y.P., Xiao, C.B., Wu, Y.F., Ding, Z.M.: A convolutional neural network for aspect-level sentiment classification. Int. J. Pattern Recogn. Artif. Intell. **33**(14), 46–59 (2019)
9. Zhang, M., Zhang, Y., Vo, D.T.: Neural networks for open domain targeted sentiment. In: Marquez, L., Callison-Burch, C. (eds.) Proceedings of the 2015 Conference on Empirical Methods in Natural Language Processing, EMNLP, Portugal, vol. 1073, pp. 612–621. Association for Computational Linguistics (2015). https://doi.org/10.18653/v1/D15-1073
10. Li, X., Bing, L., Li, P.J., Lam, W.: A unified model for opinion target extraction and target sentiment prediction. In: Pascal, V.H., Zhang, Z.H. (eds.) Proceedings of the AAAI Conference on Artificial Intelligence, AAAI, USA, vol. 33, pp. 6714–6721. Association for the Advancement of Artificial Intelligence (2019). https://doi.org/10.1609/aaai.v33i01.33016714
11. Li, X., Bing, L., Li, P.J., Lam, W.: Exploiting BERT for end-to-end aspect-based sentiment analysis. In: 5th Workshop on Noisy User-Generated Text (W-NUT 2019), Hong Kong, pp. 34–41. Association for Computational Linguistics (2019). https://doi.org/10.18653/v1/D19-5505

12. Strubell, E., Verga, P., Belanger, D.: Fast and accurate entity recognition with iterated dilated convolutions. In: Proceedings of the 2017 Conference on Empirical Methods in Natural Language Processing, Denmark, pp. 2670–2680. Association for Computational Linguistics (2017)

13. Yang, B., Li, J., Wang, D.F.: Context-aware self-attention networks. In: Pascal, V.H., Zhihua, Z. (eds.) Proceedings of the AAAI Conference on Artificial Intelligence, AAAI, USA, vol. 33, pp. 387–394. Association for the Advancement of Artificial Intelligence (2019). https://doi.org/10.1609/aaai.v33i01.3301387

14. Shang, F.J., Ran, C.F.: An entity recognition model based on deep learning fusion of text feature. Inf. Process. Manag. **59**(2), 16–30 (2022)

15. Yu, F., Koltun, V.: Multi-scale context aggregation by dilated convolutions. In: ICLR (2016). https://doi.org/10.48550/arXiv.1511.07122

16. Elman, J.L.: Finding structure in time. Cogn. Sci. **14**(2), 179–211 (1990)

17. Hochreiter, S., Schmidhuber, J.: Long short-term memory. Neural Comput. **9**(8), 1735–1780 (1997)

18. Bahdanau, D., Cho, K., Bengio, Y.: Neural machine translation by joint learning to align and translate. Comput. Sci. (2014). https://doi.org/10.48550/arXiv:1409.0473

19. Vaswani, A., Shazeer, N., Parmar, N.: Attention is all you need. In: 31st International Conference on Neural Information Processing Systems. Curran Associates Inc., Long-Beach (2017)

20. He, K., Zhang, X.Y., Ren, S.H., Sun, J.: Deep residual learning for image recognition. In: CVPR, pp. 770–778 (2015)

21. Lafferty, J.D., Mccalum, A., Pereira, F.C.N.: Conditional random fields: probabilistic models for segmenting and labeling sequence data. In: 18th International Conference on Machine Learning, pp. 282–289. Morgan Kaufmann Publishers Inc., San Francisco (2001)

22. Huang, Z.H., Xu, W., Yu, K.: Bidirectional LSTM-CRF models for the sequence tagging. Comput. Sci. (2015). https://doi.org/10.48550/arXiv.1508.01991

23. Lample, G., Ballesteros, M., Subramanian, S., Kawakami, K.: Neural architectures for named entity recognition. In: Kevin, K., Ani, N. (eds.) Proceedings of the 2016 Conference of the North American Chapter of the Association for Computational Linguistics: Human Language Technologies, NAACL, California, vol. 1030, pp. 260–270. Association for Computational Linguistics (2016). https://doi.org/10.18653/v1/N16-1030

24. Ma, X.Z., Hovy, E.: End-to-end sequence labeling via bi-directional LSTM-CNNs-CRF. In: Proceedings of the 54th Annual Meeting of the Association for Computational Linguistics, Berlin, pp. 1064–1074. Association for Computational Linguistics (2016)

How to Quantify Perceived Quality from Consumer Big Data: An Information Usefulness Perspective

Tong Yang[(⊠)] [iD], Yanzhong Dang, and Jiangning Wu

Dalian University of Technology, Dalian, China
yangt014@mail.dlut.edu.cn

Abstract. Perceived quality reflects consumers' subjective perceptions of a product and is important for manufacturers to improve quality. In recent years, social media becomes a new channel for consumers to share perceived quality, but existing studies overlooked the information usefulness of each piece of data, which creates barriers for manufacturers to process impactful information. This paper proposes a two-stage approach to quantifying perceived quality based on information usefulness. First, the usefulness categories of perceived quality are identified through a combination of deep learning and the knowledge adoption model; then, multiple usefulness categories are considered to quantify the perceived quality information. In the empirical study, the method was validated using an automobile dataset from Autohome. Results show that the method obtains more effective perceived quality information. The proposed method contributes to the research on both perceived quality quantification and information usefulness.

Keywords: Perceived quality quantification · Social media data · Information usefulness · Knowledge adoption model

1 Introduction

Building a mass prestige can help manufacturers succeed [1], and one important concern is consumer perceived quality, which is the consumer's perceptual under-standing of the product's attributes [2]. This subjective perception is affected by personal characteristics [3], and does not directly reflect the objective quality of the product [4, 5], but it influences consumers' evaluation of the product and their satisfaction with it [6, 7]. Accordingly, by focusing on the perceived quality, manufacturers can better understand consumers' perceptions of their products, and thus develop and adjust their production and marketing strategies [3].

Driven by this practical need, more and more studies are devoted to the quantification of perceived quality [8, 9]. For example, Duraiswamy et al. proposed a survey-based method combining direct attribute evaluation and choice experiment, by which to evaluate the perceived quality of vehicle body split gaps and found that consumers have a preference for smaller gaps [8]. Such methods typically use structured data (e.g., interviews, questionnaires), which suffer from lags, inadequacies, and incompleteness [10].

J. Chen et al. (Eds.): KSS 2023, CCIS 1927, pp. 63–77, 2023.
https://doi.org/10.1007/978-981-99-8318-6_5

With the rapid growth of social media, the pattern of consumer feedback on products has changed and social media data is considered comprehensive and timely by industry and academia [10, 11]. Based on this view, He, Zhang, and Yin proposed a fuzzy-based approach to mining attribute-level perceived quality from social media, offering new ideas for manufacturers to improve their products [12].

While the explosion of social media data has facilitated access to information, the value of information has become weak and fragmented, leading to cognitive overload [13]. To this end, existing studies on the quantification of perceived quality mainly average the consumers' ratings to obtain perceived quality, without considering the usefulness of each piece of data [9, 12, 14]. The information usefulness reflects the extent to which people find it valuable, informative, and helpful [15, 16]. Ignoring the information usefulness of the data would set barriers for manufacturers to process and judge impactful information [17]. Thus, the current study treats information usefulness as the basis for quantifying perceived quality.

To measure the usefulness of information, most websites have built-in voting mechanisms that allow consumers to vote on the value of published data, and examples are shown in Fig. 1. However, since voting is voluntary, it leads to a systematic problem of "error of omission". As evidenced by Liu et al. [18] and Cao, Duan, and Gan [19], the consumer voting rate for usefulness is typically low, meaning that the majority of social media data cannot be evaluated by voting, which makes effective usefulness evaluation methods important. Traditional usefulness evaluation methods mainly rely on features such as linguistic features (e.g., length, readability), lexical features (e.g., n-gram vectors), and reviewer's information (e.g., engagement, expertise) [17, 18], disregarding the semantic feature embedded in content, which is also a vital factor that makes the reviews appealing [20, 21]. Thereby, to fill the above gaps, we propose our research question: how to construct a method for quantifying perceived quality based on the information usefulness?

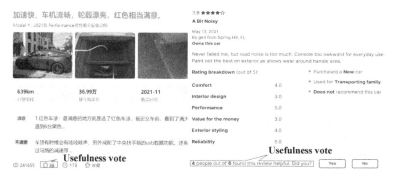

Fig. 1. Examples of usefulness votes from Autohome.com and car.com

The knowledge adoption model (KAM) is a widely used theory of modeling information usefulness [15], which describes usefulness as an important step of information adoption, influenced by the different elements of information. As KAM fits well with our research questions, we use it as a theoretical guide to propose a two-stage approach to quantifying perceived quality based on the information usefulness. Specifically, we first follow the idea [22] of combining CNN and SVM to estimate consumer usefulness votes by fusing semantic word vectors in the framework of KAM. Based on the perceived quality usefulness categories obtained in the previous stage, the perceived quality quantification is achieved by adjusting the quantitative weight of each category.

The proposed methodology is validated using automobile consumers data from the Autohome online community. The proposed method provides a new approach to perceived quality research and is relevant for automobile manufacturers to improve their data processing capabilities.

2 Related Works

2.1 Perceived Quality

Zeithaml defined perceived quality as the customer's personal evaluation of the product's overall strengths [23]. Following this, Mitra and Golder [24] described perceived quality as an overall subjective quality judgment relative to the desired quality. From an engineering perspective, Stylidis, Wickman, and Söderberg regarded perceived quality as an experience for the consumer, which results from the interaction between product quality and the consumer [14]. In general, perceived quality is the evaluation information people generate through their senses (sight, touch, hearing, smell) about a product or its attributes, and thus high perceived quality indirectly reflects products' attractiveness [14]. So perceived quality has an impact on consumer satisfaction [25], and purchase intention [26], which indirectly affects the sales of manufacturers [4, 27, 28].

The increasing complexity of products today has led consumers to perceive different qualities for different attribute categories [29]. Thus, while traditional research has considered perceived quality as too subjective to measure, more and more research is now focusing on the quantification of attribute-level perceived quality [9, 12]. Stylidis, Wickman, and Söderberg established a framework for the terminology of perceived quality, including value based perceived quality and technological perceived quality which embody consumer perception and engineering approach respectively [30]. Based on this framework, Stylidis et al. [31] and Stylidis, Wickman, and Söderberg [14] used a survey-based method called Best-Worst Scaling to further rank perceived quality attributes. Ma, Chen, and Chang processed the perceived quality of the service attributes of vehicles and used the KANO model to classify the attributes into three categories [32]. Unlike the above studies which used structured questionnaire data, He et al. evaluated attribute-level perceived quality from social media data using a fuzzy-based method [12].

Cognitive load theory suggests that humans have a limited cognitive capacity to store only 5–9 pieces of basic information or chunks of information at a time [33], which creates a barrier to manufacturers using social media data. As mentioned above, however, current studies typically average all data to obtain the perceived quality due to the impact of information overload. Despite its simplicity, such a method overlooks the fact that the value and usefulness of each piece of data are different [16, 34], which could leave the truly valuable information overwhelmed, especially when dealing with social media data, the impact of which is exacerbated by the proliferation of data volumes. Hence the current study proposes incorporating information usefulness analysis to quantify consumer perceived quality more accurately.

2.2 Information Usefulness Analysis

In recent years, emerging social media has provided an easy way to share information and the sophistication of products has made potential consumers more dependent on information from social media. But while social media generates massive amounts of data, it also creates a problem of cognitive overload for those who access the information. Consumers' votes of usefulness based on whether the information is useful or not can help manufacturers obtain more valuable information.

Regarding information usefulness research, scholars have proposed many computable models. Among them is the representative text mining approach proposed by Cao, Duan, and Gan [19], which extracted features from social media data on CNET Download.com and investigated the effect of various features (basic, stylistic, semantic) on the number of votes received from online data. Some studies also explored the information usefulness evaluation by more models, such as text regression, SVM, conditional random fields, neural networks and integrated learning. These computational models are generally heavily influenced by the task context; that is, most of the features in these studies are based on experience or an intuitive understanding of what features are important [18]. A key problem with this approach is the lack of a unified theoretical framework.

KAM argues that perceived information usefulness is influenced by the different elements of the received information, it takes the information recipient's perceived usefulness of the information as a direct determinant of information adoption [15], a function that other models cannot achieve. The influencing factors of perceived information usefulness include the argument quality and the source credibility, as shown in Fig. 2. Consequently, the KAM is used in the current study, mainly because it provides a clear pathway for quantifying perceived quality.

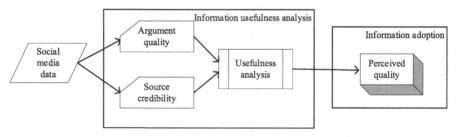

Fig. 2. KAM

The usefulness analysis based on the KAM can derive more valuable information about perceived quality than direct averaging of data [9, 12, 30]. However, such KAM approaches typically ignored the hidden semantic information in the text. Drawing on deep learning, an efficient feature extraction technique, this work integrates the semantic features into KAM theory to improve the effects of information usefulness analysis.

3   Methodology

In this section, we propose a method for capturing perceived quality from social media data. Such data are huge in quantity but not high in valuable information content, so the theoretical framework constructed needs to take this characteristic into account. Therefore, this study constructs a two-stage perceived quality acquisition method guided by KAM theory. First, following the idea of combining CNN and SVM [22], we evaluate the usefulness categories of the data by fusing deep learning with KAM, and then, the quantitative weights of the perceived quality of different usefulness categories are determined, to acquire the consumer perceived quality with the value of information considered. The specific process is shown in Fig. 3.

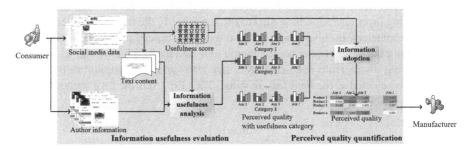

Fig. 3. Method of acquisition of perceived quality based on KAM

3.1   Information Usefulness Evaluation

KAM extracts argument quality features and source credibility features from the data, and the effectiveness of usefulness classification mainly relies on whether these explicit features are constructed comprehensively and accurately [18]. Whereas, social media data contain rich information on how consumers evaluate a product's various quality dimensions [35], and these implicit semantics can be used as an effective complement to KAM features.

Among the deep learning algorithms for processing semantic information of text, Convolutional Neural Network (CNN) shows good performance by extracting high-dimensional semantic features. When dealing with text classification tasks, CNN mainly consists of a convolutional layer, a pooling layer, and a task layer. Assuming that there are k texts in the dataset and the longest length is m (the shorter ones are zeroed), the $m*n$

dimensional matrix X_i of the i-th text is shown in Eq. (1), where x_{ij} is the n dimensional word vector for the j-th word of the i-th text.

$$X_i = \{x_{i1}, x_{i2}, \cdots, x_{im}\}, \quad i = 1, 2, \cdots, k \tag{1}$$

The convolution layer is an important component of the CNN implementation of dimensionality reduction, which is achieved by a sliding scan of the filter with parameters denoted as W, height denoted as h, width denoted as k, and the bias matrix noted as B. The convolution operation is shown in Eq. (2).

$$c_i = f(W \otimes X_{i:i+h-1} + B) \tag{2}$$

where $\otimes$ represents the convolution operator, c_i represents the new feature obtained by convolving the word vectors located at $i : i + h - 1$. $f(\cdot)$ represents the activation function, as the ReLU function can mitigate the gradient disappearance problem to a certain extent, so the ReLU function is chosen here as the activation function, which is expressed as shown in Eq. (3).

$$f(x) = \max(x, 0) \tag{3}$$

The dimensionality of the features after the convolution operation is often still large, and the pooling layer operation is needed to further simplify the features and reduce the number of parameters. In this section, the Max-pooling strategy is chosen, as shown in Eq. (4). Setting the height of the pooled region to h' and the width to w, the whole new feature region is divided into y sub-regions of size $w*h'$, and the maximum value within each region is selected as the final feature value. Then Eq. (5) is the final output feature vector of the CNN.

$$z_i = \max\{c_i\} \tag{4}$$

$$Z_i = \{z_{i1}, z_{i2}, \cdots, z_{iy}\}, \quad i = 1, 2, \cdots, k \tag{5}$$

By incorporating the semantic word vectors extracted by CNN in KAM, we integrate deep learning into KAM theory. The hidden information in the semantic word vector is used as semantic features, which together with the traditional linguistic features constitute the argument quality features. After adding the source reliability features, the features are input to the SVM to perform the final perceived quality usefulness classification, the specific structure of which is shown in Fig. 4.

With AQ_i and SC_i denoting the argument quality linguistic feature vector and the source reliability feature vector of the i-th text after normalization, respectively, and $\oplus$ denoting the vector splice operator, the final input feature vector CK_i to the SVM is generated by the splicing process shown in Eq. (6). After processing by the Support Vector Machine (SVM), which is a classic machine learning classifier, the perceived quality usefulness classification results are obtained.

$$CK_i = Z_i \oplus AQ_i \oplus SC_i \tag{6}$$

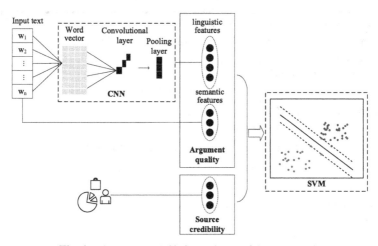

Fig. 4. The structure of information usefulness analysis

3.2 Perceived Quality Quantification

Following the idea of combining CNN and SVM [22], the massive perceived quality information is classified into information categories with different levels of usefulness after usefulness analysis. The high information usefulness perceived quality has greater value, but not always, and information in other usefulness categories is also important [16]. Quantifying the perceived quality with reasonable weights of multiple usefulness categories can, to some extent, assist manufacturers in adopting a more comprehensive perceived quality.

Suppose $CI = \{CI_{j1}, CI_{j2}, \cdots, CI_{jm_j}\}$ denotes the set of perceived quality ratings for product j, and m_j is the number of ratings for that product; $\Phi = \{\phi_1, \phi_2, \cdots, \phi_k\}$ denotes the set of quantitative weights for different usefulness categories, where k is the number of usefulness categories for the consumer data. Consumer perceived quality is expressed as a set of product attribute-rating pairs (A, S) with l product attributes. Then, the set of g products represents $P = \{P_1, P_2, \cdots, P_g\}$, and $RK = \{RK_1, RK_2, \cdots, RK_g\}$ represents the quantified perceived quality information. The process of quantifying perceived quality considering multiple usefulness categories is shown in Algorithm 1.

Algorithm 1 perceived quality quantification method

Input: Set of consumer perceived quality information CI, Set of products P, Set of quantitative weights Φ.

Output: Set of perceived quality RK.

1: Create rationalized perceived quality lists $RK_1, RK_2, \cdots, RK_g$;

2: **for** each product P_x in P **do**

3:   **for** each set I_y in $CI_1, CI_2, \cdots, CI_k$ **do**

4:     Create set P_x_all and put every information into it that talks about P_x ;

5:     Create set $P_x\_I_y$;

6:     **if** P_x is $\varnothing$ **then**

7:       Go to Step 3;

8:     **end if**

9:     **else then**

10:       **for** z in $(1,\ l)$ **do**

11:         $S_z \leftarrow average(S_z)$;

12:         Put (A_z, S_z) into $P_x\_I_y$;

13:       **end for**

14:     **end else**

15:   **end for**

16:   $RK_x = \sum_{i=1}^{k} P_x\_CI_i * \phi_i$;

17: **end for**

18: **return** $RK = \{RK_1, RK_2, \cdots, RK_g\}$;

---

4 Results

The validity and performance of the proposed method need to be critically evaluated. For this purpose, the AutoHome online community, which is currently the largest platform for the exchange of perceptions on the use of automobiles in China, with 45 million daily mobile users as of early 2022, was selected as the data source.

4.1 Data Collection and Description

A Python program was used to crawl 26,051 consumer perceived quality posts from three vehicle categories: mini, midsize, and large in the AutoHome online community. Consistent with existing research [18], this study used vote numbers as an evaluation indicator of information usefulness, with a maximum value of 1082 and a minimum value of 0. The distribution of vote numbers in the dataset is shown in Fig. 5.

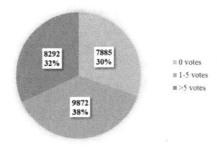

Fig. 5. Distribution of usefulness votes in the data set

4.2 Results of Information Usefulness Analysis

A KAM-based usefulness classification model was constructed to analyze the usefulness of consumer perceived quality in 26,051 data items. Referring to the work of Liu et al. [18], the features in the KAM were identified and used as the basic model, as shown in Table 1.

Table 1. Features of the baseline model

Category	Feature	Details	Category	Feature	Details
Argument quality	F1	Number of characters in the title	Argument quality	F439	Interval between purchase and posting
	F2	Number of characters in the content		F440	Vehicle category
	F3	Number of words in the title		F441	Number of replies
	F4	Number of words in the content		F442	Number of views
	F5	Number of sentences in the content		F443	Number of videos
	F6–F405	n-gram vector		F444–F456	Purchase purpose
	F406–F411	The lexical ratio of the title	Source credibility	F457	Engagement

(continued)

Table 1. (*continued*)

Category	Feature	Details	Category	Feature	Details
	F412–F417	The lexical ratio of the content		F458	Historical posting count
	F418	Average characters per sentence		F459	Number of followers
	F419	Average words per sentence		F460	Number of following
	F420	Number of punctuation marks		F461	Is there a nickname
	F421	Relevance of the title to the content		F462	Region
	F422	Overlapping words in title and content		F463	Is a certified owner of this vehicle or not
	F423–F430	Perceived quality of each attribute		F464	Number of historical best posts
	F431–F438	Information density of each attribute			

The Z-Score method can convert different features to the same scale, making them comparable and increasing the classification effect and interpretability. Thus, before classification, the pre-processed feature set is first normalized by Eq. (7):

$$X_i = \frac{(x_i - \mu)}{\sigma} \tag{7}$$

where X_i represents the normalized feature value, x_i represents the feature value before normalization, μ represents the mean value of the feature value, and σ represents the standard deviation of the feature value.

Considering the superiority of CNN-SVM in text classification [22], we construct this deep learning-based model to quantify the information usefulness. Specifically, BERT was used to vectorize the semantic representation of the text, and then we extracted its high-dimensional features based on CNN, which were subsequently merged with the previous KAM main features and fed into SVM. Parameters of the constructed CNN are shown in Table 2. Meanwhile, Table 3 introduces the experimental environment and configuration.

Table 2. Parameters of the constructed CNN

Key layers by order	Parameters
Convolutional layer	Conv1D (filter = 64, window size = 4)
Pool layer	MaxPool1D
Dense layer	Dense (node = 128, activation = 'relu')
Dropout layer	Dropout (0.4)

Table 3. Experimental environment and configuration

Experimental environment	Configuration
Programming language	Python 3.6
Deep learning framework	TensorFlow 2.4.1
Machine learning framework (SVM)	sklearn.svm
Local development environment	IntelliJ IDEA 2019.1.4
Memory	64G
CPU	32
GPU	Tesla V100-SXM2-32 GB:2

4.3 Results of Perceived Quality Quantification

The value of perceived quality data varies for different degrees of usefulness, but adopting only high usefulness perceived quality would make the information incomplete. Consequently, we attempt to quantify perceived quality by combining high usefulness information, medium usefulness information, and low usefulness information with the weights of ϕ_1, ϕ_2, and ϕ_3 respectively, with reference to Algorithm 1.

Perceived quality, as subjective consumer perception of a product, has a positive correlation with consumer satisfaction [2, 25]. To verify the effectiveness of the proposed method for quantifying perceived quality, we supplementally crawled the overall satisfaction of consumers with different vehicle models on Autohome. As this indicator reflects the acceptance of a vehicle by consumers and is somewhat indicative, we used the Pearson correlation coefficient between the quantified perceived quality information and satisfaction (hereafter referred to as the "correlation coefficient") to evaluate the validity of the quantification method. The higher the value of the correlation coefficient, the closer the perceived quality information is to the true perception of the consumer group. Figure 6 shows the correlation coefficients for various quantitative weights, with $\phi_1 \geq \phi_2 \geq \phi_3$ and $\phi_1 + \phi_2 + \phi_3 = 1$ as rule restrictions to be more realistic.

In Fig. 6, "averaging" is the result of averaging all data directly without classifying usefulness categories; "weighting" is the result of quantifying the information in multiple usefulness categories according to the quantitative weights. As can be seen from Fig. 6, the perceived quality quantified by the weight of "10/0/0" is not most relevant

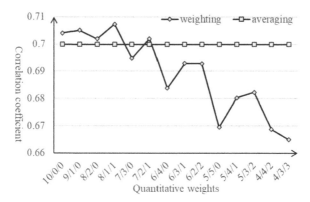

Fig. 6. The quantification effect of different quantitative weights

to satisfaction, so albeit that high usefulness information is recognized by relatively more consumers, it is not fully representative of consumer perceived quality. Perceived quality is best quantified when the quantitative weights for each usefulness category are $\phi_1 = 0.8$, $\phi_2 = 0.1$, $\phi_3 = 0.1$. Accordingly, we quantified the perceived quality by the weights and obtained the perceived quality information.

5 Discussion and Conclusion

Capturing consumers' perceived quality from massive social media data can help manufacturers understand how consumers feel about using their products and improve them in a targeted manner. In this study, a KAM-based perceived quality quantification method is designed to reduce the cognitive load on manufacturers through two stages: information usefulness evaluation and perceived quality quantification. In the information usefulness evaluation stage, a usefulness classification method combining deep learning and KAM is proposed by fusing semantic word vectors in the traditional KAM method, which mines the implicit semantic information to obtain the multiple usefulness categories of the data. In the perceived quality quantification stage, the quantitative weights of the perceived quality with different usefulness categories are determined to obtain perceived quality information. By using Autohome data, the optimal weights for quantification were determined. After obtaining the perceived quality information, it is applied in a competitive analysis.

Compared with previous studies and applications, the present study has several theoretical implications. First, most previous research on quantifying perceived quality directly averaged the data, without considering the varying usefulness of each piece of data. This is mainly because most traditional studies have used structured data such as questionnaires, which on the one hand have small volumes and have little impact without filtering for usefulness, and on the other hand such data lack a usefulness evaluation method. With the development of social media, researchers begin to use social media data to quantify perceived quality, and the massive volume and sparse value of social media data complicate this issue. The proposed method regards the usefulness of each piece of data to quantify perceived quality information that more closely matches group satisfaction. As the first study of perceived quality considering information usefulness, this work is thus significant for studies on perceived quality quantification, especially those using consumer big data, and future studies should pay more attention to the usefulness of the data in their analysis. Furthermore, traditional studies of information usefulness evaluate usefulness based on explicit features in the data, for example, in studies applying KAM, usefulness is classified mainly by the argument quality and the source credibility. However, the text of social media data is rich in hidden semantic information, which contains consumer perceived quality. Based on the idea of fusion of deep learning and machine learning, we further integrate these methods within the KAM framework. As such, this study thus extends information usefulness research by providing a new idea for extracting semantic information using deep learning in usefulness modeling. Subsequent research can build on this foundation for more discussions on the integration of deep learning and usefulness-related theories.

Meanwhile, our research also has practical implications for manufacturers. Social media data, notwithstanding being rich in information about consumer perceptions, is so voluminous and rapidly growing that manufacturers often face cognitive overload when processing it. The proposed framework for quantifying perceived quality enhances manufacturers' consumer big data processing capabilities, which can help them capture consumers' real feelings from it and thus design and improve product attributes in a more targeted manner.

Despite the implications above, some limitations exist in our work that we hope to improve in the future. Firstly, with the continuous development of social media, consumer information distribution is gradually changing towards multi-channel and multi-modal. In the future, we hope to add multi-source and multi-modal data to the overall framework for analysis, especially the comprehensive consideration of reviewers with different identities. In addition, this paper focuses on the application of well-established CNN and SVM for the combination of deep learning with the classical usefulness analysis theory so as to supplement semantic information in usefulness evaluation, but there is no discussion on the efficiency of the methods. More research on the effectiveness and efficiency of different methods is desired in the future.

References

1. Paul, J.: Toward a 'masstige' theory and strategy for marketing. Eur. J. Int. Manag. **12**(5–6), 722–745 (2018)
2. Golder, P.N., Mitra, D., Moorman, C.: What is quality? An integrative framework of processes and states. J. Mark. **76**(4), 1–23 (2012)
3. Harju, C.: The perceived quality of wooden building materials—a systematic literature review and future research agenda. Int. J. Consum. Stud. **46**(1), 29–55 (2022)
4. Akdeniz, M.B., Calantone, R.J.: A longitudinal examination of the impact of quality perception gap on brand performance in the US Automotive Industry. Mark. Lett. **28**(1), 43–57 (2017)
5. Chowdhury, H.K., Ahmed, J.U.: An examination of the effects of partitioned country of origin on consumer product quality perceptions. Int. J. Consum. Stud. **33**(4), 496–502 (2009)
6. Yieh, K., Chiao, Y., Chiu, Y.: Understanding the antecedents to customer loyalty by applying structural equation modeling. Total Qual. Manag. Bus. Excell. **18**(3), 267–284 (2007)

7. Yu, C.J., Wu, L., Chiao, Y., Tai, H.: Perceived quality, customer satisfaction, and customer loyalty: the case of lexus in Taiwan. Total Qual. Manag. Bus. Excell. **16**(6), 707–719 (2007)
8. Duraiswamy, V., Campean, F., Harris, S., Munive-Hernandez, J.E.: Development of a methodology for robust evaluation of perceived quality of vehicle body panel gaps. In: The Proceedings of the DESIGN 2018 15th International Design Conference, (2018)
9. Hazen, B.T., Boone, C.A., Wang, Y., Khor, K.S.: Perceived quality of remanufactured products: construct and measure development. J. Clean. Prod. **142**, 716–726 (2017)
10. Sun, B., Mao, H., Yin, C.: How to identify product defects and segment consumer groups on an online auto forum. Int. J. Consum. Stud. **46**(6), 2270–2287 (2022)
11. Danner, H., Thøgersen, J.: Does online chatter matter for consumer behaviour? A priming experiment on organic food. Int. J. Consum. Stud. **46**(3), 850–869 (2021)
12. He, L., Zhang, N., Yin, L.: The evaluation for perceived quality of products based on text mining and fuzzy comprehensive evaluation. Electron. Commer. Res. **18**(2), 277–289 (2018)
13. Bawden, D., Robinson, L.: The dark side of information: overload, anxiety and other paradoxes and pathologies. J. Inf. Sci. **35**(2), 180–191 (2009)
14. Stylidis, K., Wickman, C., Söderberg, R.: Perceived quality of products: a framework and attributes ranking method. J. Eng. Des. **31**(1), 37–67 (2020)
15. Sussman, S.W., Siegal, W.S.: Informational influence in organizations: an integrated approach to knowledge adoption. Inf. Syst. Res. **14**(1), 47–65 (2003)
16. Wu, R., Wu, H.H., Wang, C.L.: Why is a picture 'worth a thousand words'? Pictures as information in perceived helpfulness of online reviews. Int. J. Consum. Stud. **45**(3), 364–378 (2020)
17. Filieri, R.: What makes online reviews helpful? A diagnosticity-adoption framework to explain informational and normative influences in e-WOM. J. Bus. Res. **68**(6), 1261–1270 (2015)
18. Liu, X., Wang, G.A., Fan, W., Zhang, Z.: Finding useful solutions in online knowledge communities: a theory-driven design and multilevel analysis. Inf. Syst. Res. **31**(3), 731–752 (2020)
19. Cao, Q., Duan, W., Gan, Q.: Exploring determinants of voting for the 'helpfulness' of online user reviews: a text mining approach. Decis. Support Syst. **50**(2), 511–521 (2011)
20. Liu, Z., Park, S.: What makes a useful online review? Implication for travel product websites. Tour. Manage. **47**, 140–151 (2015)
21. Schindler, R.M., Bickart, B.: Perceived helpfulness of online consumer reviews: the role of message content and style. J. Consum. Behav. **11**(3), 234–243 (2012)
22. Chen, Y., Zhang, Z.: Research on text sentiment analysis based on CNNs and SVM. In: 2018 13th IEEE Conference on Industrial Electronics and Applications (ICIEA), pp. 2731–273 (2018)
23. Zeithaml, V.A.: Consumer perceptions of price, quality, and value: a means-end model and synthesis of evidence. J. Mark. **52**(3), 2–22 (1988)
24. Mitra, D., Golder, P.N.: How does objective quality affect perceived quality? Short-term effects, long-term effects, and asymmetries. Mark. Sci. **25**(3), 230–247 (2006)
25. Yoon, B., Jeong, Y., Lee, K., Lee, S.: A systematic approach to prioritizing R&D projects based on customer-perceived value using opinion mining. Technovation **98**, 102164 (2020)
26. Gottlieb, U.R., Brown, M.R., Drennan, J.: The influence of service quality and trade show effectiveness on post-show purchase intention. Eur. J. Mark. **45**(11/12), 1642–1659 (2011)
27. Singh, A., Jenamani, M., Thakkar, J.J., Rana, N.P.: Propagation of online consumer perceived negativity: quantifying the effect of supply chain underperformance on passenger car sales. J. Bus. Res. **132**, 102–114 (2021)
28. Wang, Y.-Y., Guo, C., Susarla, A., Sambamurthy, V.: Online to offline: the impact of social media on offline sales in the automobile industry. Inf. Syst. Res. **32**(2), 582–604 (2021)
29. Slotegraaf, R.J., Inman, J.J.: Longitudinal shifts in the drivers of satisfaction with product quality: the role of attribute resolvability. J. Mark. Res. **41**(3), 269–280 (2004)

30. Stylidis, K., Wickman, C., Söderberg, R.: Defining perceived quality in the automotive industry: an engineering approach. Procedia CIRP **36**, 165–170 (2015)
31. Stylidis, K., Dagman, A., Almius, H., Gong, L., Söderberg, R.: Perceived quality evaluation with the use of extended reality. In: Proceedings of the Design Society: International Conference on Engineering Design, vol. 1, no. 1, pp. 1993–2002 (2019)
32. Ma, M.-Y., Chen, C.-W., Chang, Y.-M.: Using Kano model to differentiate between future vehicle-driving services. Int. Ind. Ergon. **69**, 142–152 (2019)
33. Sweller, J.: Cognitive load during problem solving effects on learning. Cogn. Sci. **12**(2), 257–285 (1988)
34. Moradi, M., Zihagh, F.: A meta-analysis of the elaboration likelihood model in the electronic word of mouth literature. Int. J. Consum. Stud. **46**(5), 1900–1918 (2022)
35. Oh, Y.K., Yi, J.: Asymmetric effect of feature level sentiment on product rating: an application of bigram natural language processing (NLP) analysis. Internet Res. **32**(3), 1023–1040 (2022)

Chinese Medicinal Materials Price Index Trend Prediction Using GA-XGBoost Feature Selection and Bidirectional GRU Deep Learning

Ye Liang ⓘ and Chonghui Guo$^{(\boxtimes)}$ ⓘ

Institute of Systems Engineering, Dalian University of Technology, Dalian, China
dlutguo@dlut.edu.cn

Abstract. Predicting future Chinese medicinal materials price index (CMMPI) trend plays a significant role in risk prevention, cultivation, and trade for farmers and investors. This study aims to design a high precision model to predict the future trend of the CMMPI. The model incorporates environmental factors such as weather conditions and air quality that have a greater impact on the growth of Chinese medical plants and Chinese medicinal materials market supply. In this study, we constructed informative features using the multi-source heterogeneous data. In addition, we proposed a feature selection method based on the genetic algorithm and XGBoost to select features. Finally, we transferred the selected features to the bidirectional GRU deep learning to realize the accurate prediction of the CMMPI trend. We collected 46 CMMPIs datasets to test the proposed model. The results show that the proposed model obtained more superior prediction compared to the state-of-the-art methods, and specialized in predicting long-term goal (90 days). Results also show the weather and air quality data can improve the prediction performance.

Keywords: Chinese medicinal material price index · Genetic algorithm · XGBoost · Feature selection · Deep learning · Prediction

1 Introduction

Medical plants in China are essential for human health, traditional Chinese medicine (TCM), and COVID-19 treatment [1], which has raised global awareness of the advantages of TCM in the medical and health industry. TCM is gradually being accepted and sought after by people inside and outside China. According to China's General Administration of Customs, China exported 70,965 tons of Chinese medicinal materials (CMM) and Chinese patent medicine from

This research is supported by the National Natural Science Foundation of China (Grant No. 71771034), the Liaoning Province Applied Basic Research Program Project (Grant No. 2023JH2/101300208), and the Science and Technology Program of Jieyang (2017xm041).

January to June 2022, the value of which has increased to \$690 million. The World Health Organization estimates that the global trade in medicinal plants and their derivatives will reach \$830 million by 2050 [2]. Under such rapidly developing market circumstances, cultivating CMM has become essential for increasing farmers' incomes.

Chinese Medicinal Material Price Indices (CMMPI) provides valuable insights for TCM stakeholders. Predicting price movements helps manage risk and supports the industry's growth. Since most Chinese medicine materials are from plants [3], environmental factors, including geography, weather, soil, biological factors and human development activities, can affect medical plant growth and product prices, but research on specific price-influencing factors is limited [4]. Weather influences agriculture and the wider economy, but its role in causing economic fluctuations is often overlooked [5]. Changing weather conditions, including rising temperatures, cold periods, droughts, and changes in rainfall patterns, have significant effects on the living environments of Chinese medicinal plants and the sustainable development of the TCM industry [6,7]. Moreover, the State Administration of Traditional Chinese Medicine of the People's Republic of China has formulated the Production Quality Management Standards for Chinese Medicinal Materials. Living conditions are crucial for CMM production, requiring pollution-free environments and meeting air quality standards, making weather and air quality vital for predicting CMMPI trends.

The rapid development of disruptive technologies has led to an explosion of data in various fields, which have brought many different challenges to various organizations [8–10]. Machine learning [11] and its application in TCM research has attracted increasing attention [12,13]. Feature selection plays a vital role in machine learning, providing critical information to solve the given problem and obtain more accurate decisions [14]. CMMPI features contain more valuable information than the price itself. Moreover, features that are not related to price but that act as factors affecting price are helpful for prediction tasks. Thus, this work constructed informative features from multisource heterogeneous data using feature engineering and predicting CMMPI trend from a data-driven point of view.

Deep learning revolutionizes machine learning and data mining, automating processes without manual features or expert experience [15,16]. As one of the variants of Recurrent Neural Network (RNN), Gate Recurrent Unit (GRU) can extract temporal patterns from sequence data and organize information flow, so it can remember temporal patterns in a longer period of time, which has gained wide attention in market price index prediction [17,18]. However, limited research has explored the CMMPI trend using deep learning models with weather data. This study uses a deep learning model to capture the nonlinear relationship between input features and CMMPI prediction, leading to improved performance.

This study proposes a model based on GA-XGBoost feature selection and bidirectional GRU deep learning, called GA-XGB-biGRU, to predict the CMMPI trend. Firstly, the multi-source heterogeneous data, including CMMPI data, weather data, and air quality data, were used to construct the features. Secondly, we proposed a feature selection method based on the genetic algorithm and XGBoost to select features. Then, we transferred the selected features to the bidirectional GRU deep learning as input to capture the data pattern from two directions and finally realize the prediction of the CMMPI. Compared with advanced methods, the proposed model has certain advantages in forecasting the trend of CMMPI.

The structure of this study is as follows: Sect. 2 summarizes the critical related works. Section 3 describes GA-XGB-biGRU in detail, including data retrieval, problem formalization, feature construction, feature selection, model generation, and model evaluation. Section 4 presents the experimental results and analysis. Section 5 discussed the implications of this study. Finally, Sect. 6 concludes the paper and indicates future work.

2   Related Works

2.1   Chinese Medicinal Material Price Prediction

Chang and Mao [19] used the Chengdu Chinese medical material price index from December 2010 to October 2013 to forecast an index based on the ARMA model and established an early-warning model. The same two authors [20] constructed the gray GM(1, 1) model via a differential equation in a discrete form to predict Chinese herbal medicine prices. Li et al. [21] used correlation analysis to analyze factors affecting Panax notoginseng price fluctuations. They trained an LSTM neural network with factor vectors to forecast the price in 2019. Ma et al. [22] adopted a rolling forecast method based on a genetic algorithm and BP neural network to predict the price of Radix notoginseng. The first two consecutive months' price data were used as inputs to train the model, and the next month's price was forecasted. The work of Yu and Guo [23] used news data and a domain dictionary to predict CMMPI movements, improving accuracy significantly. Their prediction results showed that news retrieved by the domain dictionary significantly affects the price of a given material and significantly improves the prediction accuracy. Yao et al. [24] used HP filtering and the ARCH model to analyze monthly price index data of primary indigenous medicinal materials in Guizhou (2010–2019) and explore their fluctuation characteristics. Li et al. [25] proposed a mixed model of Hodrick-Prescott filter, LSTM, and MLP to predict the price index of 30 kinds of CMM. Compared with LSTM and GRU models, the model has certain advantages. Chen et al. [2] predicted a slight increase in sand ginseng market price in Chifeng, Inner Mongolia, and Anguo, Hebei with ARIMA model. This study can guide future medicinal material cultivation decisions.

Fluctuations in the prices of CMM reflect the supply and demand situation in the TCM market. We aimed to predict future price trend using factors that affect supply or demand. Notably, research on CMMPI prediction is insufficient, and most of the available studies predicted CMMPI trend based on price data. In particular, weather and air quality factors that significantly influence CMM' living environments were not considered in these studies.

2.2   Effects of Weather Factors on Markets

Weather contains important factors that affect nearly every economic activity directly or indirectly [26]. Weather effects have been documented in the literature, including its influence on financial markets, investors' behaviors, economic outcomes, agricultural operations, and retail supply chains.

In previous studies, researchers in behavioral finance put their efforts into investigating mood fluctuations induced by weather effects, which influence investors' evaluations of equities. Frühwirth and Sögner [27] studied weather factors and seasonal affective disorder on financial markets, finding varying effects across segments. Barometric pressure affects corporate bonds, while cloud cover and humidity impact stock returns. Yang et al. [28] presented evidence of the relationships between weather factors and investor sentiment in the TAIEX options market using daily stock data between 2008 and 2012 from the Taiwan Future Exchange and daily weather data from the Central Weather Bureau of Taiwan. Shahzad [29] studied the impact of weather on stock returns and volatility using four weather numerically-measured variables and five weather forecast dummy variables and evaluated the effects of both returns and volatility simultaneously using the ARCH model.

Brown et al. [10] explored climate change effects on pasture growth in Charters Towers Region, emphasizing the importance of a qualified climate prediction system and predicting pasture growth days for better grazing management. Bisbis et al. [30] reviewed climate change impacts on vegetables include production benefits from increased carbon dioxide but potential alterations to product quality. Jørgensen et al. [31] analyzed a climate risk insurance scheme and explored under what conditions farmers would prefer to purchase crop insurance in the market or adapt to climate change via farm management. Le Gouis et al. [32] found that climate changes and extreme years significantly affect wheat production. Warm late autumn, wet spring, and low solar radiation reduce grain yields.

Based on the abovementioned literature review and analysis, the following conclusions can be drawn: (1) Inadequate research on CMMPI trend prediction, especially in combining weather and air quality data. (2) Limited literature on using deep learning with weather data to predict CMMPI trend. (3) This study lays the foundation for future research on CMMPI trend prediction using heterogeneous data and deep learning methods.

3   Methodology

In this study, we proposed a model based on GA-XGBoost feature selection and bidirectional GRU deep learning, named GA-XGB-biGRU, to predict the CMMPI trend. We illustrated the main workflows of GA-XGB-biGRU in Fig. 1. There are five typical and necessary fundamental tasks: data input, feature construction, feature selection, model generation, and prediction.

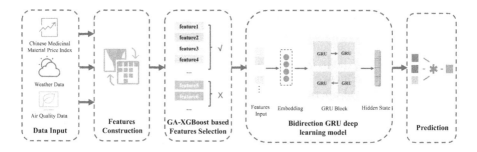

Fig. 1. The workflows of GA-XGB-biGRU

3.1 Materials

Data was collected from Chinese medicinal material markets(cnkmprice.kmzyw.com.cn/), weather stations(www.ncdc.noaa.gov), and air quality monitoring platforms(www.aqistudy.cn/) spanning different time periods. Weather data from 33 meteorological stations (2000–2019) and air quality data from 365 cities (2013–2019) were retrieved. CMMPI data, including 46 price indices, was obtained from the Kangmei China CMMPI website (2014–2019). The specific indicators of all data and the details of the collected CMMPI are as shown in Table 1 and Tabel 2, respectively.

Table 1. Indicator categories and specific indicators

Indicator categories	Specific indicators
Chinese medicinal materials price index	Price index
Weather indicators	Precipitation, Average temperature, Maximum temperature, Minimum temperature, Temperature, Wind direction, Wind speed, Dewpoint temperature, Height, Sea level pressure, Visibility, Relative humidity
Air quality indicators	Air quality index, PM2.5, PM10, Sulfur dioxide (SO_2), Carbon monoxide (CO), Nitrogen dioxide (NO_2), Ozone (O_3), Air quality level

3.2 Problem Formalization

We aimed to predict CMMPI trend with given multisource data. Specifically, given CMMPI data $X_p = [p_{i,j}]_{T \times M}$, weather data $X_s = [s_{i,j}]_{T \times N}$, and air quality data $X_q = [q_{i,j}]_{T \times K}$ as three kinds of time series, where T, M, N, and K represents the length of each time series, the number of features of CMMPI, weather data and air quality data, respectively.

Table 2. Indicator categories and specific indicators

Index category	Index code(Index detail)
General price index	ALL (The superlative price index the company compiled)
Position index	A-000, B-000, C-000, D-000, E-000, F-000, G-000, H-000, I-000, J-000, K-000, I-000
Origin index	CD-001, CD-002, CD-003, CD-004, CD-005, CD-006, CD-007, CD-008, CD-009, CD-010, CD-011, CD-012, CD-013, CD-014, CD-015, CD-016, CD-017, CD-018, CD-019, CD-020, CD-021, CD-022, CD-023, CD-024, CD-025, CD-026, CD-029, CD-033
Market index	Market 121, Market 122, Market 123, Market 141, Market 201

Future returns were calculated over different time periods to determine the price index trend label. Positive returns indicated a positive label, while negative returns indicated a negative label, considering investors' varying risk preferences. We limited our calculation of future arithmetic returns to classical formulas:

$$R_t = p_{t+l,j} - p_{t,j}, (t \in [1, T - l], j \in (1, 2, \cdots, M)) \tag{1}$$

where R_t denotes the change in the associated CMMPI and is used as the arithmetic return of the next l days in the future, and $p_{t,j}$ and $p_{t+l,j}$ denote the price index of features j on the current date t and that over the next l days in the future. In this study, we used original CMMPI values to construct R_t. The labeling methodology described was calculated as follows:

$$y_t = \begin{cases} 1 \text{ , if } R_t > v \\ 0 \text{ , if } R_t \le v \end{cases} \tag{2}$$

3.3 Feature Construction

We constructed statistical features for historical weather, air quality, and price data, considering the influence of the environment on CMM growth cycles and investors' expectations: (1) Maximum value in the previous k days. (2) Minimum value in the previous k days. (3) Average value in the previous k days.

Given a time series $X = \{x_1, x_2, \cdots, x_T\}$ and a parameter k, the general form of applying the moving average (or maximum or minimum) for a centered date t is represented by the following equation:

$$C_t = f(x_{(t-k):(t-1)}), t \in [k+1, T] \tag{3}$$

where C_t is the result of the moving average (maximum or minimum) at date t, $x_{(t-k):(t-1)}$ denotes interval data contains the first k values, f denotes the function mean (max or min), and T is the length of the given time series.

Two more simple features that we constructed are as follows: (1) The value over the previous k days. (2) The location relative to the day before the first of k days. The aim of constructing relative location features is to make the most of the relationship between current and historical values rather than just their difference. The general form of the relative location equation is:

$$Level_t = \frac{x_t - mean(x_{(t-k):(t-1)})}{mean(x_{(t-k):(t-1)})}, t \in [k+1, T] \tag{4}$$

where $Level_t$ is the value of the relative location at date t, $mean(x_{(t-k):(t-1)})$ denotes the moving average of the interval data over the previous k days, and x_t is the value at date t.

We constructed features for the CMMPI, weather, and air quality data depending on the set of empirical values for k mentioned above. Since the original data directly reflect the market supply and demand situation, we also used the original data directly as inputs. We list the features and corresponding implications in Table 3.

Table 3. Feature naming and meaning

ID	Features ($k = 10$, 30, 60)	Implication	Numbers of features
1	Fea_value	Original indicators	26
2	Fea_ma_k	Average value in previous k days	60
3	Fea_max_k	Maximum value in previous k days	3
4	Fea_min_k	Maximum value in previous k days	3
5	Fea_shift_k	The value of previous k days	3
6	Fea_level_k	The relative location to the day before k days	60

3.4   GA-XGBoost Feature Selection

Feature selection is crucial to simplify tasks and avoid complexity caused by a large number of features. Genetic Algorithm (GA) is a meta-heuristic method using natural selection to find optimal solutions. It employs mathematical operators such as crossover, mutation, fitness, and survival of the fittest. Therefore, GA is widely used for optimal selection [33,34]. GA solves complex problems without standardized optimization formulas. It uses chromosomes, mutation, and crossing to evolve and select based on the fitness function. In this study, XGBoost, known for its fast learning and high accuracy, was used for machine learning. The initial population had ten chromosomes, with binary genes representing relevant features. Genetic algorithm operations were applied, and the XGBoost classification AUC was computed as the fitness measure. After 30 iterations, the chromosome with the highest fitness represented the optimal feature set.

3.5   Bidirectional GRU Deep Learning and Evaluation

After obtaining the optimal feature combination, we used bidirectional GRU deep learning to predict price index movements, treating it as a classification problem. Since the features of price index, weather, and air quality are not in the same vector space, a one-dimensional feedforward neural network FNN is used to transform the feature vectors into the same feature space:

$$\dot{x} = FNN(X_f) = \text{ReLU}(W_f X_f + b)(\dot{x} \in \mathbb{R}^d) \tag{5}$$

where $X_f \in \mathbb{R}^L$ denote the feature vector which contain selected L features, $W_f \in \mathbb{R}^{d \times L}$, $b \in \mathbb{R}^d$ are the parameters of FNN, d is the dimension of the dense vector $\dot{x}$, and ReLU is the activate function. The feature vectors X_f are transformed into dense vector $\dot{x}$. After the embedding operation, dense vector $\dot{x}$ was transmitted into the GRU [17] to learn hidden representation.

$$h_i = GRU(\dot{x}, h_{i-1}) \tag{6}$$

We used $\overrightarrow{h_i}$ and $\overleftarrow{h_i}$ to represent the output of directional GRU and concat $\overrightarrow{h_i}$ and $\overleftarrow{h_i}$ to form hidden representation $h_c = [\overrightarrow{h_i}, \overleftarrow{h_i}] \in \mathbb{R}^{2d}$. Finally, we adopt a logistic regression as the classifier for CMMPI trend:

$$\hat{y} = \sigma(W_c h_c + b_c) \tag{7}$$

where $\hat{y}$ is the prediction, $\sigma(x) = 1/(1 + \exp(-\text{x}))$ denoted the Sigmoid function, and $W_c \in \mathbb{R}^{2d}$, $b_c \in \mathbb{R}$ are learnable weights. With the training set $\mathcal{D}$, binary cross-entropy loss $\mathcal{L}$ between the ground truth y and the prediction probabilities $\hat{y}$ to train the model and get the learned parameters θ,

$$\mathcal{L} = -\frac{1}{|\mathcal{D}|} \sum_{i=1}^{|\mathcal{D}|} (y_i \log(\hat{y}_i) + (1 - y_i) \log(1 - \hat{y}_i)) \tag{8}$$

where i is the index of samples and $|\mathcal{D}|$ is the number of samples.

In order to verify the validity of proposed model, several commonly used benchmark methods are selected for comparison: Decision Trees (DT), Random Forest (RF) [35], Gradient Boosting Machine (GBM) [36,37]. Artificial neural networks (NN) [21], Support vector machine (SVM) [38], GRU [17], and Gaussian mixture model (GMM) [39]. We evaluated the models based on the confusion matrix:

$$Accuracy = \frac{TP + TN}{N} \tag{9}$$

$$Precision = \frac{TP}{TP + FP} \tag{10}$$

$$Recall = \frac{TP}{TP + FN} \tag{11}$$

$$F1 = \frac{2 \times Precision \times Recall}{Precision + Recall} \tag{12}$$

where TP and TN denote the true positive and true negative data, FP and FN represent the false positive and false negative data, respectively, and N denotes the total number of data records. Besides, we adopted AUC (area under ROC curve) [40] as an overall performance measure.

4 Experiment and Results

4.1 Experimental Process

We used nested cross validation [41,42], to evaluate each model's performance comprehensively. We evaluated model performance using shuffled data to avoid bias and averaged test prediction results. We compared GA-XGB-biGRU's classification effects with the above methods on all constructed features. For the labels, we tested nine values of l ($l = 10, 20, \cdots, 90$) and three values of v ($v = 0, 2, 5$). For instance, $l = 10$ and $v = 5$ mean predicting whether the given CMMPI will increase by 5 yuan in the next 10 days. We applied the SMOTE [43] to address the class imbalance problem in our study. Finally, we used the averaged prediction result in the test step to evaluate the model performance. Optimal parameters for the model were obtained (Table 4) through parameter tuning before testing.

Table 4. List of classification algorithms and their optimal parameters

Algorithm	Optimal parameters
Decision Trees (DT), GMM	–
Random Forests (RF)	The number of trees in the forest, $ntrees = 500$ The function to measure the quality of a split: $impurity =$ "$gini$" The maximum depth of the tree, $maxDepth = 50$ The number of features to consider: $max\_features = sqrt$(number of features)
Gradient Boosting Machine (GBM)	Based classifier: decision trees The number of trees, $ntrees = 500$ Learning rate, $learning\_rate = 0.001$
Neural Networks (NN)	The number of hidden layers: $layers = 2$ The number of hidden layer units: $size = 128$ Learning rate: $learning\_rate = 0.0001$ Activation function: $activation =$ "$relu$"
Support Vector Machines (SVM)	Cost of constraints violation: $C = 10$ Kernel type: $kernel =$ "rbf"
GRU	Learning rate: $learning\_rate = 0.0001$ Optimizer: Adam The number of embedding layer units: $size = 128$ The number of hidden layer units, $size = 128$

4.2 Results and Discussion

Table 5 shows the classification performances of six algorithms, including "Ours" (GA-XGB-biGRU), for CMMPI under various label parameters ($l = 10, 20, \cdots, 90$) and $v = 5$. In Table 5, deep learning outperforms traditional shallow methods significantly in classification results. The proposed model showed the best performance, followed by basic GRU. SVM and GBM also demonstrated advantages over traditional methods, with SVM having the highest accuracy, followed by GBM. DT's performance was the worst, while NN performed better but was complex and less interpretable.

We conducted ablation experiments on feature construction and feature selection separately to observe their impact on the model's performance. The results are shown in Table 6 and Table 7, respectively. In Table 6, it can be seen that constructed features add valuable information to the classification model, improving its performance over using only CMMPI. In Table 7, GA-XGB-biGRU performs similarly to the basic GRU model (AUC 0.76), but GA-XGBoost's feature selection improves classification by eliminating redundancies and selecting relevant features.

Table 5. Average evaluation results of comparative classification algorithms

Evaluate indicators	DT	RF	NN	GBM	SVM	GMM	GRU	Ours
Accuracy	0.64 ± 0.11	0.70 ± 0.08	0.70 ± 0.24	0.76 ± 0.06	0.77 ± 0.10	0.72 ± 0.04	0.82 ± 0.06	$\mathbf{0.83} \pm 0.09$
Precision	0.53 ± 0.10	0.56 ± 0.10	0.58 ± 0.07	0.60 ± 0.16	0.56 ± 0.15	0.56 ± 0.13	0.76 ± 0.15	$\mathbf{0.80} \pm 0.08$
Recall	0.51 ± 0.18	0.56 ± 0.10	0.56 ± 0.09	0.60 ± 0.11	0.60 ± 0.10	0.60 ± 0.11	0.74 ± 0.07	$\mathbf{0.78} \pm 0.12$
F1	0.52 ± 0.09	0.57 ± 0.09	0.56 ± 0.08	0.59 ± 0.10	0.58 ± 0.10	0.58 ± 0.12	0.74 ± 0.11	$\mathbf{0.77} \pm 0.11$
AUC	0.52 ± 0.08	0.57 ± 0.08	0.60 ± 0.07	0.58 ± 0.10	0.58 ± 0.07	0.57 ± 0.04	0.76 ± 0.07	$\mathbf{0.78} \pm 0.10$

Table 6. AUC results in comparison between with/without feature construction

Feature construction	DT	RF	NN	GBM	SVM	GMM	GRU	Ours
With	0.52 ± 0.08	0.57 ± 0.08	0.60 ± 0.07	0.58 ± 0.10	0.58 ± 0.07	0.57 ± 0.04	0.76 ± 0.07	$\mathbf{0.78} \pm 0.10$
Without	0.49 ± 0.12	0.53 ± 0.11	0.56 ± 0.10	0.55 ± 0.11	0.54 ± 0.13	0.53 ± 0.16	0.65 ± 0.13	$\mathbf{0.67} \pm 0.14$

Table 7. AUC results in comparison between with/without GA-XGBoost feature selection

	GRU	Ours (without GA-XGBoost)	Ours
Average AUC	0.76 ± 0.07	0.76 ± 0.15	$\mathbf{0.78} \pm 0.10$

According to the CMMPI category, we calculated statistics for the classification results of each method and obtained the average classification results in each evaluation index, as shown in Table 8. Deep learning excels in predicting price indexes, with the general price index showing the best performance for AUC and the position index performing less effectively due to CMM's location and supply-demand variations.

Table 8. Average evaluation results by categories of CMMPI

Evaluate indicator	Index category	DT	RF	NN	GBM	SVM	GMM	GRU	Ours
Accuracy	General price index	0.65 ± 0.25	0.72 ± 0.12	0.69 ± 0.24	0.74 ± 0.06	0.77 ± 0.15	0.76 ± 0.13	**0.82 ± 0.05**	**0.82 ± 0.06**
	Market index	0.59 ± 0.04	0.66 ± 0.05	0.69 ± 0.05	0.71 ± 0.03	0.77 ± 0.04	0.72 ± 0.06	**0.78 ± 0.06**	**0.78 ± 0.08**
	Origin index	0.60 ± 0.05	0.64 ± 0.06	0.63 ± 0.08	0.67 ± 0.08	0.71 ± 0.11	0.69 ± 0.09	**0.81 ± 0.06**	**0.81 ± 0.09**
	Position index	0.64 ± 0.08	0.69 ± 0.10	0.70 ± 0.10	0.71 ± 0.09	0.73 ± 0.10	0.71 ± 0.10	**0.79 ± 0.07**	**0.79 ± 0.11**
Precision	General price index	0.52 ± 0.08	0.63 ± 0.10	0.57 ± 0.04	0.67 ± 0.11	0.60 ± 0.06	0.62 ± 0.06	0.80 ± 0.15	**0.82 ± 0.06**
	Market index	0.50 ± 0.09	0.52 ± 0.09	0.61 ± 0.06	0.57 ± 0.11	0.59 ± 0.08	0.57 ± 0.07	0.75 ± 0.14	**0.78 ± 0.08**
	Origin index	0.54 ± 0.12	0.53 ± 0.11	0.55 ± 0.10	0.56 ± 0.22	0.53 ± 0.23	0.55 ± 0.13	0.75 ± 0.14	**0.77 ± 0.09**
	Position index	0.54 ± 0.12	0.55 ± 0.11	0.57 ± 0.08	0.58 ± 0.22	0.53 ± 0.22	0.57 ± 0.22	0.74 ± 0.15	**0.79 ± 0.09**
Recall	General price index	0.49 ± 0.24	0.67 ± 0.25	0.60 ± 0.23	0.69 ± 0.25	0.61 ± 0.18	0.65 ± 0.15	0.78 ± 0.05	**0.81 ± 0.08**
	Market index	0.51 ± 0.04	0.55 ± 0.03	0.54 ± 0.03	0.57 ± 0.03	0.64 ± 0.06	0.59 ± 0.05	**0.76 ± 0.06**	0.75 ± 0.11
	Origin index	0.53 ± 0.05	0.55 ± 0.05	0.55 ± 0.05	0.57 ± 0.07	0.59 ± 0.09	0.58 ± 0.07	0.75 ± 0.09	**0.78 ± 0.13**
	Position index	0.54 ± 0.05	0.57 ± 0.09	0.56 ± 0.08	0.59 ± 0.08	0.56 ± 0.07	0.56 ± 0.07	0.74 ± 0.09	**0.76 ± 0.15**
F1	General price index	0.50 ± 0.24	0.65 ± 0.10	0.58 ± 0.11	0.67 ± 0.08	0.61 ± 0.12	0.63 ± 0.11	0.79 ± 0.11	**0.81 ± 0.07**
	Market index	0.51 ± 0.04	0.53 ± 0.05	0.57 ± 0.08	0.57 ± 0.10	0.61 ± 0.04	0.58 ± 0.09	0.75 ± 0.10	**0.76 ± 0.09**
	Origin index	0.53 ± 0.05	0.54 ± 0.11	0.55 ± 0.07	0.56 ± 0.10	0.56 ± 0.11	0.56 ± 0.08	0.75 ± 0.10	**0.78 ± 0.11**
	Position index	0.54 ± 0.05	0.56 ± 0.09	0.56 ± 0.09	0.58 ± 0.11	0.54 ± 0.13	0.56 ± 0.11	0.74 ± 0.11	**0.76 ± 0.13**
AUC	General price index	0.52 ± 0.06	0.64 ± 0.07	0.66 ± 0.04	0.65 ± 0.08	0.60 ± 0.04	0.63 ± 0.07	0.80 ± 0.05	**0.81 ± 0.08**
	Market index	0.54 ± 0.09	0.55 ± 0.08	0.61 ± 0.06	0.57 ± 0.09	0.61 ± 0.06	0.60 ± 0.07	0.77 ± 0.06	**0.77 ± 0.11**
	Origin index	0.52 ± 0.10	0.53 ± 0.09	0.55 ± 0.10	0.55 ± 0.11	0.56 ± 0.09	0.55 ± 0.11	0.74 ± 0.07	**0.78 ± 0.11**
	Position index	0.51 ± 0.09	0.54 ± 0.09	0.58 ± 0.08	0.54 ± 0.10	0.54 ± 0.08	0.56 ± 0.10	0.73 ± 0.08	**0.76 ± 0.12**

Table 9. The p-values of the Wilcoxon rank sum test

$H0$: Indicator's values are equal	DT vs Ours	RF vs Ours	NN vs Ours	GBM vs Ours	SVM vs Ours	GMM vs Ours	GRU vs Ours
Accuracy	1.0425e−9***	1.5373e−9***	1.9876e−8***	1.0176e−8***	1.0526e−8***	1.4137e−8***	1.5879e−6***
F1	1.2378e−9***	2.3579e−9***	2.4757e−9***	2.4757e−9***	2.4757e−9***	2.4757e−9***	0.8979
AUC	1.0338e−9***	2.4757e−9***	2.6413e−9***	2.6413e−9***	2.6413e−9***	2.4757e−9***	6.2884e−5***

Significance levels: ***< 0.001.

Wilcoxon rank sum tests were used to compare the classification results of GA-XGB-biGRU with other methods, considering label parameters $l = 10$ and $v = 5$. As shown in Table 9, GA-XGB-biGRU shows significant differences in Accuracy and AUC compared to traditional shallow methods, while F1-Score has no significant difference between GA-XGB-biGRU and GRU.

The effect of the period length l on detecting the predictive accuracy is shown in Fig. 2. Longer length periods lead to more accurate classification performance. Deep learning exhibits stable advantages in classification. Traditional shallow methods have varying prediction results for different periods. Short-term prediction is challenging due to volatility, while long-term prediction is more accurate but faces feature validity challenges. GA-XGB-biGRU performs similarly to GRU but saves storage by eliminating redundant features through GA-XBoost selection.

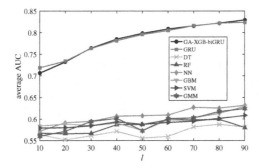

Fig. 2. AUC values of methods under different period lengths l

We take the average AUC of parameters $l = 10$ to 90 and $v = 0, 2$, and 5 to observe the classification results of GA-XGB-biGRU on different labels in each dataset, which is shown in Fig. 3. The overall prediction results of $v = 5$ in Fig. 3 are better and more stable than the classification results of $v = 0, 2$. It is worth noting that a greater value v is not necessarily better. An appropriate price threshold improves prediction results and aligns with business scenarios.

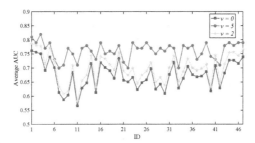

Fig. 3. Average AUC values of each data sets

We validated weather and air quality data in Yunnan Province using CD-001 index. Combined with price data, we assessed their impact on predictions. The comparison results are shown in Fig. 4. As can be seen from Fig. 4, under different values l, the prediction results with weather and air quality features are better than that without weather and air quality features.

A total of 71 features were selected by GA-XGBoost. To facilitate the analysis, we counted the frequency of features of the same type, and the detailed results are shown in Fig. 5. Out of the 71 features, 7 are related to price index, 9 to weather (including precipitation, height, maximum temperature, visibility, and minimum temperature), and 6 to air quality (PM2.5 and NO_2). It is indicated that the informative additional data provide effectiveness for predicting the CMMPI trend.

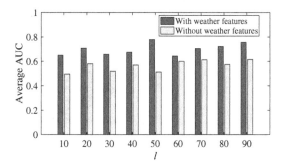

Fig. 4. CD-001 average AUC values with/without weather and air quality features

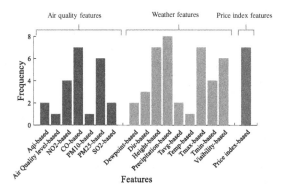

Fig. 5. Selected features of CD-001

5   Implication

Predicting CMMPI trend is essential due to the rapid growth of the TCM industry. An advanced model is needed to uncover latent influencing factors and aid decision-making. From the methodology perspective, This study proposes GA-XGB-biGRU, a model combining GA-XGBoost feature selection and bidirectional GRU deep learning, for CMMPI trend prediction. The model shows advantages compared to advanced methods and offers diverse price index prediction solutions for different users. To maintain stable TCM prices and avoid frequent sharp fluctuations, government departments should implement policies that balance long-term development and short-term market health. Such measures are vital to benefit supply chain stakeholders and the entire CMM industry chain. From a marketing planning point of view, using the proposed model for CMMPI prediction benefits farmers' planting plans, assists investors in managing investment amounts, and reduces capital risk. Government and market regulators can use the predictive results for price management and planned medicine production, fostering TCM industry development.

6   Conclusions

We proposed a GA-XGBoost feature selection and bidirectional GRU model for predicting 46 CMMPIs' price trends, demonstrating its superiority over six classification algorithms. Our work's main contributions are as follows: (1) GA-XGB-biGRU achieved superior performance with an average classification AUC of 0.78, outperforming comparative methods significantly. Feature construction and selection improved prediction performance. (2) We explored label parameter thresholds' effects and found long-term predictive goals (90 days) yielded better results. (3) Weather and air quality data validity were verified using Yunnan's original index, showing improved prediction performance with aggregated or corresponding provincial data. This study has limitations. Firstly, the model may not fully grasp policy-related information. Relevant policies impact prices, but constructed features might struggle to capture policy nuances. Secondly, deep learning's opacity is a drawback, requiring interpretable techniques to analyze feature contributions. Future research should improve CMMPI trend prediction by incorporating more data, developing an interpretable model based on economic rationale, and investigating diverse factors' influence on CMMPI trends using data-driven methods.

References

1. Wang, W., Xie, Y., Zhou, H., Liu, L.: Contribution of traditional Chinese medicine to the treatment of COVID-19. Phytomedicine **85**, 153279 (2021). https://doi.org/10.1016/j.phymed.2020.153279
2. Chen, Y., et al.: Quality control of Glehniae Radix, the root of Glehnia Littoralis Fr. Schmidt ex Miq., along its value chains. Front. Pharmacol. **12**, 2752 (2021). https://doi.org/10.3389/fphar.2021.729554
3. Chan, K.: Chinese medicinal materials and their interface with western medical concepts. J. Ethnopharmacol. **96**(1–2), 1–18 (2005). https://doi.org/10.1016/j.jep.2004.09.019
4. Cunningham, A., Long, X.: Linking resource supplies and price drivers: lessons from Traditional Chinese Medicine (TCM) price volatility and change, 2002–2017. J. Ethnopharmacol. **229**, 205–214 (2019). https://doi.org/10.1016/j.jep.2018.10.010
5. Gallic, E., Vermandel, G.: Weather shocks. Eur. Econ. Rev. **124**, 103409 (2020). https://doi.org/10.1016/j.euroecorev.2020.103409
6. Chi, X., et al.: Threatened medicinal plants in China: distributions and conservation priorities. Biol. Cons. **210**, 89–95 (2017). https://doi.org/10.1016/j.biocon.2017.04.015
7. Gupta, A., et al.: Medicinal plants under climate change: impacts on pharmaceutical properties of plants. In: Climate Change and Agricultural Ecosystems, pp. 181–209. Elsevier (2019). https://doi.org/10.1016/B978-0-12-816483-9.00008-6

8. Katal, A., Wazid, M., Goudar, R.H.: Big data: issues, challenges, tools and good practices. In: 2013 6th International Conference on Contemporary Computing, pp. 404–409. IEEE, India (2013). https://doi.org/10.1109/IC3.2013.6612229

9. Canito, J., Ramos, P., Moro, S., Rita, P.: Unfolding the relations between companies and technologies under the big data umbrella. Comput. Ind. **99**, 1–8 (2018). https://doi.org/10.1016/j.compind.2018.03.018

10. Brown, J.N., Ash, A., MacLeod, N., McIntosh, P.: Diagnosing the weather and climate features that influence pasture growth in Northern Australia. Clim. Risk Manag. **24**, 1–12 (2019). https://doi.org/10.1016/j.crm.2019.01.003

11. Huck, N.: Large data sets and machine learning: applications to statistical arbitrage. Eur. J. Oper. Res. **278**(1), 330–342 (2019). https://doi.org/10.1016/j.ejor.2019.04.013

12. Yu, T., et al.: Knowledge graph for TCM health preservation: design, construction, and applications. Artif. Intell. Med. **77**, 48–52 (2017). https://doi.org/10.1016/j.artmed.2017.04.001

13. Han, N., Qiao, S., Yuan, G., Huang, P., Liu, D., Yue, K.: A novel Chinese herbal medicine clustering algorithm via artificial bee colony optimization. Artif. Intell. Med. **101**, 101760 (2019). https://doi.org/10.1016/j.artmed.2019.101760

14. Huber, J., Müller, S., Fleischmann, M., Stuckenschmidt, H.: A data-driven newsvendor problem: from data to decision. Eur. J. Oper. Res. **278**(3), 904–915 (2019). https://doi.org/10.1016/j.ejor.2019.04.043

15. Zhong, L., Hu, L., Zhou, H.: Deep learning based multi-temporal crop classification. Remote Sens. Environ. **221**, 430–443 (2019). https://doi.org/10.1016/j.rse.2018.11.032

16. Khan, A., Vibhute, A.D., Mali, S., Patil, C.: A systematic review on hyperspectral imaging technology with a machine and deep learning methodology for agricultural applications. Ecol. Inform. 101678 (2022). https://doi.org/10.1016/j.ecoinf.2022.101678

17. Gupta, U., Bhattacharjee, V., Bishnu, P.S.: StockNet-GRU based stock index prediction. Expert Syst. Appl. **207**, 117986 (2022). https://doi.org/10.1016/j.eswa.2022.117986

18. Jiang, W.: Applications of deep learning in stock market prediction: recent progress. Expert Syst. Appl. **184**, 115537 (2021). https://doi.org/10.1016/j.eswa.2021.115537

19. Chang, F., Mao, Y.: Study on early-warning of Chinese materia medica price base on ARMA model. China J. Chin. Materia Med. **39**(9), 1721–1723 (2014)

20. Mao, Y., Chang, F.: Prediction of Chinese herbal medicine price index based on gray GM(1,1) prediction model. China Pharm. **25**(23), 2200–2202 (2014)

21. Li, F., Song, Q., Chen, C., Liu, J., Gao, X.: Forecast of Panax notoginseng price index based on LSTM neural network. Mod. Chin. Med. **21**(4), 536–541 (2019)

22. Ma, G., Ma, D., Shao, X.: On price forecasting of radix Notoginseng based on genetic BP neural network. J. Tianjin Norm. Univ. (Nat. Sci. Edn.) **37**(6), 76–80 (2019)

23. Yu, M., Guo, C.: Using news to predict Chinese medicinal material price index movements. Industr. Manag. Data Syst. **118**(5), 998–1017 (2018). https://doi.org/10.1108/IMDS-06-2017-0287

24. Yao, Q., Huang, Y., Lu, D., Tian, W.: Study on price index fluctuation of China's authentic medicinal materials-taking Guizhou Ainaxiang, Uncaria and other eight high-quality traditional Chinese medicinal meterials as examples. Price: Theory Pract. **2**, 87–90 (2021)

25. Li, Y., Yang, H., Liu, J., Fu, H., Chen, S.: Predicting price index of Chinese herval medicines in China. J. Huazhong Agric. Univ. **40**(6), 50–59 (2021). https://doi.org/10.13300/j.cnki.hnlkxb.2021.06.007

26. Lazo, J.K., Lawson, M., Larsen, P.H., Waldman, D.M.: U.S. economic sensitivity to weather variability. Bull. Am. Meteorol. Soc. **92**(6), 709–720 (2011). https://doi.org/10.1175/2011BAMS2928.1

27. Frühwirth, M., Sögner, L.: Weather and SAD related mood effects on the financial market. Q. Rev. Econ. Financ. **57**, 11–31 (2015). https://doi.org/10.1016/j.qref.2015.02.003

28. Yang, C.Y., Jhang, L.J., Chang, C.C.: Do investor sentiment, weather and catastrophe effects improve hedging performance? Evidence from the Taiwan options market. Pac. Basin Financ. J. **37**, 35–51 (2016). https://doi.org/10.1016/j.pacfin.2016.03.002

29. Shahzad, F.: Does weather influence investor behavior, stock returns, and volatility? Evidence from the Greater China region. Phys. A **523**, 525–543 (2019). https://doi.org/10.1016/j.physa.2019.02.015

30. Bisbis, M.B., Gruda, N., Blanke, M.: Potential impacts of climate change on vegetable production and product quality-a review. J. Clean. Prod. **170**, 1602–1620 (2018). https://doi.org/10.1016/j.jclepro.2017.09.224

31. Jørgensen, S.L., Termansen, M., Pascual, U.: Natural insurance as condition for market insurance: climate change adaptation in agriculture. Ecol. Econ. **169**, 106489 (2020). https://doi.org/10.1016/j.ecolecon.2019.106489

32. Le Gouis, J., Oury, F.X., Charmet, G.: How changes in climate and agricultural practices influenced wheat production in Western Europe. J. Cereal Sci. **93**, 102960 (2020). https://doi.org/10.1016/j.jcs.2020.102960

33. Chung, H., Shin, K.S.: Genetic algorithm-optimized multi-channel convolutional neural network for stock market prediction. Neural Comput. Appl. **32**(12), 7897–7914 (2020). https://doi.org/10.1007/s00521-019-04236-3

34. El-Rashidy, M.A.: A novel system for fast and accurate decisions of gold-stock markets in the short-term prediction. Neural Comput. Appl. **33**(1), 393–407 (2021). https://doi.org/10.1007/s00521-020-05019-x

35. Yin, L., Li, B., Li, P., Zhang, R.: Research on stock trend prediction method based on optimized random forest. CAAI Trans. Intell. Technol. 1–11 (2021). https://doi.org/10.1049/cit2.12067

36. Shrivastav, L.K., Kumar, R.: An ensemble of random forest gradient boosting machine and deep learning methods for stock price prediction. CAAI Trans. Intell. Technol. 1–11 (2021). https://doi.org/10.4018/JITR.2022010102

37. Lu, H., Mazumder, R.: Randomized gradient boosting machine. SIAM J. Optim. **30**(4), 2780–2808 (2020). https://doi.org/10.1137/18M1223277

38. Cortes, C., Vapnik, V.: Support-vector networks. Mach. Learn. **20**(3), 273–297 (1995). https://doi.org/10.1007/BF00994018

39. Aljohani, H.M., Elhag, A.A.: Using statistical model to study the daily closing price index in the Kingdom of Saudi Arabia (KSA). Complexity **2021**, 1–5 (2021). https://doi.org/10.1155/2021/5593273

40. Sokolova, M., Lapalme, G.: A systematic analysis of performance measures for classification tasks. Inf. Process. Manage. **45**(4), 427–437 (2009). https://doi.org/10.1016/j.ipm.2009.03.002

41. Varma, S., Simon, R.: Bias in error estimation when using cross-validation for model selection. BMC Bioinform. **7**(1), 1–8 (2006). https://doi.org/10.1186/1471-2105-7-91

42. Parvandeh, S., Yeh, H.W., Paulus, M.P., McKinney, B.A.: Consensus features nested cross-validation. Bioinformatics **36**(10), 3093–3098 (2020). https://doi.org/10.1093/bioinformatics/btaa046

43. Maldonado, S., Vairetti, C., Fernandez, A., Herrera, F.: FW-SMOTE: a feature-weighted oversampling approach for imbalanced classification. Pattern Recogn. **124**, 108511 (2022). https://doi.org/10.1016/j.patcog.2021.108511

Complex Systems Modeling, Decision Analysis and Knowledge Management

Maximum Effort Consensus Modeling Under Social Network Considering Individual Emotions

Mengru Xu[1], Fuying Jing[2], Xiangrui Chao[3]($\boxtimes$), and Enrique Herrera-viedma[4]

[1] School of Management Science and Engineering, Chongqing Technology and Business University, Chongqing 400067, China
[2] National Research Base of Intelligent Manufacturing Service, Chongqing Technology and Business University, Chongqing 400067, China
[3] School of Business, Sichuan University, Chengdu 610065, China
chaoxr@scu.edu.cn
[4] Andalusian Research Institute in Data Science and Computational Intelligence, Department of Computer Science and Artificial Intelligence, University of Granada, 18071 Granada, Spain

Abstract. It's beyond disputed that everyone has their own unique understandings of words, which induces personalized individual semantics (PISs) attached to linguistic expressions. Social network illustrates trust relationships among group members. In linguistic social network decision making (SNGDM), social network analysis usually aids in determining the importance weights of decision makers (DMs). Actually, emotions may have an impact on trust propagation. For instance, positive emotions would strengthen trust whereas negative emotions work in a reverse way. Thus, a social network with individual emotions is firstly developed. Generally, DMs are frequently required to modify their opinions in consensus reaching process (CRP). To some extent, emotions also reflect their adjusted probabilities. Meanwhile, DM's willingness and attitude to making necessary modifications for better agreement, that is measured by effort degree, will bring to different consensus results. Under a limited cost budget, a maximum effort consensus model driven by maximizing all DMs' effort degree is proposed for instructing feedback regulation.

Keywords: Linguistic social network group decision making · Personalized individual semantics · Emotions · Maximum effort

1 Introduction

Linguistic social network group decision making (SNGDM) involves a group of decision makers (DMs) reaching an agreement from several alternatives based on linguistic terms under social network [1, 2]. Personalized individual semantics (PISs), showing people's diverse meanings of words, is used to transform linguistic expressions to numerical values [3]. Obviously, DMs will behave differently concerning their present emotions, which might strengthen or weaken their trust from others. For instance, people tend to

J. Chen et al. (Eds.): KSS 2023, CCIS 1927, pp. 97–104, 2023.
https://doi.org/10.1007/978-981-99-8318-6_7

believe that when they are in a positive mood, they can assess situations more accurately than when they are down. Besides, effort degree, that is DM's adjusted willingness and attitude, will lead to various consensus results. The DMs contribute more to consensus, the more effort they put forth. Without loss of generality, the consensus reaching process (CRP) consists of two phases [4, 5]: a consensus measurement phase to compute the consensus level of the group and a feedback regulation phase to provide adjustment advice for DMs. One approach in the latter phase is to solve optimization-based consensus models [15–18].

With regard to SNGDM [6], emotions in group [11–14] and optimization-based consensus models [15–18], many studies have concentrated on different perspectives.

Social Network Group Decision Making. The Trust relationships among DMs plays a key role in the SNGDM problem [7, 8]. There might be over one trust path available between two DMs [8]. Some approaches have been developed to estimate the unknown trust values [6, 9]. The weights of those DMs trusted more by other DMs can be considered higher [10].

Emotions in Group. In cooperative contexts, people have a tendency to read each other's emotions [11]. Individuals respond differently to another's expressions of anger as a function of the nature of the situation [12, 13]. Happier people are more willing to integrate themselves and respond supportively and cooperatively in group [12, 14].

Optimization-Based Consensus Models. Due to practical resource constraints, it is imperative to meet cost expectations. Hence, the feedback mechanism based on minimum cost consensus models (MCCM) have drawn a lot of attention [15], especially in classical and complex GDM problems (e.g., social network and large scale GDM problems) [16]. Asymmetric cost [17], individual tolerance [18, 19], and satisfaction level [18] of DMs are also widely discussed.

However, emotions are not be fused in trust propagation. In addition to restricted cost, it is also important to consider how much efforts that DMs are prepared to put in during real-word CRP.

In this study, we aim to identify the influence of individual emotions in social network for revaluating DMs' weights. A distributed trust-emotion socio-matric is designed. After obtaining numerical values solving by PIS model and computing current collective consensus, a maximum effort consensus model considering cost limitation is established to provide adjustment suggestions.

The structure of this paper is the following: Preliminaries section introduces basic concepts about linguistic model, personalized individual semantics, and social network. Next, trust propagation and aggregation with individual emotions is presented. Then, a maximum effort consensus model based on PIS is constructed. Finally, we end this paper in Conclusion section.

2 Preliminaries

In this section, we first review some fundamental knowledge of 2-tuple linguistic model, numerical scale model, linguistic preference relations and PIS model. Then, some related work of social network is briefly introduced.

In the context of linguistic preference relation, a GDM consist in finding the best alternative from a set of alternatives $X = \{x_1, x_2, ..., x_n\}$ according to a group of DMs $D = \{d_1, d_2, ..., d_m\}$ $(m, n \geq 2)$ based on a set of linguistic terms $S = \{s_0, s_1, ..., s_g\}$.

2.1 2-Tuple Linguistic and Numerical Models

For computing with words, the 2-tuple linguistic model is available without any loss of information [20].

Definition 1 [20] : Let $\beta \in [0, g]$ be a value representing the result of a symbolic aggregation operation, where $s_i \geq s_j$, if $i \geq j$. The conversion function to acquire the 2-tuple expressing the consistent information of β is defined as

$$\Delta : [0, g] \to S \times [-0.5, 0.5),$$
$$\Delta(\beta) = (s_i, \alpha), \text{ with } \begin{cases} s_i, i = \text{round}(\beta), \\ \alpha = \beta - i, \alpha \in [-0.5, 0.5). \end{cases} \tag{1}$$

Definition 2 [21] : Let R be the set of real numbers. The function $NS: S \to R$ is called a numerical scale of S and $NS(s_i)$ is referred to as the numerical index of s_i. If $NS(s_{i+1}) > NS(s_i)$, for $i = 0, 1, ..., g - 1$, then the NS on S is ordered. The numerical scale $NS(s_i, \alpha)$ is presented as

$$NS(s_i, \alpha) = \begin{cases} NS(s_i) + \alpha \times (NS(s_{i+1}) - NS(s_i)), \alpha \geq 0, \\ NS(s_i) + \alpha \times (NS(s_i) - NS(s_{i-1})), \alpha < 0. \end{cases} \tag{2}$$

Then the inverse operator of NS [22] can further be attained.

2.2 Linguistic Preference Relations and PIS Model

In linguistic GDM problems, the most popular preference representation structure is linguistic preference relation. And PIS model assists in translating linguistic preferences into precise numerical preferences.

Definition 3 [23] : A linguistic preference relation is described as a square matrix $L = (l_{ij})_{n \times n}$, where $l_{ij} \in S$ indicates the favor degree of x_i over x_j, and $l_{ij} = Neg(l_{ji})$, $l_{ii} = s_{g/2}$, for $i, j = 1, 2, ..., n$. If $l_{ij} \in S \times [-0.5, 0.5)$, then L is called a 2-tuple linguistic preference relation. Suppose that $l_{ij} = null$, for $\forall i > j$.

Let NS^k be an ordered numerical scale of DM d_k on S. The first four constraints define the overall range for numerical scales, and the last condition guarantees that the numerical scales are ordered. Then, the PIS model [3], which aims to maximize the consistency index CL to deduce the personalized numerical scales (PNS), is formulated as

$$\max CI(L^k) = \max \left[1 - \frac{4}{n(n-1)(n-2)} \sum_{i,j,\varsigma=1; i<j<\varsigma}^{n} \left| NS(l_{ij}^k) + NS(l_{j\varsigma}^k) - NS(l_{i\varsigma}^k) - 0.5 \right| \right]$$

$$s.t. \begin{cases} NS^k(s_0) = 0, \\ NS^k(s_i) \in \left[\frac{i-1}{g}, \frac{i+1}{g}\right], i = 1, 2, \ldots, g-1 \text{ and } i \neq g/2, \\ NS^k(s_{g/2}) = 0.5, \\ NS^k(s_g) = 1, \\ NS^k(s_{i+1}) - NS^k(s_i) \geq \lambda, i = 1, 2, \ldots, g-1. \end{cases} \tag{3}$$

2.3 Social Network

Social network analysis explores the connections and relationships between individuals or communities, mainly consisting of the set of actors, relations among them, and their properties [7, 8, 10]. To unify the important network concepts, three notational schemes are available.

- Sociometric: in which the information about DMs' relationships is often displayed in a two-dimensional matrix called sociometric or adjacency matrix. In a sociometric matrix, the value of t_{kh} is 1 if DM d_k directly trusts d_h, else it is 0.
- Graph theory: in which the network is regarded as a graph with nodes linked by edges.
- Algebraic: this representation distinguishes several notable relations and shows combinations of relations.

3 Trust Propagation and Aggregation with Emotions

3.1 Social Network with Emotions

As is shown in the above sociometric which does not notice the effect that DMs' emotions have on trust relations. However, DMs with positive emotions are more inclined to provide supportive feedback than DMs with negative emotions [12–14]. Additionally, negative emotions may cause DMs to question their judgments and weaken the trust value of others. Therefore, we adopt trust information with emotions to accurately reflect relationships among DMs in social network.

Definition 4. Let EMO_{kh} be the emotion assessment of the DM d_k over d_h, then the 2-tuple trust-emotion sociometric TE can be defined as

$$TE = \begin{bmatrix} - & (t_{12}, EMO_{12}) & \cdots & (t_{1m}, EMO_{1m}) \\ (t_{21}, EMO_{21}) & - & \cdots & (t_{2m}, EMO_{2m}) \\ \vdots & \vdots & \ddots & \vdots \\ (t_{m1}, EMO_{m1}) & \cdots & (t_{mm-1}, EMO_{mm-1}) & - \end{bmatrix}. \tag{4}$$

3.2 Trust Propagation and Aggregation Process

Hence et al. [24] developed a trust propagation approach based on t-norms to deduce indirect trust relationships.

Definition 5 [24] **:** Let $d_k \xrightarrow{1} d_{\sigma(1)} \xrightarrow{2} d_{\sigma(2)} \xrightarrow{3} \cdots \xrightarrow{q} d_{\sigma(q)} \xrightarrow{q+1} d_h$ be a path from DM d_k to d_h with the length of $q + 1$, then the trust value t_{kh} can be estimated as.

$$t_{kh} = \frac{2t_{k,\sigma(1)} \cdot t_{\sigma(b),h} \prod_{b=1}^{q-1} t_{\sigma(b),\sigma(b+1)}}{(2 - t_{k,\sigma(1)})(2 - t_{\sigma(q),h}) \prod_{b=1}^{q-1}(2 - t_{\sigma(b),\sigma(b+1)}) + t_{k,\sigma(1)} \cdot t_{\sigma(q),h} \prod_{b=1}^{q-1} t_{\sigma(b),\sigma(b+1)}}. \tag{5}$$

The aggregated value of EMO_{kh} is calculated similarly.

3.3 The Importance Weights of DMs in Social Network

Without loss of generality, DMs' importance weights can be imputed by their trust relationships in social network.

Definition 6. Let $TE = (te_{kh})_{m \times m}$ be a complete trust-emotion sociometric, then in-degree centrality index of DM d_h is formulated as.

$$C(d_h) = \frac{1}{m-1} \sum_{k=1,h\neq k}^{m} (t_{kh} \cdot EMO_{kh}). \tag{6}$$

Then the relative importance weight w_h can be computed as

$$w_h = \frac{C(d_h)}{\sum_{k=1}^{m} C(d_k)}. \tag{7}$$

4 Maximum Effort Consensus Model Based on PIS

4.1 PIS Learning

After gathering the individual linguistic preference relations $L^k = (l_{ij}^k)_{n \times n}$ provided by each DM, we transformed them into the individual fuzzy preference relations $F^k = (f_{ij}^k)_{n \times n}$, where $f_{ij}^k = NS^k(l_{ij}^k)$.

4.2 Consensus Measurement

Based on the aggregation function, the moderator integrates the individual fuzzy preference relations into a collective preference relation $F^c = (f_{ij}^c)_{n \times n}$, where $f_{ij}^c = \sum_{k=1}^{m} (w_k \times f_{ij}^k), W = (w_1, w_2, \ldots, w_m)$ is the weight vector of DMs and $\sum_{k=1}^{m} w_k = 1$. Then, the level of agreement in the group, that is, collective consensus level $CL = \sum_{k=1}^{m} (w_k \times CL^k)$, can be calculated by aggregating individual consensus level

$CL^k = 1 - \sum_{i,j=1;i\neq j}^{n} \frac{|f_{ij}^k - f_{ij}^c|}{n(n-1)}$ associated with DM d_k.

4.3 Feedback Regulation

It is certain that DMs are often asked to try their best to align their preferences in a way that contributes to consensus when confronted with a given cost budgets. We define adjusted willingness and attitude that DMs produce for a better consensus as effort degree.

Definition 7. DMs with more positive emotions are more likely to make changes. Let p_k be the probability that the k-th DM is willing to adjust, then it can be defined as.

$$p_k = \frac{\sum_{h=1,h\neq k}^{m} EMO_{hk} - \min_{\kappa=1}^{m} \sum_{h,\kappa=1,h\neq\kappa}^{m} EMO_{h\kappa}}{\max_{\kappa=1}^{m} \sum_{h=1,h\neq\kappa}^{m} EMO_{h\kappa} - \min_{\kappa=1}^{m} \sum_{h,\kappa=1,h\neq\kappa}^{m} EMO_{h\kappa}}. \tag{8}$$

Definition 8. Let o_k and o'_k be the k-th DM's original and adjusted opinion, and $\bar{o}'$ be the collective opinion, then the effort degree of the k-th DM can be described as

$$e_k = \operatorname{sgn}\left(\left|o_k - \bar{o}'\right| - \left|o'_k - \bar{o}'\right|\right) \cdot \frac{\left|p_k \cdot o'_k - o_k\right|}{o_k}. \tag{9}$$

where $\operatorname{sgn}\left(\left|o_k - \bar{o}'\right| - \left|o'_k - \bar{o}'\right|\right) = \begin{cases} -1, & \left|o_k - \bar{o}'\right| < \left|o'_k - \bar{o}'\right| \\ 0, & \left|o_k - \bar{o}'\right| = \left|o'_k - \bar{o}'\right| \\ 1, & \left|o_k - \bar{o}'\right| > \left|o'_k - \bar{o}'\right| \end{cases}$.

The group efforts can be acquired by adding the effort degree of all DMs with their importance weights based on social network. Then the maximum effort consensus model (MECM) is established as

$$\max \sum_{k=1}^{m} w_k e_k = \max \sum_{k=1}^{m} \left[w_k \cdot \operatorname{sgn}\left(\left|o_k - \bar{o}'\right| - \left|o'_k - \bar{o}'\right|\right) \cdot \frac{\left|p_k \cdot o'_k - o_k\right|}{o_k} \right]$$

$$s.t. \begin{cases} \left|o'_k - \bar{o}'\right| \leq \varepsilon, & (10\text{--}1) \\ \bar{o}' = \sum_{k=1}^{m} w_k o'_k, & (10\text{--}2) \\ CL \geq \gamma, & (10\text{--}3) \\ \sum_{k=1}^{m} c_k \cdot \left|o'_k - o_k\right| \leq B, & (10\text{--}4) \\ o'_k, \bar{o}' \geq 0, \ k = 1, 2, \ldots, m. & (10\text{--}5) \end{cases} \tag{10}$$

In model (10), ε, w_k, o_k, γ, c_k, and B are known variables, and the rest are unknown variables. The objective function is to attain the maximum efforts for all DMs to modify their opinions. Constraint (10–1) represents the maximum distance between adjusted opinion and collective opinion. Constraint (10–2) is the aggregation approach to collective opinion. Constraint (10–3) denotes that the collective consensus level is not less than threshold γ, and constraint (10–4) indicates the total cost can't exceed the limited budget B.

5 Conclusions

In reality, the CRP under linguistic SNGDM is complicated and dynamic, many factors will affect the result, efficiency and consensus level. This study incorporates emotions into social network and examines how they influence the importance weights assigned to DMs. The modified probabilities of DMs can also be determined based on individual emotions. Next, the feedback regulation phase involves the effort degree reflecting the adjusted willingness and attitude of the DM. Then, the problem of the maximum effort consensus is first discussed. When the budget cost is limited to a certain range in GDM problems, the better consensus is also what we pursue. The proposed model aims to maximize the efforts of decision-makers, that is, to maximize their contribution to consensus. The MECM solutions could offer DMs more useful feedback guidance since they satisfy both the level of collective consensus and the cost budget. Especially in relocation and resettlement decision-making issues, a greater consensus level can facilitate the project's smooth progress. Further, the CRP can function with greater effectiveness.

Funding. This research was supported in part by grants from the National Natural Science Foundation of China (#72274132, #71874023, #72002020), and the Spanish State Research Agency under Project PID 2019-103880RB-100/AEI/10.13039/501100011033.

References

1. Wu, J., Zhao, Z., Sun, Q., Fujita, H.: A maximum self-esteem degree based feedback mechanism for group consensus reaching with the distributed linguistic trust propagation in social network. Inf. Fusion **67**, 80–93 (2021)
2. Gupta, M.: Consensus building process in group decision making—an adaptive procedure based on group dynamics. IEEE Trans. Fuzzy Syst. **26**(4), 1923–1933 (2018)
3. Li, C.C., Dong, Y.C., Herrera, F., Herrera-Viedma, E., Martínez, L.: Personalized individual semantics in computing with words for supporting linguistic group decision making: an application on consensus reaching. Inf. Fusion **33**, 29–30 (2017)
4. Cao, M.S., Wu, J., Chiclana, F., Ureña, R., Herrera-Viedma, E.: A personalized consensus feedback mechanism based on maximum harmony degree. IEEE Trans. Syst. Man Cybern.: Syst. **51**(10), 6134–6146 (2020)
5. Wang, S., Wu, J., Chiclana, F., Sun, Q., Herrera-Viedma, E.: Two stage feedback mechanism with different power structures for consensus in large-scale group decision-making. IEEE Trans. Fuzzy Syst. **30**(10), 4177–4189 (2022)
6. Capuano, N., Chiclana, F., Fujita, H., Herrera-Viedma, E., Loia, V.: Fuzzy group decision making with incomplete information guided by social influence. IEEE Trans. Fuzzy Syst. **26**(3), 1704–1718 (2018)
7. Xiao, J., Wang, X., Zhang, H.: Managing classification-based consensus in social network group decision making: an optimization-based approach with minimum information loss. Inf. Fusion **63**, 74–87 (2020)
8. Wu, J., Chiclana, F., Herrera-Viedma, E.: Trust based consensus model for social network in an incomplete linguistic information context. Appl. Soft Comput. **35**, 827–839 (2015)
9. Liu, Y., Liang, C., Chiclana, F., Wu, J.: A trust induced recommendation mechanism for reaching consensus in group decision making. Knowl.-Based Syst. **119**, 221–231 (2017)
10. Wu, J., Chiclana, F.: A social network analysis trust–consensus based approach to group decision-making problems with interval-valued fuzzy reciprocal preference relations. Knowl.-Based Syst. **59**, 97–107 (2014)

11. Van Kleef, G.A., Cheshin, A., Fischer, A.H., Schneider, I.K.: Editorial: the social nature of emotions. Front. Psychol. **7** (2016)
12. Barsade, S.G.: The ripple effect: emotional contagion and its influence on group behavior. Adm. Sci. Q. **47**(4), 644–675 (2002)
13. Sy, T., Côté, S., Saavedra, R.: The contagious leader: impact of the leader's mood on the mood of group members, group affective tone, and group processes. J. Appl. Psychol. **90**(2), 295–305 (2005)
14. Clark, M.S., Pataki, S.P., Carver, V.H.: Knowledge Structures in Close Relationships: A Social Psychological Approach, 1st edn. Psychology Press, New York (1996)
15. Ben-Arieh, D., Easton, T.: Multi-criteria group consensus under linear cost opinion elasticity. Decis. Support. Syst. **43**(3), 713–721 (2007)
16. Zhang, H., Zhao, S., Kou, G., Li, C.C., Herrera, F.: An overview on feedback mechanisms with minimum adjustment or cost in consensus reaching in group decision making: research paradigms and challenges. Inf. Fusion **60**, 65–79 (2020)
17. Ji, Y., Li, H., Zhang, H.: Risk-averse two-stage stochastic minimum cost consensus models with asymmetric adjustment cost. Group Decis. Negot. **31**(2), 261–291 (2022)
18. Cheng, D., Yuan, Y., Wu, Y., Hao, T., Cheng, F.: Maximum satisfaction consensus with budget constraints considering individual tolerance and compromise limit behaviors. Eur. J. Oper. Res. **297**(1), 221–238 (2022)
19. Guo, W., Gong, Z., Zhang, W.-G., Xu, Y.: Minimum cost consensus modeling under dynamic feedback regulation mechanism considering consensus principle and tolerance level. Eur. J. Oper. Res. **306**(3), 1279–1295 (2023)
20. Herrera, F., Martínez, L.: A 2-tuple fuzzy linguistic representation model for computing with words. IEEE Trans. Fuzzy Syst. **8**(6), 746–752 (2000)
21. Dong, Y.C., Xu, Y.F., Yu, S.: Computing the numerical scale of the linguistic term set for the 2-tuple fuzzy linguistic representation model. IEEE Trans. Fuzzy Syst. **17**(6), 1366–1378 (2009)
22. Dong, Y.C., Li, C.C., Herrera, F.: Connecting the linguistic hierarchy and the numerical scale for the 2-tuple linguistic model and its use to deal with hesitant unbalanced linguistic information. Inf. Sci. **367**, 259–278 (2016)
23. Alonso, S., Chiclana, F., Herrera, F., Herrera-Viedma, E., Alcalá-Fdez, J., Porcel, C.: A consistency-based procedure to estimate missing pairwise preference values. Int. J. Intell. Syst. **23**(2), 155–175 (2008)
24. Victor, P., Cornelis, C., De Cock, M., Herrera-Viedma, E.: Practical aggregation operators for gradual trust and distrust. Fuzzy Sets Syst. **184**(1), 126–147 (2011)

Research on the Behavioral Characteristics and Performance Influence Factors of Venture Capitalists Based on Heterogeneous Networks

Zijun Liu, Zhongbo Jing, and Yuxue Chi[✉]

School of Management Science and Engineering,
Central University of Finance and Economics, Beijing 100081, China
chiy@cufe.edu.cn

Abstract. Venture capital plays an important role in enterprise value reconstruction, R&D innovation, and a range of other areas. To study the investment behavior characteristics and the influencing factors of investment performance, we construct a heterogeneous network model composed of investment institutions and invested enterprises and proposes a series of network indicators. According to the empirical analysis of 61402 Chinese venture capital records from 2001 to 2020, we found that compared with corporate venture capital institutions, state-backed investment institutions have closer internal cooperation, investment firms with greater resource diversity. Meanwhile, these institutions took on the function of the bridge in the primary stage with some CVCs gradually becoming the bridge between the two after 2016. In terms of performance influencing factors, with the intensification of market competition, the positive impact or insignificant impact of Resource Limitation, Resource Adequacy, Resource Uniqueness, Industry Diversity and Regional Diversity on investment performance gradually turns negative.

Keywords: Venture Capital · Heterogeneous Network · State-backed Capital · Venture Institution · Corporate Venture Capital Institution

1 Introduction

Venture capital plays an important role in enterprise value reconstruction, research and development innovation, and so on [1]. With the development of the venture capital industry, more and more investment institutions with different equity backgrounds have participated. Studies have shown that the factors that influence venture capital performance vary among different equity background investment institutions, and there are significant differences in investment mode choices [2]. Among them, state-backed venture capital (SVC) and corporate venture capital (CVC) are two representative categories. State-backed venture capital (SVC) refers to investment institutions established by government agencies, government capital or government-related industrial funds, and

J. Chen et al. (Eds.): KSS 2023, CCIS 1927, pp. 105–119, 2023.
https://doi.org/10.1007/978-981-99-8318-6_8

it usually has a strong policy guidance. Corporate venture capital (CVC) refers to non-financial enterprises making equity investments in start-up companies to achieve financial or strategic goals. CVC institutions generally invest based on the parent company's main business, with abundant industry experience and financial support.

Though the investment motives of state-backed investment institutions and corporate venture capital (CVC) firms are not entirely the same, both are important components of the venture capital industry, and their behavior characteristics and performance have gradually drawn the attention of researchers. Some studies have found that the production efficiency of target companies invested by state-backed venture capital is significantly lower than that of private venture capital [3]. Especially when there is a conflict of policy orientation and interests, state-backed venture capital institutions need to make a trade-off between realizing social value and obtaining higher investment returns. In this regard, corporate venture capital has similar challenges as state-backed venture capital, CVC often faces resistance from the parent company in investment practice [4]. In response to this problem, startups reduce potential risk by establishing a community of independent venture capitalists and other CVC investors [5]. For state-backed venture capital, some studies have shown that a mixed-ownership structure is more conducive than a pure-ownership structure to attracting domestic and foreign private venture capital into the market [6].

Investment performance has long been a major concern. In addition to the influence of different equity backgrounds on investment performance, scholars also studied the impact of factors such as reputation of investment institutions [7], geographical concentration [8], industry affiliation of target companies [9], network status [10], investment diversity [11], cross-community relationship [12] and other factors on investment performance. Among them, studies [10–12] applied network models. The network model can reveal the complexity and dynamics of venture capital, help researchers understand the structure and characteristics of venture capital market more comprehensively, and play an important role in predicting the future development trend of venture capital and exploring the characteristics of investor behavior. The current relevant researches mainly applied joint investment networks composed of homogeneous nodes, and there are few studies that considered the relationship between investors and invested enterprises in network model construction. However, the investors' selection of invested companies directly determines their future investment returns. If both this relationship and the cooperation between investors can be taken into account when constructing the analytical model, it would help to better study the characteristics of investor behavior. Thus, we attempt to put the relationship between investors and invested companies into concern. In this paper, we constructed a heterogeneous network model of venture capital composed of investment institutions and invested enterprises, and further proposed a series of network indicators based on the network.

The research of this article mainly focus on two aspects, the investment behavior characteristics of different types of investment institutions and the influencing factors of investment performance. The research on investment performance of SVC and CVC in different periods not only helps us to gain insight into the behavioral characteristics of them, but also helps us to understand the role they have assumed in the development of the venture capital industry.

In details, this paper divided the nodes of investment institutions into different types, and their behavioral characteristics were analyzed using the indicators proposed and community mining algorithm. Then, we analyzed the influencing factors of investment performance by constructing a regression model. In this paper, we found that the cooperation phenomenon within SVC was more obvious than within CVC, and it tended to invest in diversified resources. Meanwhile, with the intensification of market competition, the positive or insignificant impact of Resource Limitation, Resource Adequacy, Resource Uniqueness, Regional Diversity, and Industry Diversity on investment performance gradually turns into a negative impact.

2   Venture Capital Heterogeneous Network Model

2.1   Heterogeneous Network Model Construction

In order to better characterize the relationship between investors and invested enterprises, this article constructs a directed weighted heterogeneous network $G(V, E, W)$. The example of the heterogeneous network model is shown in Fig. 1.

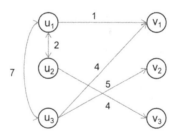

Fig. 1. Heterogeneous network diagram

Node set V The Node set V includes two types of nodes, investment institution nodes and invested enterprises nodes. $V = \{u_1, u_2, \ldots, u_a, v_1, v_2, \ldots, v_b\}$, among them, u represents investment institution nodes, a represents the number of nodes u, and v represents invested enterprises nodes with the number of b. The investment institution nodes u_i are divided into state-backed investment institutions, corporate investment institutions, and other types of investment institutions based on the nature of the investment institution. The invested nodes v_k include industry attributes and geographical attributes.

Edge Set E. The edge set contains two types of edges, which can be represented as $(u_i, u_j)(i \neq j)$ and (u_i, u_k). $(u_i, u_j)(i \neq j)$ represents the cooperative relationship between joint investment partners, while (u_i, u_k) represents the relationship between venture capital institutions and invested enterprises.

Weights Set W. The weights W of different types of edges have different meanings. The weight of edge $(u_i, u_j)(i \neq j)$ between investment institutions represents the number of times they have cooperated. The weight of edge (u_i, u_k) between investment institutions and invested enterprises represents the number of times the investment institution has invested in the company.

2.2   Network Indicators Construction

Previous studies have shown that network centrality has a significant impact on investment performance in homogeneous joint investment networks. However, network centrality indicators based on homogeneous networks are difficult to depict the impact of invested enterprises on investment institutions. Therefore, based on the heterogeneous network model of venture capital, this paper proposes a set of network resource indicators by regarding invested institutions as investment resources, as follows:

Resource Popularity. Resource Popularity is measured by the node importance of enterprises invested by investment institutions using degree centrality, as shown in formula (1). q_i represents the number of invested enterprises nodes directly connected to the investment institution i, and NC_{ij} represents the degree centrality of the enterprise j invested by investment institution i, which can be measured using degree centrality, eigenvector centrality, etc.

$$RP_i = \frac{\Sigma_{j=1}^{q_i} NC_{ij}}{q_i} \tag{1}$$

Resource Limitation. In this paper, we use two indicators to measure Resource Diversity, Resource Limitation and Resource Adequacy.

Resource Limitation. Previous studies based on joint venture capital networks measured resource constraints by the number of partners and the strength of connections between partners, with an increase in the number of partners reducing resource constraints and an increase in connectivity reinforcing resource constraints [13]. Similarly, this paper argues that resource constraints are important considerations and proposes Resource Limitation measured by the number of unreachable points of investors in heterogeneous networks. The more unreachable points, the greater the Resource Limitation.

Resource Adequacy. Some scholars used the number of cross-community movements of an investment institution to measure the investors' ability to obtain heterogeneous resources [14]. They reckoned that these resources could provide enterprises with more portfolio choices. These studies focused more on co-investors, while in heterogeneous networks, this paper argued that Resource Adequacy was not only related to co-investors but also directly related to invested companies. Therefore, this article uses the reciprocal of the path length that investors reach all invested institutions in the heterogeneous network to measure Resource Adequacy, as shown in formula (2). L_{ij} represents the shortest path distance between the investor node i and invested node j and m represents the number of invested nodes.

$$RA_i = \sum_{j=1}^{m} \left(\frac{1}{L_{ij}} \right) \tag{2}$$

Resource Uniqueness. In the heterogeneous network, we measure the investor's Resource Uniqueness by the reciprocal sum of the number of investors connected through all invested nodes, which are the neighbor nodes of the investor node. As shown in formula (3), k_i represents the number of enterprises that investor i has invested in directly, and C_{ij} represents the number of investor nodes directly connected to invested enterprise j that has been invested by the investment institution i. The larger the value of RU_i, the higher the Resource Uniqueness of investor i, indicating that this investor has invested in many companies that are not being invested by other investors.

$$RU_i = \sum_{j=1}^{k_i} \left(\frac{1}{C_{ij}} \right) \tag{3}$$

Resource Diversity. There were studies that measured Resource Diversity by the diversification of investment partners in terms of industry and geographical investment [15], This paper also considers industry and geographical factors, but since the invested enterprises are also an important part of investment firm resources, this article measures Resource Diversity basing on the distribution of invested companies in region and industry of the investors. The greater the Resource Diversity, the more inclined investment institutions are to invest in heterogeneous resources.

Regional Diversity. We use the concept of entropy to measure Regional Diversity, as shown in formula (4). n represents the total number of regions, and T_{ij} represents the number of investments invested in the region j as a percentage of the total investment of the investment institution i. When the entropy value is close to 0, it indicates that the investment is mainly concentrated in a certain region.

$$RD_i = \sum_{j=1}^{n} T_{ij} \ln\left(\frac{1}{T_{ij}}\right) \tag{4}$$

Industry Diversity. Similar to Regional Diversity, we use entropy to measure Industry Diversity. As shown in formula (5), h represents the total number of industries, and P_{ij} represents the percentage of the number of invested enterprises in industry i among all invested by investor i.

$$ID_i = \sum_{j=1}^{h} P_{ij} \ln\left(\frac{1}{P_{ij}}\right) \tag{5}$$

3   Analysis of SVC and CVC Investment Behavior Based on the Heterogeneous Network

In this article, IT Orange was used as the data source. After preprocessing, a total of 61,402 venture capital event records and 2016 IPO or M&A exit records were obtained. To better analyze the characteristics and differences in investment behavior of SVC and CVC, venture capital heterogeneous networks were constructed based on data from four time periods: 2001–2005, 2006–2010, 2011–2015, and 2016–2020.

3.1   Analysis of Institutional Cooperation Based on Community Detection

Combined with classic network indicators and community mining algorithms, the characteristics and differences in investment behavior of these two types of institutions in each time window are analyzed as follows.

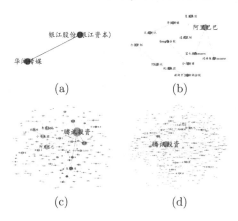

Fig. 2. (a) 01–05 CVC Cooperation Network; (b) 06–10 CVC Cooperation Network; (c) 11–15 CVC Cooperation Network; (d) 11–15 CVC Cooperation Network

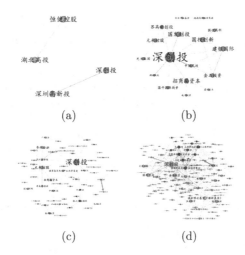

Fig. 3. (a) 01–05 SVC Cooperation Network; (b) 06–10 SVC Cooperation Network; (c) 11–15 SVC Cooperation Network; (d) 11–15 SVC Cooperation Network

Figures 2 and 3 show that the collaborations in CVCs and collaborations in SVCs have both become more frequent and complex, and have formed a clear community structure in the past decade. The communities formed within CVC networks were mainly centered around Tencent Investment, Alibaba, and Qihoo 360, with significantly higher network centrality. The communities formed within SVC networks were mainly centered around Shenzhen Capital Group Co.,Ltd., Guoxin Fund, Jinpu Investment, Suzhou Oriza Holdings Co.,Ltd., etc., with a more balanced community size compared to CVC.

In addition, by combining network density, it was found that the density of CVC collaboration networks (06–10: 0.09; 11–15: 0.017; 16–20: 0.012) was lower

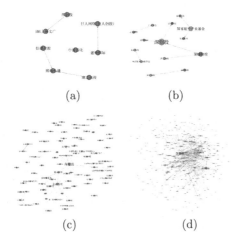

Fig. 4. (a) 01–05 CVC and SVC cooperation network; (b) 06–10 CVC and SVC cooperation network; (c) 11–15 CVC and SVC cooperation network; (d) 16–20 CVC and SVC cooperation network

than that of SVC collaboration networks during the same period (06–10: 0.176; 11–15: 0.034; 16–20: 0.053), indicating that the collaborations within SVC were closer, to a certain extent.

Figure 4 shows the cooperative network between CVC and SVC, where orange and purple nodes respectively represent SVC and CVC, and green nodes represent companies that are both SVC and CVC. As time passed, the collaboration between SVC and CVC become more frequent. In the early stage, it was mainly state-backed venture capital institutions that assumed the role of a bridge, showing a star distribution of multiple CVC institutions around state-backed institutions, and by 2016–2020, some CVCs began to assume the role of the bridge.

3.2   Analysis of SVC and CVC Network Resource Indicators

Using classical indicators proposed by previous studies and network resource indicators proposed above, the average values of each indicator for CVC and SVC were calculated in each time window, as shown in Table 1. In the first three time windows, SVC had better average investment performance than CVC, while in the 2016 - 2020 time window CVC performed better than SVC. Furthermore, the average Regional Diversity, Industry Diversity, and Resource Uniqueness of both CVC and SVC increased over time. It could be seen that compared with CVC, SVC tended to invest in more diverse areas and industries and engaged in more risky investment activities while expanding their unique resources.

Table 1. Comparison of CVC and SVC indicators in the 5-year window

	Resource Adequacy	Resource Uniqueness	Regional Diversity	Industry Diversity	Total number of investments	Eigenvector Centrality	Number of IPO/M&A
01–05 CVC	7.563	0.439	0.000	0.000	117.278	0.002	3.500
01–05 SVC	6.443	1.443	0.134	0.389	118.240	0.005	6.133
06–10 CVC	202.639	1.122	0.314	0.414	65.284	0.009	3.205
06–10 SVC	274.816	4.435	0.690	1.083	107.306	0.043	5.600
11–15 CVC	2387.730	2.698	0.662	0.789	26.481	0.007	2.328
11–15 SVC	2879.209	4.806	0.692	1.352	60.779	0.007	2.548
16–20 CVC	4729.388	3.311	0.999	1.050	24.600	0.006	1.659
16–20 SVC	5451.010	5.509	1.224	1.546	46.528	0.013	1.083

Through the above analysis, it is found that: 1) Both SVC and CVC collaboration networks had gradually formed clear community structures, with CVC communities featuring prominent core nodes and SVC communities having more balanced community sizes with more apparent internal collaboration; 2) In the collaborative investment between SVC and CVC, the bridge role was mainly played by SVC in the early stage, and some CVC companies started to play a bridge role in later stages; 3) Compared with CVC, SVC was more inclined to invest in diverse areas and industries.

4   Analysis of the Impact of Network Resources on Investment Performance

4.1   Analysis of SVC and CVC Network Resource Indicators

Network Centrality. Network centrality was often used as an explanatory variable when studying the factors affecting investment performance, such as exploring the impact of degree centrality and betweenness centrality on investment performance [10]; the influence of proximity centrality on investment performance was studied [16]. Most studies suggested a significant positive correlation between network centrality and investment performance. Therefore, the hypothesis is proposed as follows:

Hypothesis 1. *Network centrality plays a positive role in helping venture capital institutions to achieve better investment performance.*

Network Resources. Resource Popularity was mainly related to the network position of the invested enterprises, and the invested enterprises that had been invested by multiple investors were regarded as enterprises in the central position. In this part, the network centrality of an enterprise is measured by feature vector centrality. Meanwhile, companies located in central positions usually receive more attention and possess better development prospects and resources, which are more conducive for investors to achieve better investment performance. Based on this, the following hypothesis is proposed:

Hypothesis 2. *Resource Popularity plays a positive role in helping venture capital institutions to achieve better investment performance.*

Resource Quality can be divided into two parts: Resource Limitation and Resource Adequacy. The more accessible investment nodes and the closer the distance, the greater the diversity of potential resources that investment institutions can obtain. However, having access to more resources didn't necessarily mean better investment performance [17]. Based on this, the following hypothesis is proposed:

Hypothesis 3. *There is no significant association between Resource Limitation and investment performance.*

Hypothesis 4. *There is no significant association between Resource Adequacy and investment performance.*

Resource Uniqueness of an institution is affected by two factors: the number of invested enterprises and the number of venture capital institutions that invest in those enterprises. Investment institutions that can invest extensively and have access to unique resources usually have strong investment strength and rich investment experience, the following hypotheses are proposed:

Hypothesis 5. *Resource Uniqueness plays a positive role in helping venture capital institutions to achieve better investment performance.performance.*

Resource Diversity consists of Regional Diversity and Industry Diversity. In terms of geographical location, some studies focused on the geographical concentration of joint investment partners, believing that geographical concentration could help reduce costs and improve the exit probability of venture capital institutions [8]. Others focused on the spatial distance between investors and invested enterprises and believed that as the distance between the two increases, the probability of a successful transaction would decrease [18], which would also have a negative effect on the technical performance of start-up companies. Cross-industry investment requires industry experience to support it compared to regions. Studies on CVCs showed that diversification in the investment industry could negatively impact CVCs' investment performance [19]. Make assumptions based on below:

Hypothesis 6. *Regional Diversity plays a negative role in helping venture capital institutions to achieve better investment performance.*

Hypothesis 7. *Industry Diversity plays a negative role in helping venture capital institutions to achieve better investment performance.*

4.2   Variable Definitions

Table 2. Variable definitions

Variable Properties	Variable Categories	Variable Symbol
Dependent variable	Investment Performance	SE
Independent variable	Network Centrality	EC
		DC
	Resource Popularity	RP
	Resource Quality	RL
		RA
	Resource Uniqueness	RU
	Resource Diversity	RD
		ID
Control variables	Total number of investments	TI
	investment Institution Attributes	SVC
		CVC
	Lead-to-Investing Ratio	LR

Since venture investment primarily relies on exits to generate returns, the number of IPOs and M&A exits was the commonly used indicator for evaluating investment performance. This paper also uses this indicator as the dependent variable. Based on above hypothesis, the variable definitions in this article are shown in Table 2.

4.3   Analysis of Experimental Results

Since the Chinese government launched the ChiNext board in 2009 and introduced a series of tax incentives for the risk investment industry, including the "Notice on Implementation of Preferential Income Tax for Venture Capital Enterprises", the venture investment industry in China has entered a golden period of development [20]. The development of China's venture capital industry has entered a new stage since then. Thus, we divided the data into two-time windows, 2001–2010 and 2011–2020, for regression analysis. Due to the large difference in data volume between the two-time windows, a network model was constructed using only edges with a weight greater than 1 in the 2011–2020 window.

SPSSAU, a widely used online data analysis platform, was used for regression analysis. To avoid multicollinearity effects, this study employed ridge regression [21]. In this study, the minimum value of 'k' was found when the normalization coefficient of the respective variables tended to be stable on the ridge trace map. The determined value of 'k' was then input to obtain the estimation of the ridge regression model. Model 1 is a regression model that introduces control variables, while model 2 adds two network centrality variables. Models 3–6 gradually add three explanatory variables: Resource Popularity, Resource Quality, Resource Uniqueness and Resource Diversity.

Table 3. Ridge regression results (2001–2010)

	Model 1	Model 2	Model 3	Model 4	Model 5	Model 6
constant	1.519**	−0.093	−0.009	1.544	0.412	1.369
Control variable						
SVC	3.433**	0.584	0.694	0.728	0.446	0.707
CVC	−0.194	1.024	1.067	1.086	1.05	0.969
LR	2.878	4.119*	4.085*	4.259*	4.673**	4.493**
TI	0.027**	0.015**	0.015**	0.015**	0.012**	0.012**
Independent Variable						
EC		−8.824	−0.054	1.793	0.498	0.734
DC		547.797**	452.207**	449.186**	246.104**	237.484**
RP			−1.104	−0.202	−0.563	0.887
RL				−0.001	0	−0.001
RA				−0.003	0.001	−0.002
RU					0.419**	0.407**
RD						0.930**
ID						−0.379
R^2	0.544	0.781	0.78	0.78	0.805	0.809
Adjust R^2	0.54	0.779	0.776	0.776	0.8	0.804

Note: There are 467 sample data. Model 3-6 takes a K value of 0.02 and Model 1-2 takes a K value of 0. *p<0.1, **p<0.05, ***p<0.01.

The empirical results of the two phases are shown in Table 3 and 4, which corresponds to the time window of 2001-2010 and 2011-2020 for models 1-6. The conclusions of models 2 and 3 are consistent with the first stage. The difference lies in the experimental results corresponding to models 3, 4, and 5, where it is found that the Resource Limitation, Resource Adequacy, Resource Uniqueness, and Industrial Diversity have a significant negative impact on investment performance, while Regional Diversity has no significant impact on investment performance.

Table 4. Ridge regression results (2011-2020)

	Model 1	Model 2	Model 3	Model 4	Model 5	Model 6
constant	0.555**	0.229	0.284	3.123**	2.710**	0.987
Control variable						
SVC	−0.467	−0.81	−0.807	−0.854*	−1.129**	−1.429**
CVC	0.279	0.404	0.421	0.372	0.365	0.37
LR	2.532**	2.160*	2.209*	2.279**	1.774*	1.319
TI	0.022**	0.017**	0.017**	0.017**	0.018**	0.018**
Independent Variable						
EC		4.054*	4.325*	8.305**	8.453**	10.853**
DC		77.571	72.943	57.164	105.539*	88.075*
RP			−3.333	7.909	5.435	−4.882
RL				−0.001*	−0.000*	0
RA				−0.003**	−0.003**	−0.001
RU					−0.092**	−0.083**
RD						−0.352
ID						−0.448**
R^2	0.572	0.586	0.586	0.594	0.604	0.614
Adjust R^2	0.569	0.583	0.583	0.59	0.599	0.608

Note: There are 798 sample data. Model 2–6 takes a K value of 0.02 and Model 1 takes a K value of 0. *p<0.1, **p<0.05, ***p<0.01.

Comparing the two time windows, it is found that network centrality, lead investment ratio and total investment have significant positive effects on investment performance in both time windows. Resource Popularity has no significant impact on investment performance in either time window. Some studies used network size to measure the number of co-investment partners of investment institutions and found that the network size and investment performance exhibit an inverted U-shaped relationship [17], which meant the best effect was achieved when the number of joint investment partners was moderate. Therefore, we can infer that Resource Popularity had no significant effect on the investment performance in the two time windows, which might be due to the uncertain direction of the influence of Resource Popularity on the investment performance, resulting in that its impact on investment performance was not significant.

In view of the phenomenon that the positive or non-significant impact of Resource Limitation, Resource Adequate, Resource Uniqueness, Regional Diversity and Industry Diversity on investment performance gradually turned into a negative impact, we infer that it might be caused by changes of the competitive environment. From 2001 to 2010, China's venture capital was in the early stage of development, the number of investors in the venture capital market was relatively small, with low liquidity of funds and insufficient cooperation opportunities among investors or there is a high-risk aversion, which belonged to a low competitive environment. Under the 2011–2020 time window, the number of investors and joint investment events had doubled compared with 2001–2010, and the number of venture investors was large, the liquidity of funds was high, which belonged to a high competitive environment.

As for Resource Quality, Resource Limitation had a significant negative impact on investment performance in a highly competitive environment, which might be due to the fact that the venture capital market had developed relatively maturely, but investment institutions with high resource limitation still hadn't broken through the barriers to cooperation with other investment institutions, which might lead to weak investment strength or investment experience of the investment institution, resulting in lower investment performance. Resource Adequate also had a significant negative impact on investment performance in a highly competitive environment. We speculate it was because that this part of the investment institutions didn't have a clear investment tendency and investment advantages, which was not conducive to obtaining better investment performance in a highly competitive environment.

As for Resource Uniqueness, in a low-competition environment, investment institutions could diversify their investments or occupy some unique resources for investment, which helped to form resource advantages. With the gradual intensification of competition and the gradual disintegration of resource barriers, investment institutions still invested in enterprises that had no investment from other investment institutions, possibly, leading to invest in enterprises with poor development prospects or certain hidden dangers, resulting in in a decline in investment performance.

In view of diversity, researchers hold different views on the impact of regional diversity on investment performance, and some indicated that network density may have a moderating effect on this impact [22]. Therefore, we can speculate that investors who invest in multiple regions usually had strong investment capabilities, which helped them obtain heterogeneous resources and improve investment performance. Meanwhile, the multi-regional investment would not only increase costs, but also made it difficult to form a stable cooperation alliance, while this effect is particularly detrimental to investors in a highly competitive environment. This may explain why Regional Diversity has distinct effect in different competitive environment.

5   Conclusion

This study analyzes the investment behavior characteristics of state-backed institutions and corporate investment institutions, as well as the factors influencing investment performance. To address the limitations of existing research, this paper innovatively constructs a venture investment heterogeneous network model, and a set of network metrics is proposed based on the model. It is the first time that invested enterprises are regarded as resources in the research of investors' investment performance and their preference for resource selection. The research conclusions are helpful for CVC and SVC in making better investment decisions. By analyzing the cooperation networks within and between the CVC and SVC groups, it is found that SVC has closer internal cooperation, and in the cooperation between SVC and CVC, state-backed institutions mainly played a bridge role in the early stage, while some CVCs also began to play this role in the later stage. Combined with the network resource indicators, it is found that SVC is more inclined to diversify regions and industries in the selection of invested enterprises than CVC.

Furthermore, this study applies ridge regression to analyze the factors affecting investment performance and finds that investing in hot enterprises doesn't have a significant impact on investment performance. What's more, the positive impact or insignificant impact of Resource Limitation, Resource Adequate, Resource Uniqueness, Industry Diversity and Regional Diversity on investment performance gradually becomes negative as market competition intensifies.

Therefore, considering SVC's performance in terms of Resource Adequacy, Industry Diversity, Regional Diversity, and Resource Uniqueness, SVC can try to adjust its investment strategy, from multi-point investment to a planned focus on specific regions or industries, as the venture capital industry develops and competition intensifies. As for CVC, which has lower network centrality than SVC, it can improve its network status by cooperating with high-quality investment institutions to strengthen its resource and information advantages and improve competitiveness and investment performance.

Acknowledgements. The authors of the present paper are supported by the National Natural Science Foundation of China (72204283, 72271253), as well as the Program for Innovation Research in Central University of Finance and Economics.

References

1. Chemmanur, T.J., Krishnan, K., Nandy, D.K.: How does venture capital financing improve efficiency in private firms? A look beneath the surface. Rev. Financ. Stud. **24**(12), 4037–4090 (2011)
2. Tian, X., Kou, G., Zhang, W.: Geographic distance, venture capital and technological performance: evidence from Chinese enterprises. Technol. Forecast. Soc. Chang. **158**, 120155 (2020)
3. Alperovych, Y., Hübner, G., Lobet, F.: How does governmental versus private venture capital backing affect a firm's efficiency? Evidence from Belgium. J. Bus. Ventur. **30**(4), 508–525 (2015)

4. Basu, S., Phelps, C.C., Kotha, S.: Search and integration in external venturing: an inductive examination of corporate venture capital units. Strateg. Entrep. J. **10**, 129–152 (2016)
5. Hallen, B.L., Katila, R., Rosenberger, J.D.: How do social defenses work? A resource-dependence lens on technology ventures, venture capital investors, and corporate relationships. Acad. Manag. J. **57**(4), 1078–1101 (2014)
6. Dahaj, A.S., Cozzarin, B.P.: Government venture capital and cross-border investment. Glob. Financ. J. **41**, 113–127 (2019)
7. Nahata, R.: Venture capital reputation and investment performance. J. Financ. Econ. **90**(2), 127–151 (2008)
8. Kang, J.-K., Li, Y., Oh, S.: Venture capital coordination in syndicates, corporate monitoring, and firm performance. J. Financ. Intermediation **50**, 100948 (2022)
9. Li, B., Liang, K.: Performance comparison between lead investors and follow-on investors in co-investment: an empirical study based on the sinovation ventures network in China's venture capital market. Management World (1), 2 (2017)
10. Hochberg, Y.V., Ljungqvist, A., Lu, Y.: Whom you know matters: venture capital networks and investment performance. J. Financ. **62**, 251–301 (2007)
11. Wang, Y.X.: Study on the moderating effect of network embedding on network capability and exit performance of venture capital firms. Soft Sci. **11**, 29–33 (2018)
12. Xue, C., Jiang, P., Dang, X.: The dynamics of network communities and venture capital performance: evidence from China. Financ. Res. Lett. **28**, 6–10 (2019)
13. Hu, X., Wang, J., Wu, B.: Venture capital firms' lead orientation, network position, and selection of familiar syndicate partners. North Am. J. Econ. Financ. **62**, 101757 (2022)
14. Gu, J., Zhang, F., Xu, X., Xue, C.: Stay or switch? The impact of venture capitalists' movement across network communities on enterprises' innovation performance. Technovation **125**, 102770 (2023)
15. Gao, Y., Xie, Y., Yang, Y.: Network collaboration capability and venture capital self-network dynamics: the moderating effects of knowledge attributes. Manage. Rev. **35**(05), 29–41 (2023)
16. Zhou, L., Shan, J., Zhang, J.: The impact of joint investment network position on investment performance: an empirical study from venture capital. Manage. Rev. (12), 160–169+181 (2014)
17. Yang, Y.P., Gao, Y.G.: The impact of network size and 2-step reachability on venture capital performance: the moderating effect of knowledge attributes. Manage. Rev. **06**, 114–126 (2020)
18. Li, Z.P., Luo, G.F., Long, D., An, R.: Geographic proximity in venture capital: an empirical study on China's venture capital market. Manage. Sci. **3**, 124–132 (2014)
19. Wang, S.Y., Chen, S., Qiao, H., Lu, Q.Y.: Diversification and overinvestment behavior of corporate venture capital: performance evaluation based on inter-industry mutual investment network. Nankai Bus. Rev. **5**, 128–140 (2021)
20. Zhang, J., Guo, R.: Evolution, current situation, and future prospects of China's venture capital development. Global Technol. Econ. Outlook **31**(09), 34–43 (2016)
21. Hoerl, A.E., Kennard, R.W.: Ridge regression: biased estimation for nonorthogonal problems. Technometrics **12**(1), 55–67 (1970)
22. Wang, Y.X., Zhang, C., Wang, X.: Network capability and investment performance of venture capital firms: the interactive effect of network position and relationship strength. Mod. Financ. Econ. (J. Tianjin Univ. Financ. Econ.) **2**, 91–101 (2018)

Social Capital, Market Motivation and Knowledge Sharing: A Multiplex Network Perspective on Blockchain-Based Knowledge Communities

Zhihong Li, Yongjing Xie[ID], and Xiaoying Xu[✉][ID]

School of Business and Management, South China University of Technology,
Guangzhou 510000, Guangdong, China
bmxyxu@scut.edu.cn

Abstract. With the development of blockchain technology and token economies, blockchain-based knowledge communities (BKC) have emerged. One unique feature of these communities is that the blockchain tokens used to incentivize users can be freely traded on external markets. However, the mechanism by which market motivations in token transactions affect users' knowledge-sharing behavior is not yet clear. Therefore, taking Steemit, the typical BKC, as the research object, this study uses a multiplex network perspective and draws on social capital theory and limited attention theory to empirically investigate the impact of social capital and market motivations on users' knowledge-sharing behavior in BKC. The results show that social capital has a significant positive impact on users' knowledge-sharing behavior. In addition, users' market motivations play a negative moderating role in the impact of social capital on knowledge-sharing behavior. The research findings enrich the understanding of the factors that influence users' knowledge-sharing behavior and provide important insights for how BKC can effectively incentivize users to make sustained contributions.

Keywords: Blockchain · Knowledge Communities · Social Capital · Market Motivation

1 Introduction

Online knowledge community refers to a novel knowledge production model that integrates "knowledge sharing" and "online social networking" [1]. Sustained knowledge sharing is considered a pivotal factor in ensuring the sustainable development of online knowledge communities [2]. However, the lack of effective incentive mechanisms often leads to a dearth of motivation among users to generate content, resulting in a significant prevalence of free-riding behavior. A recent study discovered that within the majority of knowledge communities, only 1% of users contribute original content, 9% contribute by curating and amalgamating content from others, while 90% of users merely browse the content without actively contributing [3].

© The Author(s), under exclusive license to Springer Nature Singapore Pte Ltd. 2023
J. Chen et al. (Eds.): KSS 2023, CCIS 1927, pp. 120–135, 2023.
https://doi.org/10.1007/978-981-99-8318-6_9

Blockchain technology, as an emerging field, has attracted considerable attention from both academia and industry [4]. With the growing maturity of blockchain technology, token-based incentives offer a fresh approach to tackling the aforementioned challenges. The fusion of blockchain and token economy facilitates the "value conversion" and "value transfer" between distinct value systems [5], thereby fostering the emergence of novel organizational models. Blockchain-based knowledge communities (BKC) serve as prominent exemplars of such models.

BKC refers to the online community that leverages the underlying technology of blockchain to provide decentralized social networking services [6], exemplified by platforms like Steemit and BiHu. The disparities between BKC and conventional knowledge communities primarily manifest in two facets. First, BKC implements a proof-of-brain incentive mechanism, which rewards users who actively engage in content creation and exploration, effectively addressing the challenge of insufficient motivation among users to generate content. Second, tokens, a distinctive feature of BKC, represent tradable digital assets and serve as proof of ownership [7]. Unlike traditional monetary incentives, tokens can be freely exchanged on external markets [8].

The introduction of token incentive mechanisms enables users to profit from the community in two ways. On the one hand, the presence of refined incentive mechanisms transforms users' knowledge sharing based on interests and enthusiasm into knowledge production driven by the goal of earning profits, giving rise to a novel "knowledge monetization" model [9]. On the other hand, the market attributes of tokens, such as value appreciation, liquidity, and convertibility [5], have fostered market motivations to profit from participation in token transactions. This profit-seeking model, detached from knowledge production, falls within the realm of "non-knowledge monetization." Taking the example of Steemit, the largest globally recognized BKC, the market motivations embedded in token transactions within the community encompass profit-seeking incentives through services such as token trading, leasing, and vote buying and selling [10]. These market motivations can significantly impact knowledge sharing. However, existing research fails to encompass the characteristics of market motivations and neglects consideration of their potential effects, thereby inadequately elucidating the mechanisms underlying blockchain token incentives.

Due to the complexity, dynamics, and interdependence of the interaction between motivations for knowledge monetization and market motivations for non-knowledge monetization, traditional modeling approaches struggle to effectively capture these relationships. Research on social networks has indicated that organizational activities are influenced by multiple types of network relationships [11]. For instance, in a recent study by Li et al. [12], the factors influencing innovation performance in enterprises were investigated using a multiplex network theory perspective. The study revealed the mechanisms through which different characteristics of multiplex network embedding affect innovation activities at various levels. Drawing inspiration from this research approach, this paper adopts a multiplex network perspective based on social capital theory and limited attention theory to explore the impact of different types of network relationships in BKC on knowledge-sharing behavior.

The remaining sections of this paper are organized as follows: Sect. 2 presents the theoretical foundations and research hypotheses, Sect. 3 describes the research design, Sect. 4 presents the empirical analysis, and finally, the conclusion is provided.

2 Literature Review and Hypothesis Development

Building upon social capital theory and limited attention theory, this study employs a multiplex network perspective to establish the follower network, vote network, governance network, and transaction network. Centrality indicators are extracted from each network layer to measure the structural capital, relational capital, cognitive capital, and market motivations of users within the community. Based on these factors, the study examines their influence on users' knowledge-sharing behavior, culminating in the development of the theoretical model presented in Fig. 1.

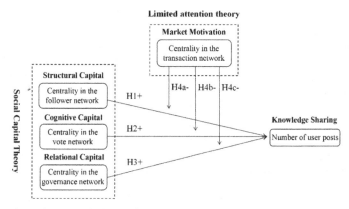

Fig. 1. The theoretical model of the influencing factors of knowledge-sharing behavior of users in BKC.

2.1 Social Capital and Knowledge Sharing

In social capital theory, social capital is conceptualized as the accumulation of assets or resources embedded within individuals, communities, and social relationship networks [13]. Moreover, existing research has commonly categorized social capital into three dimensions: structural, cognitive, and relational [14]. In recent years, with the rapid development of online knowledge communities, the impact of social capital on knowledge-sharing behavior has garnered extensive attention from scholars both domestically and internationally. Although previous studies have demonstrated the positive influence of social capital on users' knowledge-sharing behavior in online communities [15], further in-depth research is warranted due to the substantial disparities between BKC and traditional knowledge communities, particularly in the areas of community governance and incentive mechanisms [16].

Structural Capital and Knowledge Sharing. Structural social capital refers to the social connections that reflect the structural characteristics of a social network, such as node positions and connectivity [13]. The relationships between individuals or the structural links established through social interactions are important antecedents to collective action and cooperation among individuals [17].

In previous studies, researchers often used the number of followers in a network as an indicator of structural capital. For example, Wang et al. (2013) found that microblog users with a larger number of followers possess more structural social capital, resulting in higher readership and likes for their posted content and making it easier for them to gain attention from other users [18]. In BKC, users can acquire relevant knowledge by actively following others and engage in knowledge sharing and dissemination through their followers. This leads to the formation of a follower network where users act as nodes and the following relationships represent the edges. The centrality of an individual in a network measures the extent of their connections to other actors and their level of resource acquisition and control [10]. Previous social network research has shown that centrality reflects an individual's proximity to central hubs in the network and has a significant influence on knowledge transfer and sharing [19, 20]. Thus, in this study, the centrality of the follower network is used to assess users' structural capital. For knowledge providers, a higher centrality in the follower network, indicating a larger number of connected follower nodes, signifies a higher level of structural social capital. To maintain this sense of superiority (structural capital), these individuals are more inclined to engage in knowledge sharing with other members of the network [21]. Based on this, the following hypothesis is proposed:

H1: Centrality in the follower network has a positive impact on the number of posts by users.

Cognitive Capital and Knowledge Sharing. Cognitive social capital represents the embodiment of shared interests, values, and modes of expression among individuals in information exchange [13]. Typical research variables for cognitive capital include shared vision and shared language [13]. Shared vision refers to the common goals, interests, and values held by members of an organization, connecting individuals who may be unfamiliar with each other within a network and leading to interactive behaviors [13].

Existing research suggests that the shared vision between users and knowledge creators is often expressed through voting behavior [22]. In BKC, knowledge seekers express their shared vision with knowledge providers through voting actions, resulting in a vote network where users serve as nodes and voting relationships represent the edges. Diverging from conventional knowledge communities, voting relationships within BKC can significantly influence the token earnings for both the voters and the knowledge providers. Consequently, for knowledge providers, higher centrality in the vote network signifies that they have obtained greater expectations and recognition, reflecting a higher level of cognitive social capital. Previous studies have indicated that shared vision enhances the willingness to share knowledge and improves the perception of knowledge quality [23]. Based on this, the following hypothesis is proposed:

H2: Centrality in the vote network has a positive impact on the number of posts by users.

Relational Capital and Knowledge Sharing. Relational social capital refers to the interpersonal relationships among individuals in social interactions, such as trust and reciprocity [13]. Trust is regarded as one of the fundamental elements of social capital [24]. It is a social perception that arises from emotional interactions among group members, leading to trust in the group as a whole [25].

In the Steemit community, witnesses are responsible for creating and verifying transaction blocks [26]. Unlike traditional knowledge communities with centralized management, Steemit is operated by a team composed of 21 witnesses. Witnesses are dynamically elected by all users, and each user has the right to vote and be voted for [26]. Community members engage with one another through participation in witness elections, expressing emotional trust in other users rather than being solely driven by economic incentives. Consequently, a governance network is formed, with users serving as nodes and election relationships representing the connections. Higher centrality in the governance network indicates greater support from other users, indicating a higher level of trust within the governance network and thus possessing greater relational capital. Previous research has indicated that trust is considered a "prerequisite" for effective knowledge exchange [23]. Trust fosters a relaxed and open environment for knowledge sharing among community members and facilitates the exchange and sharing of valuable knowledge [27]. Based on this, the following hypothesis is proposed:

H3: Centrality in the governance network has a positive impact on the number of posts by users.

2.2 The Moderating Role of Market Motivation

Previous research has indicated that when the evaluation mechanism for content shifts from a "sharing is rewarding" mechanism characterized by invitations and votes to a payment mechanism, the scarcity of knowledge is replaced by the scarcity of attention [28]. Attention is a finite resource that cannot be replenished in the short term [29] and is susceptible to various external factors [30]. Limited attention theory, a research focus on behavioral finance, suggests that individuals tend to prioritize tasks associated with physiological and behavioral costs over other tasks [31]. Recent research has shown that when individuals are faced with numerous attention-diverting stimuli during task execution, they may become distracted, leading to compromised outcomes [32]. In the Steemit community, market motivations stemming from the market attributes of tokens have the potential to divert users' attention away from knowledge sharing. This market motivation refers to the profit-seeking incentives embedded in token transactions. Therefore, the role of users' market motivation in this process should not be overlooked when examining its impact on knowledge-sharing behavior.

In the Steemit community, the primary economic participants in token transactions include cryptocurrency exchange accounts, users offering token leasing and vote-selling services, as well as regular users engaged in transactions [10]. The structural characteristics of their transaction behaviors can be described by the network structure formed through connections between these participants, known as the transaction network [12]. Moreover, with blockchain-based token economies, each transaction activity is recorded in a public database, replicating real-world economic systems and providing data support for constructing the transaction network [10]. Thus, this study utilizes the centrality of users in the transaction network to indicate their level of market motivation. A higher centrality in the transaction network suggests a more significant role played by the user in the token trading network, indicating a stronger market motivation to profit through various market services. According to the limited attention theory [33], as market motivation increases, these users tend to focus their attention on transactional relationships, thereby weakening the influence of social capital on users' knowledge-sharing behavior. Based on these observations, the following hypotheses are proposed:

H4a: The centrality in the transaction network negatively moderates the positive relationship between the centrality in the follower network and the number of posts by users.

H4b: The centrality in the transaction network negatively moderates the positive relationship between the centrality in the vote network and the number of posts by users.

H4c: The centrality in the transaction network negatively moderates the positive relationship between the centrality in the governance network and the number of posts by users.

3 Research Design

3.1 Sample Selection and Data Collection

BKC is an online community that provides decentralized social services based on distributed platforms [6]. Steemit, as a prominent example of BKC, operates as a decentralized blockchain content platform with token incentive mechanisms. It enables participants to derive value from the community's effects through user autonomy [26]. Since its establishment in July 2016, the Steemit community has attracted over 1.59 million registered users, and the total value of distributed tokens has exceeded 250 million USD. All user activities within the community are permanently recorded on the blockchain, and the community offers an API interface to access blockchain data, providing a rich data foundation for this study.

The empirical study utilizes a dataset comprising blockchain data from June 1, 2017, to April 1, 2018. The selection of this observation period is primarily based on the high user activity observed in the community during this timeframe. The dataset includes various data points such as users' posting activities, votes received, following relationships, witness voting, transaction records, token rewards for posts, and token power-up data. The data is aggregated on a monthly basis, resulting in a panel dataset spanning 10 months. During the data processing stage, to ensure that the research sample consists of active users, only users who posted at least once per month were retained, resulting in a final sample of 3,692 users and 36,920 observations.

3.2 Variable Selection

Dependent Variable. The main variables involved in this study are defined as shown in Table 1. Following the practice of previous studies [34, 35], the number of user posts ($Post_{i,t}$) is used as a measure of users' knowledge-sharing behavior and serves as the dependent variable.

Table 1. Variable definitions

Type	Variable names	Abbreviations	Measures
Dependent variable	Number of user posts	$Post_{i,t}$	The number of posts made by user i in month $t/1000$
Independent variable	Centrality in the follower network	$Follow_{i,t}$	The centrality of user i's network influence in the follower network during month t
	Centrality in the vote network	$Vote_{i,t}$	The centrality of user i's network influence in the vote network during month t
	Centrality in the governance network	$Witness_{i,t}$	The centrality of user i's network influence in the governance network during month t
Moderating variable	Centrality in the transaction network	$Transfer_{i,t}$	The centrality of user i's network influence in the transaction network during month t
Control variable	Number of token rewards received for posts	$Reward_{i,t}$	Ln (The number of tokens rewarded to user i for posting in month $t + 1$)
	Number of tokens powered up	$Powerup_{i,t}$	Ln (The number of tokens powered up by user i in month $t + 1$)

Key Explanatory and Moderating Variables. The relationships between the multiplex networks are summarized in Fig. 2. In this study, the centrality in the follower network ($Follow_{i,t}$) is used to measure users' structural capital, the centrality in the vote network ($Vote_{i,t}$) represents users' cognitive capital, and the centrality in the governance network ($Witness_{i,t}$) reflects users' relational capital. These variables are considered as the key explanatory variables in this study. The centrality in the transaction network ($Transfer_{i,t}$) is used to measure the market motivation embedded in users' transactional behaviors and serves as the moderating variable. In this study, the degree centrality (DC) of the aforementioned variables is calculated, taking into account the directedness and weights of the relationships between network nodes. Weighted in-degree is used to calculate the relationship between nodes. The specific calculation process is described in Eq. (1).

$$DC(i) = \frac{\sum_{j=1}^{g-1} a_{j,i}}{g - 1} \tag{1}$$

where g represents the number of users in the network, i represents a specific user in the network, j represents any other user in the network except for i, and $a_{j,i}$ represents the weighted in-degree of individual j towards individual i in the network. In practice, the network constructed in this study may vary in size across different periods. Therefore, the normalization algorithm used for degree centrality helps mitigate the effects of size differences [19].

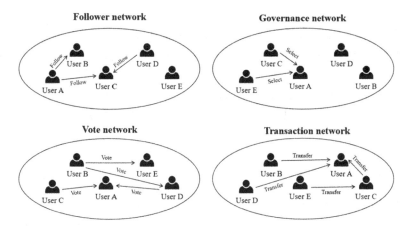

Fig. 2. The relationship between the multiplex networks.

Control Variables. Based on existing literature [16], the empirical model in this study includes the following control variables: (1) the number of token rewards received for posts ($Reward_{i,t}$); (2) the number of tokens powered up ($Powerup_{i,t}$). The inclusion of these variables as control variables is motivated by the understanding that variations in the number of token rewards received for posts can influence users' expectations regarding the outcomes of their knowledge-sharing behavior [36]. Additionally, the number of tokens held by users in the community affects their rights in incentive allocation and community governance, and an increase in the number of powered-up tokens can enhance a user's influence within the community, thereby affecting their self-efficacy in knowledge sharing [36].

3.3 Research Model

To test Hypotheses 1, 2, and 3, this study constructs the following fixed-effects panel model:

$$Post_{i,t} = \alpha_1 + \alpha_1 Follow_{i,t} + \alpha_2 Vote_{i,t} + \alpha_3 Witness_{i,t} + Controls_{i,t} + \eta_i + \eta_t + \varepsilon_{i,t}$$
$$(2)$$

where i represents the user, t represents the time measured in months, *Follow*, *Vote*, and *Witness* represent the independent variables, *Controls* represents the control variables,

η_i and η_t represent individual fixed effects and time fixed effects, and $\varepsilon_{i,t}$ represents the random disturbance term.

Based on this, to further test Hypotheses 4a, 4b, and 4c, this study constructs the following moderation effect models:

$$
\begin{aligned}
Post_{i,t} = {} & \beta_0 + \beta_1 Follow_{i,t} + \beta_2 Vote_{i,t} + \beta_3 Witness_{i,t} + \beta_4 Transfer_{i,t} \\
& + \beta_5 Transfer_{i,t} \times Follow_{i,t} + Controls_{i,t} + \eta_i + \eta_t + \varepsilon_{i,t}
\end{aligned}
\tag{3}
$$

$$
\begin{aligned}
Post_{i,t} = {} & \gamma_0 + \gamma_1 Follow_{i,t} + \gamma_2 Vote_{i,t} + \gamma_3 Witness_{i,t} + \gamma_4 Transfer_{i,t} \\
& + \gamma_5 Transfer_{i,t} \times Vote_{i,t} + Controls_{i,t} + \eta_i + \eta_t + +\varepsilon_{i,t}
\end{aligned}
\tag{4}
$$

$$
\begin{aligned}
Post_{i,t} = {} & \delta_0 + \delta_1 Follow_{i,t} + \delta_2 Vote_{i,t} + \delta_3 Witness_{i,t} + \delta_4 Transfer_{i,t} \\
& + \delta_5 Transfer_{i,t} \times Witness_{i,t} + Controls_{i,t} + \eta_i + \eta_t + \varepsilon_{i,t}
\end{aligned}
\tag{5}
$$

4 Empirical Analysis Results

4.1 Baseline Regression

Table 2. Baseline regression

Variable	Model 1	Model 2	Model 3	Model 4	Model 5
$Follow_{i,t}$		25.2591***			16.2083***
		(42.24)			(24.68)
$Vote_{i,t}$			11.2697***		8.3384***
			(46.89)		(31.39)
$Witness_{i,t}$				1.5087***	1.2130***
				(5.22)	(4.36)
$Reward_{i,t}$	0.0355***	0.0304***	0.0250***	0.0354***	0.0245***
	(45.37)	(39.49)	(31.75)	(45.26)	(31.28)
$Powerup_{i,t}$	0.0066***	0.0056***	0.0058***	0.0067***	0.0054***
	(13.19)	(11.40)	(11.82)	(13.30)	(11.09)
Cons	0.0041	−0.0056**	−0.0011	0.0028	−0.0071***
	(1.60)	(−2.24)	(−0.45)	(1.08)	(-2.84)
User FE	Yes	Yes	Yes	Yes	Yes
Year FE	Yes	Yes	Yes	Yes	Yes
R^2	0.783	0.794	0.796	0.783	0.800
N	36,920	36,920	36,920	36,920	36,920

Note: *** $p < 0.01$, ** $p < 0.05$, * $p < 0.1$.

When estimating Eq. (1), a stepwise regression method is employed in this study, and the basic regression results are summarized in Table 2. Model 1 in Table 2 represents the basic model, including only the control variables. Models 2–4 sequentially introduce the explanatory variables, namely, the centrality in the follower network ($Follow_{i,t}$), the centrality in the vote network ($Vote_{i,t}$), the centrality in the governance network ($Witness_{i,t}$), and all control variables. It can be observed that the centrality in the follower network, the centrality in the vote network, and the centrality in the governance network are significantly and positively correlated with the number of user posts ($Post_{i,t}$) at the 0.01 significance level. Model 5 represents the full model, which yields similar regression results. These results indicate that users' possession of structural capital, cognitive capital, and relational capital has a positive impact on the number of user posts, providing empirical support for Hypotheses 1, 2, and 3.

4.2 Robustness Tests

Endogeneity Test. To account for the potential presence of reverse causality between the explanatory variables and the dependent variable, an endogeneity test is conducted in this study. Drawing on existing research [37, 38], we incorporate $Follow_{i,t-1}$, $Vote_{i,t-1}$, and $Witness_{i,t-1}$ as explanatory variables in a regression analysis. The outcomes of this analysis are displayed in Model 1 of Table 3, which indicate that structural capital, cognitive capital, and relational capital continue to exert a significant positive influence on users' knowledge-sharing behavior. Moreover, an instrumental variable approach is employed to alleviate endogeneity concerns that may arise from omitted variables. Following prior research [39], we utilize $Follow_{i,t-1}$, $Vote_{i,t-1}$, and $Witness_{i,t-1}$ as instrumental variables. The results of the second-stage regression are presented in Model 2 of Table 3.

Table 3. Robustness test

Variable	Model 1	Model 2	Model 3	Model 4
$Follow_{i,t}$		20.5419***	13.0511***	176.5627***
		(10.88)	(11.38)	(14.28)
$Vote_{i,t}$		8.9959***	9.1472***	57.4018***
		(11.53)	(18.94)	(9.31)
$Witness_{i,t}$		3.9940***	1.2674***	9.1808***
		(7.46)	(3.00)	(9.14)
$Follow_{i,t-1}$	9.1295***			
	(14.04)			
$Vote_{i,t-1}$	2.8192***			
	(10.77)			

(*continued*)

Table 3. (*continued*)

Variable	Model 1	Model 2	Model 3	Model 4
$Witness_{i,t-1}$	1.5540***			
	(5.34)			
$Reward_{i,t}$	0.0268***	0.0205***	0.0310***	
	(31.93)	(21.61)	(16.06)	
$Powerup_{i,t}$	0.0051***	0.0044***	0.0064***	
	(10.01)	(8.71)	(6.39)	
Cons	0.0068**		−0.0183**	0.0008
	(2.43)		(−2.33)	(0.32)
User FE	Yes	Yes	Yes	Yes
Year FE	Yes	Yes	Yes	Yes
Anderson LM		3822.362***		
Cragg-Donald Wald F		1462.876		
R^2	0.810	0.126	0.810	0.787
N	33,228	33,228	14,000	36,920

Note: *** $p < 0.01$, ** $p < 0.05$, * $p < 0.1$.

Adjustment of Sample Size. To mitigate the potential impact of outliers on the research findings during the robustness test, it is essential to exclude individual outliers and assess the continued robustness of the conclusions. Thus, this study further refined the sample by selecting users from the original dataset who received at least one transfer per month. This refined sample consisted of a total of 1400 users, and a re-estimation was conducted. The results are presented in Model 3 of Table 3.

Substitution of Independent Variables. The PageRank algorithm takes into account the entire network structure and connectivity patterns, allowing for a more accurate reflection of node influence and status. Therefore, in this study, the PageRank algorithm is further employed to re-evaluate the study's independent variables. The results are summarized in Model 4 of Table 3.

4.3 Moderation Analysis

The moderation analysis, as demonstrated in Models 1–3 of Table 4, examines the influence of market motivation on the relationship between social capital and user posting quantity, focusing on Hypotheses 4a-c. Model 1 of Table 4 extends the baseline Model 5 from Table 2 by including the cross-term of transfer network centrality ($Transfer_{i,t}$), and follow network centrality ($Follow_{i,t}$). The results reveal a statistically significant negative correlation between the cross-term and the number of user posts ($Post_{i,t}$) at a significance level of 0.01. This suggests that with an increase in market motivation, the positive impact of users' structural capital on knowledge-sharing behavior weakens,

Table 4. Results of the moderating effect

Variable	Model 1	Model 2	Model 3
$Follow_{i,t}$	16.4737***	16.1656***	16.1699***
	(24.89)	(24.63)	(24.63)
$Vote_{i,t}$	8.2667***	8.3366***	8.2935***
	(31.11)	(31.30)	(31.22)
$Witness_{i,t}$	1.0929***	1.1453***	1.1389***
	(3.93)	(4.10)	(4.06)
$Transfer_{i,t}$	12.0825***	10.5614***	9.7783***
	(7.11)	(6.07)	(5.30)
$Reward_{i,t}$	0.0242***	0.0242***	0.0242***
	(30.90)	(30.95)	(31.01)
$Powerup_{i,t}$	0.0053***	0.0054***	0.0053***
	(11.02)	(11.08)	(11.03)
$Transfer_{i,t} \times Follow_{i,t}$	−336.7442***		
	(−3.25)		
$Transfer_{i,t} \times Vote_{i,t}$		−75.2089*	
		(−1.78)	
$Transfer_{i,t} \times Witness_{i,t}$			−10.5335
			(−0.99)
$Cons$	−0.0079***	−0.0078***	−0.0077***
	(−3.15)	(−3.14)	(−3.09)
User FE	Yes	Yes	Yes
Year FE	Yes	Yes	Yes
R^2	0.800	0.800	0.800
N	36,920	36,920	36,920

Note: *** $p < 0.01$, ** $p < 0.05$, * $p < 0.1$.

thereby confirming Hypothesis 4a. Similarly, the results in Model 2 of Table 4 imply that as market motivation increases, the positive influence of users' cognitive capital on knowledge-sharing behavior diminishes, thereby supporting Hypothesis 4b. In addition, in Model 3 of Table 4, the cross-term of transfer network centrality ($Transfer_{i,t}$) and governance network centrality ($Witness_{i,t}$) is added to the baseline Model 5 from Table 2. The results show a negative correlation between the cross-term and the number of user posts ($Post_{i,t}$), but it is not statistically significant. Hence, Hypothesis 4c lacks substantial support.

In general, Hypotheses 1, 2, 3, 4a, and 4b receive confirmation, while Hypothesis 4c does not obtain significant support. This may be attributed to the unique nature of the community's decentralized management, where users engage with other members through participation in witness elections. The demonstrated loyalty, maintenance, and effort primarily arise from a deep emotional connection to the community, rather than economic incentives, illustrating users' emotional identification with the community. Therefore, when users have a strong emotional attachment, their behavior prioritizes community interests over personal gain. Consequently, the moderating effect of market motivation on the relationship between relational capital and knowledge sharing is not deemed significant.

5 Conclusion

The research findings present compelling evidence of the positive impact of social capital on users' knowledge-sharing behavior. Specifically, the study reveals that higher levels of structural capital, cognitive capital, and relational capital correspond to a greater tendency for sustained knowledge sharing. Notably, among the three dimensions of social capital, structural capital emerges as the most influential factor in driving knowledge-sharing behavior. This finding stands in contrast to previous studies conducted in traditional knowledge communities, where the effects of different elements of social capital on knowledge sharing were found to be relatively equal. Moreover, the research results shed light on a significant negative moderating role of market motivation in the relationship between social capital and knowledge-sharing behavior.

The study makes significant theoretical contributions in several areas. Firstly, it bridges the gap between behavioral finance research on investment decision-making, which has extensively employed limited attention theory, and information systems research, where its application has been limited. By integrating limited attention theory with social capital theory and applying it to BKC, this research expands the theoretical application of limited attention theory. Secondly, unlike previous literature that explored the effects of different dimensions of social capital and network structures separately, this study takes a novel approach. Leveraging the heterogeneous characteristics of different network layers, it proposes constructing multiple networks with distinct relationships. Through this approach, the study extracts network structure indicators to measure users' social capital across various dimensions. Additionally, by considering the unique features of BKC, such as governance networks, the research perspective is enriched. Lastly, the study addresses the lack of research on the factors influencing knowledge-sharing behavior in the context of BKC. By focusing on the prominent characteristic of market motivation embedded in token transactions within BKC, the empirical analysis using real user data reveals the impact of this characteristic on users' knowledge-sharing behavior. These findings provide valuable insights for effectively incentivizing users' continuous contributions in BKC.

The research offers valuable practical implications for community managers and decision-makers. First, the findings underscore the significant positive impact of users' structural capital in the follow network on their knowledge-sharing behavior. As a result, communities should implement and enhance user recommendation mechanisms. By capturing user preferences and integrating interests and social factors into predictive models, targeted recommendations can be provided to users with similar interests and close social connections. This approach will facilitate content creators in accumulating more structural capital within the follow network. Second, the study highlights the significant positive influence of users' cognitive capital in the vote network on their knowledge-sharing behavior. Consequently, managers should focus on improving content recommendation mechanisms. Effective post recommendations can benefit content creators by generating substantial income and enable content consumers to access content that genuinely interests them. To achieve these objectives, an appropriate social recommendation framework can be employed to accurately suggest posts to users likely to be interested, thus increasing post exposure and helping content creators accumulate more cognitive capital. Additionally, the results demonstrate a significant positive relationship between users' relational capital in the governance network and knowledge sharing. Hence, managers should devise suitable incentive mechanisms to encourage greater user participation in witness elections. By assisting witness candidates in garnering more support, users can accumulate more relational capital. Lastly, considering the negative moderating effect of market motivation on the relationship between users' social capital and knowledge sharing, managers should adopt measures to limit the profit potential derived from market service provision. Implementing restrictions on transfers, similar to the risk control measures employed by banks, can reduce users' profit potential from offering various market services, thereby mitigating their market motivation.

Acknowledgment. This work was supported by grants No. 72171089 and 72071083 from the National Natural Science Foundation of China, grant No. 2021A1515012003 from the Guangdong Natural Science Foundation of China, and grant No. 2021GZQN09 from the Project of Philosophy and Social Science Planning of Guangzhou in 2021.

References

1. Chen, L., et al.: Why do participants continue to contribute? Evaluation of usefulness voting and commenting motivational affordances within an online knowledge community. Decis. Support Syst. **118**, 21–32 (2019)
2. Zhang, Y., et al.: Understanding the formation mechanism of high-quality knowledge in social question and answer communities: a knowledge co-creation perspective. Int. J. Inf. Manage. **48**, 72–84 (2019)
3. Kwon, S.: Understanding user participation from the perspective of psychological ownership: The moderating role of social distance. Comput. Hum. Behav. **105**, 106207 (2020)
4. Raza, A., et al.: GPSPiChain-blockchain and AI based self-contained anomaly detection family security system in smart home. J. Syst. Sci. Syst. Eng. **30**, 433–449 (2021)
5. Drasch, B.J., et al.: The token's secret: the two-faced financial incentive of the token economy. Electron. Mark. **30**, 557–567 (2020)
6. Guidi, B.: When blockchain meets online social networks. Pervasive Mob. Comput. **62**, 101131 (2020)
7. Qin, R., et al.: A novel hybrid share reporting strategy for blockchain miners in PPLNS pools. Decis. Support Syst. **118**, 91–101 (2019)
8. Xie, Y., et al.: Addressing wealth inequality problem in blockchain-enabled knowledge community with reputation-based incentive mechanism. In: ICEB (2021)
9. Xie, R., Zhang, W.: Online knowledge sharing in blockchains: towards increasing participation. Manage. Decis. (2023)

10. Liu, X.F., et al.: Characterizing key agents in the cryptocurrency economy through blockchain transaction analysis. EPJ Data Sci. **10**(1), 21 (2021)
11. Brass, D.J., et al.: Taking stock of networks and organizations: a multilevel perspective. Acad. Manag. J. **47**(6), 795–817 (2004)
12. Li, J., Yu, Y.: Structural holes in collaboration network, cohesion of knowledge network and exploratory innovation performance: an empirical study on the Chinese automakers. Nankai Bus. Rev. **21**(6), 121–130 (2018)
13. Nahapiet, J., Ghoshal, S.: Social capital, intellectual capital, and the organizational advantage. Acad. Manag. Rev. **23**(2), 242–266 (1998)
14. Cummings, J., Dennis, A.R.: Virtual first impressions matter: the effect of enterprise social networking sites on impression formation in virtual teams. MIS Q. **42**(3), 697–718 (2018)
15. Yan, J., et al.: Social capital and knowledge contribution in online user communities: one-way or two-way relationship? Decis. Support Syst. **127**, 113131 (2019)
16. Chang, M., et al.: Understanding members' active participation in a DAO: an empirical study on steemit. In: PACIS. p. 197 (2019)
17. Wasko, M.M., Faraj, S.: Why should I share? Examining social capital and knowledge contribution in electronic networks of practice. MIS Q. 35–57 (2005)
18. Wang, G., et al.: Wisdom in the social crowd: an analysis of quora. In: Proceedings of the 22nd International Conference on World Wide Web, pp. 1341–1352 (2013)
19. Liu, S., et al.: Human capital network and firm innovation: a study based on the resume data of LinkedIn. Manage. World **7**, 88–98+119+188 (2017)
20. Li, W., et al.: Research on emotional polarization mechanism of knowledge community from the perspective of social network structure—an empirical study on 'Zhihu' question and answer learning community. Front. Phys. **11**, 193 (2023)
21. Allameh, S.M.: Antecedents and consequences of intellectual capital: the role of social capital, knowledge sharing and innovation. J. Intellect. Cap. **19**(5), 858–874 (2018)
22. Moser, C., et al.: Communicative genres as organising structures in online communities–of team players and storytellers. Inf. Syst. J. **23**(6), 551–567 (2013)
23. Moser, C., Deichmann, D.: Knowledge sharing in two cultures: the moderating effect of national culture on perceived knowledge quality in online communities. Eur. J. Inf. Syst. **30**(6), 623–641 (2021)
24. Zhao, L., Detlor, B.: Towards a contingency model of knowledge sharing: interaction between social capital and social exchange theories. Knowl. Manag. Res. Pract. **21**(1), 197–209 (2023)
25. Hashim, K.F., Tan, F.B.: The mediating role of trust and commitment on members' continuous knowledge sharing intention: a commitment-trust theory perspective. Int. J. Inf. Manage. **35**(2), 145–151 (2015)
26. Guidi, B., et al.: Analysis of witnesses in the steem blockchain. Mob. Netw. Appl. 1–12 (2021)
27. Lu, Y., Yang, D.: Information exchange in virtual communities under extreme disaster conditions. Decis. Support Syst. **50**(2), 529–538 (2011)
28. Zeng, Q., et al.: What factors influence grassroots knowledge supplier performance in online knowledge platforms? Evidence from a paid Q&A service. Electron. Mark. **32**(4), 2507–2523 (2022)
29. Tversky, A., Kahneman, D.: Availability: a heuristic for judging frequency and probability. Cogn. Psychol. **5**(2), 207–232 (1973)
30. Zhang, Y.: Analyst responsiveness and the post-earnings-announcement drift. J. Account. Econ. **46**(1), 201–215 (2008)
31. Dukas, R.: Behavioural and ecological consequences of limited attention. Philos. Trans. R. Soc. London Ser. B: Biol. Sci. **357**(1427), 1539–1547 (2002)
32. Scalf, P.E., et al.: Competition explains limited attention and perceptual resources: implications for perceptual load and dilution theories. Front. Psychol. **4**, 243 (2013)

33. Rochanahastin, N.: Assessing axioms of theories of limited attention. J. Behav. Exp. Econ. **84**, 101499 (2020)

34. Jin, J., et al.: Why users contribute knowledge to online communities: an empirical study of an online social Q&A community. Inf. Manage. **52**(7), 840–849 (2015)

35. Wang, N., et al.: From knowledge seeking to knowledge contribution: a social capital perspective on knowledge sharing behaviors in online Q&A communities. Technol. Forecast. Soc. Chang. **182**, 121864 (2022)

36. Hsu, M.-H., et al.: Knowledge sharing behavior in virtual communities: the relationship between trust, self-efficacy, and outcome expectations. Int. J. Hum Comput Stud. **65**(2), 153–169 (2007)

37. Lin, B., Xie, Y.: Driving green technology innovation in renewable energy: does venture capital matter? IEEE Trans. Eng. Manage. (2023)

38. Lin, B., Xie, Y.: Positive or negative? R&D subsidies and green technology innovation: evidence from China's renewable energy industry. Renewable Energy **213**, 148–156 (2023)

39. Kim, S., et al.: Divergent effects of external financing on technology innovation activity: Korean evidence. Technol. Forecast. Soc. Chang. **106**, 22–30 (2016)

Exploring Knowledge Synthesis Enablers for Successful Research Projects

Siri-on Umarin[1]([✉]), Takashi Hashimoto[1], Thanwadee Chinda[2],
and Yoshiteru Nakamori[1]

[1] School of Knowledge Science, Japan Advanced Institute of Science and Technology (JAIST),
1-1 Asahidai, Nomi, Ishikawa 923-1292, Japan
aymsris@yahoo.com, {s2120403,hash}@jaist.ac.jp,
nakamori0212@gmail.com

[2] School of Management Technology (SIIT), Sirindhorn International Institute of Technology
(SIIT), Thammasat University, 99 Moo 18, Km. 41 On Paholyothin Highway Khlong Luang,
Pathum Thani 12120, Thailand
thanwadee@siit.tu.ac.th

Abstract. Research and Development (R&D) is one crucial factor for economic growth and social development as it raises the national competitiveness level and the quality of life through innovation. In Thailand, the National Science and Technology Development Agency (NSTDA), the biggest statutory government R&D institute is responsible for the National Research and Development (R&D) projects and activities that require research assessment to ensure their quality fitting for the needs, solutions for problems, and current concerns. This study proposes the idea of Knowledge Management and the concept of Knowledge Construction Methodology to observe the R&D from the start adopt the technique of semi-structured interviews for the data collection with selected respondents from NSTDA and use the confirmatory factor analysis to confirm the knowledge domains as a significant factor for successful R&D projects to introduce another approach inspired by system thinking and knowledge management called knowledge synthesis as an alternative tool for R&D quality improvement.

Keywords: Knowledge Synthesis · Assessment · R&D Project · Knowledge Construction Methodology

1 Introduction

Research and Development (R&D) is one of the crucial factors for economic growth and social development as it raises the national competitiveness level together with the quality of life [1] through innovation, thus Developing Countries must support their R&D activities [2]. As one of the Developing Countries, Thailand also established its national R&D institute called the National Science and Technology Development Agency (NSTDA), the statutory government R&D institute under Thailand's Ministry of Higher Education, Science, Research and Innovation, as the biggest National R&D body.

With the prime status, NSTDA's R&D project requires a research assessment to ensure its quality fitting for the needs, solutions for problems, and concerns. Assessment is important to guarantee the expected results. Still, it is the study from the outside perspective, observing the result after the action, operation, or project emerged. Speaking of the research assessment, there are several approaches from previous studies including the impact assessment [3] and quality assessment [4] which are significant for NSTDA.

As the biggest governmental research and development institute, NSTDA's mission is to "accelerate science, technology, and innovation development in Thailand to respond to the needs of the industry and enhance the country's competitiveness in the global economy, and as a result, contributing to national economic and social development" through "Research and Development, Design and Engineering, Technology Transfer, Science, and Technology Human Resource Development, and Infrastructure Development" [5] raising Thais' quality of life, lifting technological advancement nationally, and transferring advanced knowledge to the country. Such an ambiguous mission requires enormous support from both personnel and non-personnel, i.e., funding, data, etc., likewise, the applicable assessment ensures the essential result of its research.

Just as stated before, impact and quality assessments are two active methods to measure research interest. Impact refers to the impact factor or journal academic impact factor, using a scientometric index provided to reflect the mean number of citations. It highlights a substantial work by the journal ranking, the higher the ranking the journal is, the more important the published research inside it is. While quality refers to the "performance upon expectations and fit for function" as stated by Tech Quality Pedia [6], quality reflects satisfaction from the customers' point of view covering performance, look, objectified fulfillment, etc. There is a contradiction between these two definitions shaping the perspective of research operators (all involved stakeholders) as one focuses on the 'content of work' itself, while the other focuses on 'creating results from the work'.

In addition, either impact or quality is not the absolute assessment as the first gains negative criticism from the dependent on the editors and political issue, together with the attachment of field-dependent factors limiting the cross-disciplinary or multidisciplinary application [7–9]. When speaking of quality, Total Quality Management (TQM) is often mentioned due to its effectiveness for business operations, but this worldwide strategy also possesses downsides i.e., requiring an absolute commitment from the whole organization, time, and cost, etc., as it is the long-term strategy for customers' satisfaction [10].

Acknowledging these, both definitions shall not be adapted to research directly as the component of research exceeds any satisfaction and concrete result but covers new academic or theoretical findings craving for further study. In the process of national needs, the National Research and Development Institute could not solely give heed to the new business product, at the same time, it could not strictly explore the unseen technological discoveries. With such a dilemma, Knowledge Synthesis was introduced concerning a new possible assessment that combines the identity of system thinking and knowledge management that allows capturing the essence of work and monitoring the whole process.

2 Knowledge Synthesis

Research work is different from product development even if some parts align. Still, every product requires research work as a supportive component. Good research needs an objective-oriented purpose, efficiency of research personnel, active and good relationships among stakeholders, and supportive information. All of the previously mentioned aspects could be summarized as the 'selected knowledge' that could emerge from the synergy of the knowledge domains. The concept of the knowledge domain is new compared to the well-known theory of knowledge management, i.e., the SECI knowledge creation model, etc., as most of them emphasize the creation and emergence of knowledge in business. For this paper, the knowledge synthesis is adapted from the concept of the knowledge construction methodology by Nakamori [11]. This concept combines the basic idea of system thinking and knowledge management. The knowledge construction methodology has 3 important components;

- **Knowledge construction system model:** It proposes five nodes of knowledge creation including Intelligence, Involvement, Imagination, Intervention, and Integration. The first 3 nodes capture the knowledge existing domain in the scientific-actual domain, the social-relational domain, and the cognitive-mental domain. While the rest two nodes are in the initiative-creative domain and manage knowledge creation activities. The knowledge construction system model is shown in Fig. 1.
- **Knowledge Construction Diagram:** As stated by Nonaka et al. [12] tacit knowledge and explicit knowledge co-exist with each other continuously; hence the two types of knowledge have different natures and an efficient mutual conversion process is needed for new knowledge creation. Thus, the knowledge construction system model supports new knowledge creation through the knowledge construction diagram. The knowledge construction diagram is shown in Fig. 2

- **Constructive Evolutionary Objectivism:** It portrays action guidelines for knowledge construction using the knowledge construction system model with the following four principles. First, the intervention principle helps to define and scope the boundary of data collection for the problem solution by considering all restrictions. Second, the multimedia principle guides the practitioner to seek all possible knowledge from data and information in as many media as possible, from securing information system management, human dialogue and conversation, and ideas. Third, the emergence principle stimulates new knowledge creation through the combination of existing knowledge (information and wisdom) or intuitive inspiration. Such circumstances could happen in the interaction of participants via questions at the same time as the systematization of the collected knowledge. Last, the fourth principle is the evolutionary falsification principle, which verifies the effectiveness of newly emerged ideas. If scientific verification is difficult, the subjective consensus among the participants in the problem could be used.

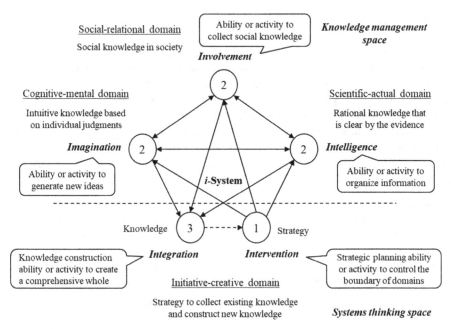

Fig.1. Knowledge construction system model

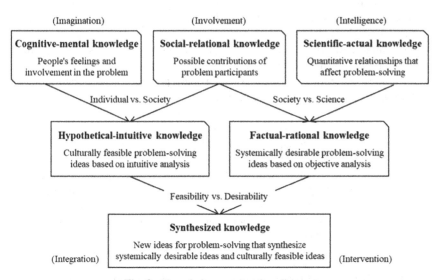

Fig. 2. Knowledge construction diagram

3 Research Method

This paper has adopted both quantitative and qualitative approaches for data collection. They were used to discover the unrevealed information from the target respondents about the R&D project environment in Thailand, another example of a Developing country. A questionnaire survey and a semi-structured interview were implemented to gather significant information about the research work, the relationship among research stakeholders, and influences from non-person factors influencing the research work. The target respondent is selected from the NSTDA representing each National Center specializing in Electronics and Computer Technology, Materials Technology, Nanotechnology, and the research support unit under the Central Officer of NSTDA a total of 26 respondents.

Questionnaire Survey: The five-point Likert Scale [13] survey was deployed to the respondents to determine their concerns with the R&D project and its components including personnel thoughts, beliefs, and experiences. The question set was constructed from the concept of Knowledge Construction Methodology [11]. These questions correspond to the observed variables in the covariance structure analysis and the variables of the five nodes of the model in Fig. 1 are the latent variables as shown in Table 1 below.

(K1) Intelligence: Activities to collect, maintain, and understand existing knowledge and related data/information
(K2) Involvement: Activities to collect, maintain, and understand external information and implicit knowledge
(K3) Imagination: Activities to create ideas through systematic knowledge and member interaction
(S1) Intervention: Activities to define problems, establish research systems including research funding, and clarify research contents
(S2) Integration: Activities to achieve research goals, publish research results, and then develop research by discovering new issues

Table 1. Corresponding questions

Knowledge Domains	Question List of the conscious activities in each Knowledge Domain
Scientific-Actual Domain	SA1 Have you collected sufficient existing knowledge from previous research surveys?
	SA2 Have you collected sufficient data and information to promote the research?
	SA3 Have you mastered enough technologies to analyze data and information?
Social-Relational Domain	SR1 Have you communicated well with the government, stakeholders, and users?
	SR2 Have you fully cooperated with domestic and foreign researchers?
	SR3 Have you established trust relationships with the managers, supervisors, and colleagues?

(continued)

Table 1. (*continued*)

Knowledge Domains	Question List of the conscious activities in each Knowledge Domain
Cognitive-Mental Domain	CM1 Have you effectively used the systematized data, information, and knowledge?
	CM2 Have you effectively used the knowledge of experts, stakeholders, and policymakers?
	CM3 Have you effectively used the knowledge of group members?
Initiative-Creative Domain	IC1 Have you been confident in the problem settings?
	IC2 Have you been confident in the research funding and implementation system?
	IC3 Have you been confident in the approach to knowledge collection?
	IC4 Have you synthesized various knowledge and got satisfactory research results?
	IC5 Have you written a satisfactory research report and academic papers?
	IC6 Have you found promising ideas for continuous research development?

- **Semi-structured interview** [14]: Considering the time effectiveness and availability of the respondents, the semi-structured interview through Face to Face (FtF) was used to save the social cues [15] of interviewees linking to the essential concealed data for further qualitative analysis.

4   Result

Analysis

A total of 26 respondents consisted of 16 personnel from the National Electronics and Computer Technology Center (NECTEC), 7 personnel from the National National Metal and Materials Technology Center (MTEC), 2 personnel from the National Nanotechnology Center (NECTEC), and one from the NSTDA Characterization and Testing Service Center, one of the Research Support Units under the Central Officer of NSTDA, corporated with the interview. The question scores were cleaned and analyzed by confirmatory factor analysis to certify the group of knowledge management into three activity types which could be referred to as three factors. This analysis was performed using the maximum likelihood method for estimation and the Promax method for rotation. From this first analysis, all observed variables, except CM1 and CM2, were categorized into 3 factors in Table 2.

Table 2. Result loadings and uniqueness

	Factor 1	Factor 2	Factor 3	Uniqueness
SA1	0.940	−0.133	0.149	0.110
SA2	0.931	−0.079	−0.001	0.194
SA3	0.983	0.072	−0.136	0.051
SR1	−0.094	1.033	0.008	0.005
SR2	0.389	0.354	0.084	0.538
SR3	0.005	0.003	0.994	0.005
CM3	−0.070	0.007	0.640	0.613

Since CM1 and CM2 contain content from the Scientific-Actial and Social-Relational domains, respectively, it was concluded that they should be excluded from the analysis. With such a result, Factor 1 (K1) has SA1, SA2, and SA3 as observed variables, Factor 2 (K2) has SR1 and SR2 as observed variables, and Factor 3 (K3) has SR3 and CM3 as observed variables. Next step, this information was used for a covariance structure analysis and to find the relationship between latent variables and observed variables.

As mentioned above, K1, K2, and K3 correspond to Intelligence, Involvement, and Imagination in Fig. 1, respectively, and S1 and S2 correspond to Intervention and Integration. These are considered latent variables, and SA1, SA2, SA3, SR1, SR2, SR3, and CM3 are observed variables in covariance structure analysis. The software used for analysis is "R" [16, 17], so the structural equation model is described according to its grammar. A result of covariance structure analysis is shown in Table 3, where we read the operator "=~" as "is observed by", and "~" as "explained by."

Table 3. A result of covariance structure analysis

	Standardized Solution	P-value
K1 =~ SA1	0.999	0.000
K1 =~ SA2	0.837	0.000
K1 =~ SA3	0.886	0.000
K2 =~ SR1	0.541	0.001
K2 =~ SR2	0.998	0.000
K3 =~ SR3	1.182	0.000
K3 =~ CM3	0.547	0.000
S1 =~ IC1	0.753	0.000
S1 =~ IC2	0.487	0.004
S1 =~ IC3	0.655	0.000

(continued)

Table 3. (*continued*)

	Standardized Solution	P-value
S2 =~ IC4	0.608	0.000
S2 =~ IC5	0.989	0.000
S2 =~ IC6	0.801	0.000
K1 ~ S1	0.785	0.000
K2 ~ S1	0.715	0.000
K3 ~ S1	0.398	0.016
S2 ~ K1	0.861	0.000
S2 ~ K2	0.044	0.487
S2 ~ K3	0.207	0.006

Refer to Table 3. Research plan (S1) strongly influences research activities (K1 and K2) but has less influence on K3 which means the activities on K3 (idea generation activities in a group) were unwell planned. Yet, research activity (K1) firmly affects research results (S2) while K2 and K3 have little effect on S2. This means the mutual idea in a research group has a minimum contribution to the research result and that is possibly the characteristic of the study of natural sciences, and also a feature of NSTDA. For this, the current recommendation is that the institute must strengthen and

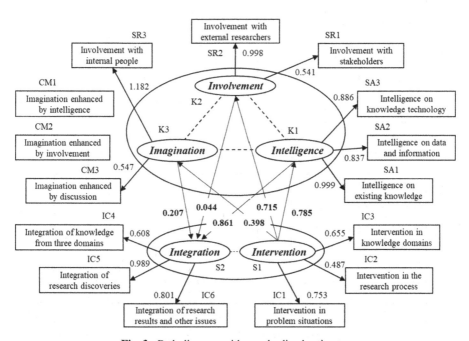

Fig. 3. Path diagram with standardized estimates

spread its cooperation with external collaborators together with the promotion of internal knowledge management. All information is shown in Fig. 3.

The correlation coefficients between observed variables are shown in Table 3. The table shows that the summary of the research (knowledge synthesis) highly relates to internal collaboration/knowledge management and problem setting (IC1 or Factor 3). Although the production of academic papers and the discovery of new themes are highly correlated with scientific knowledge/data gathering (Factor 1), exchanges of opinions with external researchers (SR2), and relationships with superiors and colleagues (SR3). The rest is the variable in SR1 which has less correlation with other variables. At this point, the distinguished characteristics of research personnel in the National R&D Institute have revealed that they have full research competency and capacity to gather precise knowledge (including data/information) in their specific or related disciplines for their research (Table 4).

Table 4. Correlation coefficients between observed variables

	SA1	SA2	SA3	SR1	SR2	SR3	CM3	IC1	IC2	IC3	IC4	IC5
SA2	0.834											
SA3	0.887	0.867										
SR1	0.295	0.271	0.374									
SR2	0.556	0.489	0.544	0.540								
SR3	0.420	0.295	0.256	0.499	0.404							
CM3	0.207	0.117	0.102	0.299	0.117	0.617						
IC1	0.596	0.609	0.390	0.188	0.533	0.361	0.346					
IC2	0.307	0.216	0.187	0.196	0.443	0.478	0.305	0.348				
IC3	0.519	0.661	0.445	0.187	0.375	0.379	0.267	0.588	0.278			
IC4	0.555	0.557	0.351	0.261	0.321	0.645	0.688	0.639	0.311	0.662		
IC5	0.945	0.783	0.765	0.312	0.589	0.589	0.219	0.631	0.431	0.550	0.587	
IC6	0.701	0.710	0.533	0.366	0.598	0.684	0.396	0.814	0.401	0.762	0.741	0.799

5   Discussion

The result from the correlation coefficients in Table 3 revealed connectivity among each observed variable which had no negative relationship. This result aligned with the information given from the semi-structured interviews with respondents. Refers to Table 3, even though all observed variables have a positive relationship, some of them have less value than others, and among all, the IC2 has less value in the relationship. Based on the respondents' perception, they have confidence in their knowledge, research skills, and technology trends, and in their team members, the research support unit officers, and even their customers, both potential and royal ones. Hence, they do not share equal positive thoughts towards other stakeholders such as policymakers and funding agencies since many of them have different knowledge, mostly of financial, political, and management. The respondents consider that the different background reflects the perspective of the research result as the researcher values the excellence and competence of each research, while the policymaker and funding agency officer values a quick result and project-based solution with a low Technology Readiness Level (TRL) which in the long term could barely improve end users' problems and create sustainability of the research environment in the country. Such concern also reflects the continuity of mistrust and unalignment of standpoints between research personnel with policymakers and funding agencies that remain unresolved [18–20] as shown on Table 5.

Table 5. Example of respondent's answer for an observed variable of IC4

Question	Answers
Have you been confident in the research funding and implementation system?	"One of the organizational challenges is the change of the executive, who is a policymaker and the organizational policy constructively. Such change leads to the reconstruction and loss of personnel. Every executive has a different background and personal interest leading to an imbalance of research selection. It is understandable, somehow, it reflects some skepticism among task operators" (researcher)
	"As the researchers, our ultimate goal is to advance the quality of life of people and create sustainable research and technological ecosystem. We recognize that the Research is a chronicled work including upstream to downstream or basic to applied research, It needs permanent financial support, together with technical, and business environmental understanding. Sadly, not everyone possesses such qualifications. So, policymakers and funding agencies often focus on the quick-win research result which might not be able to commercialize or successfully transfer to the business." (researcher)

(continued)

Table 5. (*continued*)

Question	Answers
	"It is good that the policymakers are open to feedback and comments, but it would be different if they applied the feedback and comments to reality. One of the repetitions is the revision of workflow as we face so much redundancy in our working process, that there are too many people involved with the research project proposal submission. We must pass the internal evaluation before processing to the funding agency and several times we missed the deadline for submission. Some research projects could be adjusted, but some couldn't which caused the loss of customers and even declined the customer's trust." (researcher)
	"The policymakers can improve research project operation is extending the channel for communication between the top to the bottom level and across different functional officers at the same level among the same research center and between other research centers under the same umbrella organization. I believe that our personnel have the potential to grow and nowadays they don't work with full potential and even by the function of their current position. For example, we have a business development (BD) officer who works as an additional administrator with paper works instead of being a the business facilitator or business collaboration coordinator. If such a situation has been improved, we could see different results in research project proposal submission, more trust, and recognition increasing from customers and new investors, and also higher the competitiveness level of the organization as a whole." (researcher support officer)

However, the respondents again state that the situation could be mitigated by such recommendations;

- Increase transparency of policy communication [21]: As a member of the National R&D Institute, everyone must follow the organizational guidelines interacting with the National Strategy, yet, only a few people can access the updates and changes at the policy level. Those who gain access, especially the Executives or the middle managerial officers like the unit director, must make clear and fast communication with their subordinates to establish a precise and aligned research plan responding to the satisfaction of researchers, end users, and in return, the policymakers and funding agency officers themselves.
- Effective and proactive Human Resource Management: Though the respondents give positive feedback to their supportive colleagues in the research support unit, they still consider that the overall situation possibly better if the policymakers improve the organizational human resources via an adjustment of the job description (JD) [22]

and practical function of some research support unit i.e., the business development unit, will significantly benefit the research process and reduce unnecessary tasks.

- Re-design workflow [23]: Another demanding aspect for the policymakers is the modification of the organizational work process as the respondent is facing the extra paperwork, a long internal proposal evaluation process, and redundancy. The respondents constructively confirm these issues as their extra burden and barriers.

These recommendations describe the obscure reason behind each research work which from the point of view of an operational person, affects the expected result. In addition, this remaining demand from the operational person's perspective echoes that the policymakers and funding agency officers are the key players in the result of every research project even if they might not be the **direct** end-user. The respondent agreed that the more mutual understanding between the operational officers and the policymakers arise, the more possibility for the success of the research project could increase because the sustainable research project needs full commitment, synergy, and understanding from every involved person. The respondents favor the attempt to learn new things and create more intact collaboration at both vertical and horizontal levels and both micro and macro management will reduce the noise and pressure of work, raise their focus, and increase their creativity which eventually causes a positive effect to the research work and its success.

6 Conclusion

Compared with the previous studies of scientific research assessment, many works highlight the importance of the trend of scientific research activity, research performance related to the indicated topics by involving stakeholders, and research contributions [24]. Less mentioned about the assessment of the R&D Project at the state of idea crafting as proposed by the Knowledge Construction Methodology. The result of the study pointed out the importance of knowledge domains, especially the scientific-actual (SA) and social-relational (SR) domains towards the initiative-creative domain or the R&D project. Still, this study could not capture enough support from the cognitive-mental (CM) domain as some variables went missing from the calculation. Further study with the expansion of the respondent group is suggestive of a better analysis and result.

References

1. Setkij, W.: Sustainable development in the context of national development. Burapha J. Polit. Econ. **8**(1) (2020). https://so01.tci-thaijo.org/index.php/pegbuu/article/view/242955
2. Khan, J.: The role of research and development in economic growth: a review. J. Econ. Bibliography **2**(3) (2015). https://doi.org/10.1453/jeb.v2i3.480
3. Greenhalgh, T., Raftery, J., Hanney, S., Glover, M.: Research impact: a narrative review. BMC Med. **14**, 78 (2016). https://doi.org/10.1186/s12916-016-0620-8
4. Cabral, A.P., Huet, I.: Assessment of research quality in higher education: contribution for an institutional framework. Procedia Soc. Behav. Sci. **116**, 1528–1532 (2014). https://doi.org/10.1016/j.sbspro.2014.01.429

5. National Science and Technology Development Agency (NSTDA). https://www.nstda.or.th/en/about-us/at-a-glance.html
6. Tech Quality Pedia. What is Quality? Quality Definition: Quality Meaning. (2020). https://techqualitypedia.com/quality/
7. Bornmann, L., Daniel, H.D.: What do citation counts measure? A review of studies on citing behavior. J. Doc. **64**(1), 45–80 (2008). https://doi.org/10.1108/00220410810844150
8. Anauati, V., Galiani, S., Gálvez, R.H.: Quantifying the life cycle of scholarly articles across fields of economic research. Econ. Inq. **54**(2), 1339–1355 (2016). https://doi.org/10.1111/ecin.12292
9. Callaway, E.: Beat it, impact factor! Publishing elite turns against controversial metric. Nature **535**(7611), 210–211 (2016). https://doi.org/10.1038/nature.2016.20224
10. Gillis, A.S.: Total Quality Management (TQM). https://www.techtarget.com/searchcio/definition/Total-Quality-Management
11. Nakamori, Y.: Knowledge Construction Methodology: Fusing Systems Thinking and Knowledge Management. Springer, Singapore (2020). https://doi.org/10.1007/978-981-13-9887-2
12. Nonaka, I., Hirose, A., Ishi, Y.: Management by knowledge maneuverability: synthesizing knowledge creation and maneuver warfare. Hitotsubashi Bus. Rev. **61**(3), 120–137 (2013). (in Japanese)
13. Joshi, A., Kale, S., Chandel, S., Pal, D.K.: Likert scale: explored and explained. Curr. J. Appl. Sci. Technol. **7**(4), 396–403 (2015). https://doi.org/10.9734/BJAST/2015/14975
14. Kakilla, C.: Strengths and weaknesses of semi-structured interviews in qualitative research: a critical essay. Preprints.org **2021**, 2021060491 (2021). https://doi.org/10.20944/preprints202106.0491.v1. https://www.preprints.org/manuscript/202106.0491/v1
15. Opdenakker, R.: Advantages and disadvantages of four interview techniques in qualitative research. Forum Qual. Soc. Res. **7**(4) (2006). https://doi.org/10.17169/fqs-7.4.175
16. Ihaka, R., Gentleman, R.: R: a language for data analysis and graphics. J. Comput. Graph. Stat. **5**(3), 299–314 (1996). https://doi.org/10.2307/1390807
17. R Core Team: R: a language and environment for statistical computing. R Foundation for Statistical Computing, Vienna, Austria (2016). https://www.R-project.org/
18. Alazmi, A.A., Alazmi, H.S.: Closing the gap between research and policy-making to better enable effective educational practice: a proposed framework. Educ. Res. Policy Pract. **22**, 91–116 (2023). https://doi.org/10.1007/s10671-022-09321-4
19. Gollust, S.E., Seymour, J.W., Pany, M.J., Goss, A., Meisel, Z.F., Grande, D.: Mutual distrust: perspectives from researchers and policy makers on the research to policy gap in 2013 and recommendations for the future. INQUIRY: J. Health Care Organ. Provision Financ. **54** (2017). https://doi.org/10.1177/0046958017705465
20. Uzochukwu, B., Onwujekwe, O., Mbachu, C., et al.: The challenge of bridging the gap between researchers and policymakers: experiences of a Health Policy Research Group in Engaging Policymakers to Support evidence-informed policy making in Nigeria. Glob. Health **12**, 67 (2016). https://doi.org/10.1186/s12992-016-0209-1
21. University of California Berkeley. Communication with Transparency and Integrity. https://sa.berkeley.edu/sites/default/files/images/communicatingwithtransparency.pdf
22. Ramhit, K.S.: The impact of job description and career prospect on job satisfaction: a quantitative study in Mauritius. SA J. Hum. Resour. Manag./SA Tydskrif vir Menslikehulpbronbestuur **17**, a1092 (2019). https://doi.org/10.4102/sajhrm.v17i0.1092

23. Kueng, P.: The effects of workflow systems on organizations: a qualitative study. In: van der Aalst, W., Desel, J., Oberweis, A. (eds.) Business Process Management, pp. 301–316. Springer, Heidelberg (2000). https://doi.org/10.1007/3-540-45594-9_19. https://www.res earchgate.net/publication/221585932_The_Effects_of_Workflow_Systems_on_Organizat ions_A_Qualitative_Study
24. Cheng, Z., Xiao, T., Chen, C., Xiong, X.: Evaluation of scientific research in universities based on the idea of education for sustainable development. Sustainability **14**, 2474 (2022). https://doi.org/10.3390/su14042474

Interplay Between User Coalition Voting and Knowledge Contribution in Knowledge Community Under Blockchain Token Incentive-A PVAR Approach

Zhihong Li⬤ and Jie Zhang$^{(\boxtimes)}$ ⬤

South China University of Technology, 381 Wushan Road, Tianhe District, Guangzhou 510641,
Guangdong, China
18665735607@qq.com

Abstract. This study interrogates the dynamic interaction between user coalition voting and knowledge contribution in a knowledge community, under the purview of blockchain token incentives. In analyzing user posting and voting behaviors, we discern users into three salient categories: knowledge producers, knowledge discoverers, and browsers. Following this, we utilize a Panel Vector Autoregression (PVAR) model to probe into the intricate dynamics between coalition voting and knowledge contribution among these varied user groups. Estimation outcomes reveal that the quality of knowledge proffered by knowledge producers suffers a decline when their articles are subject to coalition voting. Conversely, knowledge contributions from knowledge discoverers and browsers initially receive an uplift from coalition voting but undergo suppression in the extended term. Theoretically, our research introduces an innovative lens to the field of knowledge management by unveiling the multifaceted interplay between coalition voting and user knowledge contribution. Practically, our investigation presents robust strategies and monitoring mechanisms for community governance, with a specific focus on blockchain knowledge communities. Our findings proffer pivotal insights for designing and recalibrating incentive mechanisms, bearing substantial implications for managing blockchain platforms.

Keywords: Knowledge Contribution · Token Incentives · Panel Vector Autoregression · Coalition Voting

1 Introduction

Knowledge contribution plays a crucial role in knowledge management, involving the provision of valuable content between individuals [1]. The success of a community relies on active user participation and knowledge sharing [2]. However, not all community participants contribute knowledge, with many users posting infrequently [3]. Some tend to free ride or make insufficient contributions, leading to a tragedy of the commons [4]. Despite their significant contributions, knowledge contributors often lack corresponding economic rewards, which may affect their engagement on the platform [5]. Thus, motivating users to contribute continuously poses a significant challenge for sustainable development in online knowledge communities [6].

Blockchain token-based knowledge communities represent a new ecosystem that transparently distributes "token rewards" and "community privileges" to all contributors. The incentive mechanism aims to share the platform's value with contributing users, ensuring fair economic compensation and promoting active user participation and contribution. For instance, Steemit, a prominent blockchain-based knowledge community, had over 1.3 million registered users by January 2020 [7].

Introducing blockchain tokens transforms knowledge contribution into an investment behavior. In these communities, users typically vote on content quality, weighted by their token holdings, and receive rewards accordingly [8]. Such mechanisms may lead to coalition voting, where multiple users collude to gain higher token weight and more rewards. Li et al.'s research revealed collusion among users in voting for witnesses (blockchain data recorders) [9]. Coalition voting on posts directly impacts voting counts and token rewards, potentially influencing users' knowledge contribution behavior. However, the consequences of such collusion are uncertain. While moderate collusion can stimulate content contributions, unfair participation may disrupt the community's ecology [10]. Thus, further analysis is necessary to understand the relationship between user coalition voting and knowledge contribution.

Steemit, the earliest and largest blockchain token-incentivized knowledge community, has become a primary subject of scholarly research. This paper using data from 5764 active Steemit users and 1.16 million posts throughout 2020. Coalition voting behavior was identified through a complex network model, distinguishing between active coalition voting (users participating in voting within a coalition) and voted by coalitions (post voted on by a coalition). The study found that a user's knowledge contribution is influenced by posts voted by coalitions, while their active coalition voting doesn't directly impact contribution. For knowledge producers, being voted by coalitions hinders quality but boosts innovation. Knowledge discoverers and browsers experience decreased quality and innovation but increased contribution in the short term. Coalition voting isn't the primary driver of knowledge producers' contributions but significantly impacts discoverers and browsers. The study highlights the self-promotion effect and long-term impact of coalition voting.

This research enriches knowledge management by exploring various incentives, including traditional functions and economic incentives, unlike prior studies focusing on single factors. It addresses bias in user coalition voting's contribution to different users' knowledge and identifies dynamic relationships between coalition voting and knowledge contribution. The study's algorithm can aid user collusion discovery in blockchain platforms based on token weight, offering governance decision-making for token-incentivized blockchain communities.

2 Literature Review

2.1 User Contribution in Online Communities

Existing research on user contributions in online knowledge communities has primarily focused on several aspects. Through a literature review, this paper categorizes the factors influencing the knowledge contribution behavior of knowledge community users into two dimensions: self-interest motivation and social factors. Earlier researchers have extensively studied the self-interest factors that precipitate individual knowledge contributions [11]. Self-interest motivators include economic interest and psychological drives [12]. From the economic theory that rational individuals are typically driven by utility, we infer that virtual rewards are a vital stimulant of individuals' distribution behavior [13]. However, egoism is not confined to tangible interest pursuit; it can manifest in other forms, including intangible interests, such as innate psychological needs [14]. According to Rocha and Ghoshal [15], individuals seek a positive sense of doing good, thereby psychologically relying on others' appreciation, such as the desire to receive praise. Furthermore, numerous scholars have found that social factors (such as reputation, identity, and social relationships), technical factors (i.e., the usefulness and perceived ease of use of knowledge management systems), and personal factors influence knowledge contribution [16].

The token economy in blockchain aims to facilitate large-scale user collaboration by incentivizing continuous community contributions. It addresses challenges encountered by knowledge platforms, such as knowledge monetization difficulties and a lack of economic incentives for sustained user contributions. The integration of token economies in blockchain-based knowledge communities breaks down the centralized information barriers and value monopolies prevalent in traditional internet knowledge communities. Through token incentives, it encourages widespread participation in knowledge co-creation and sharing, offering innovative solutions to knowledge monetization issues and the lack of user economic motivation [6]. Steemit, one of the earliest and largest blockchain-based knowledge communities, has emerged as a primary subject of academic research. In the initial stages of studying blockchain-based knowledge communities, scholars generally expressed positive views on these initiatives. For instance, Larimer et al. analyzed the operational model and incentive mechanism of Steemit [17]. However, with Steemit's community token price fluctuations and user attrition, academic skepticism has arisen. For example, Chohan contends that blockchain-based knowledge communities solely rely on user autonomy, and early adopters exert significant influence, leading to the emergence of a "power center" around influential users and potential power abuse risk within the community [8]. Additionally, while economic incentives can temporarily enhance users' positive knowledge behaviors, they may, in the long run, weaken users' "altruism" and suppress their knowledge contribution efforts [18].

2.2 Coalition in Decentralized Autonomous Organizations

Collusion, a prevalent topic in economics, pertains to covert cooperation between participants, often driven by self-interest, potentially at the expense of others. Collusion entails a form of collaborative behavior that individual platform users cannot achieve independently; it necessitates user cooperation to realize outcomes unattainable through individual efforts alone [19, 20]. Decentralized Autonomous Organizations (DAOs) are entirely autonomous entities typically controlled by all platform members, unhindered by a central authority. Unlike traditional platforms, on decentralized platforms, Coalition Voting or its resultant negative consequences can prolong the governance process. In fact, such coalition behavior has been observed on decentralized blockchain platforms. Empirical research by Li and colleagues on Steemit's incentive mechanism suggests that the Delegated Proof of Stake (DPOS) consensus mechanism might not be as effective as intended. This is because alliances among miners can also emerge, resulting in the selection of specific delegates through voting behavior. However, apart from miners, ordinary users can also form alliances with the goal of enhancing their individual token earnings. Currently, research on blockchain-based knowledge community users' Coalition Voting behavior in relation to token incentives primarily concentrates on identifying such behavior through empirical studies. For instance, Li et al. utilized social network analysis to examine the voting behavior of influential users in the Steemit knowledge community, revealing the existence of Coalition Voting behavior aimed at acquiring token benefits [9].

3   Theoretical Background and Hypotheses Development

We have introduced incentive theory and community exchange theory to analyze the impact of community user participation in coalition voting and being voted on by a coalition on the knowledge contribution in the community. Users in the community can be divided into content producers, content discoverers, and viewers according to the number of posts and votings they have.

The primary proposition of the incentive theory is that human behavior is influenced by the anticipated outcomes, focusing mainly on individual behavior and motivation [21]. In simple terms, if a particular action is likely to yield positive results, people are more inclined to undertake that action. Klein and Martin discovered that once users achieve the expected incentive consequence through a particular behavior, that behavior gets reinforced [22]. By participating in coalition voting, users who post can receive higher token rewards. However, Wasko et al. found that external incentives are not effective in stimulating users' knowledge sharing [1]. Therefore, once users who were initially actively creating content are voted by the alliance, they would pay more attention to token rewards, neglecting the quality and innovation of the content.

H1a: Participation in coalition voting will reduce the quality of knowledge produced by content creators in the community.

H1b: Participation in coalition voting will reduce the innovativeness of knowledge produced by content creators in the community.

The social exchange theory primarily focuses on interactions between individuals, concentrating on relationships and interactions between people [23]. It posits that people's behaviors in social relationships depend on their comparison of anticipated returns and costs. If the expected return exceeds the anticipated cost, people will engage in social interaction. Tsai and Kang et al. found that when knowledge seekers form an intention to reciprocate, they consider the benefits they perceive. In Steemit, according to the incentive mechanism, the author monopolizes 50% of the article revenue, while all voters of the article only occupy 25%. Therefore, for voters participating in the coalition vote, these users might shift to content creation to obtain higher token earnings.

H2a: Participation in coalition voting will improve the quality of knowledge from voting users in the community.

H2b: Participation in coalition voting will enhance the innovativeness of knowledge from voting users in the community.

According to the incentive theory, users who browse content might be encouraged to participate in voting due to the token rewards from coalition voting. These additional material rewards might stimulate them to engage more actively in the community, including voting and potentially sharing and creating higher quality content. This is because they can see that their actions (such as voting, sharing, and content creation) can lead to positive outcomes (like acquiring more token rewards). Thus, they are more willing to invest in coalition voting and other community activities, which may further improve their knowledge quality and innovativeness.

H3a: Participation in coalition voting will improve the quality of knowledge from browsing users in the community.

H3b: Participation in coalition voting will enhance the innovativeness of knowledge from browsing users in the community.

4 Research Methodology

Our research process comprises four stages: data acquisition, calculation of coalition voting behavior variables, construction of knowledge contribution variables, and utilization of the Panel Vector Autoregression (PVAR) model.

4.1 Construction of Variable

Steemit, established in 2016, is a blockchain-based community with over 1.42 million users and a market value of $70 million. This study focuses on Steemit and analyzes the voting behavior and knowledge contribution of 5674 active users from 2019.

To measure coalition voting behavior, we first need to identify users' votes within coalitions, assuming only real knowledge discovery voting and coalition profit-seeking voting occur. To obtain coalition voting data, we must exclude real voting data. This paper first eliminates co-voting data related to user interests and constructs a user co-voting network. Then, using the Louvain community discovery algorithm [19], we calculate user groups for coalition voting. The number of user coalition voting behaviors is then counted by excluding non-group voting using the OTSU method [24].

First, the undirected weighted graph $G_t(V_t, E_t, W_t)$ is used refers to the relationship between users' voting together in t period. Where, $V^t = \{v_1^t, v_2^t, \ldots, v_n^t\}$ denotes the set of nodes in the network, v_n^t is the user n. E^t represents the edge between two nodes, if there is a connected edge between two users, it means that they have the behavior of jointly submitting the same article in t period. w^t is the weight of edge E_{ij}^t, which refers to the number of articles that node i and node j jointly vote in t period. If the number of active users in the community is n in the period of t, then the network of voting behavior of user coalition can use the adjacency matrix $B^t = \left(b_{ij}^t\right)$ of $n * n$, as shown in formula:

$$b_{ij}^t = \begin{cases} w_{ij}^t \\ 0 \end{cases} \tag{1}$$

Secondly, in this study, we need to divide the community of the coalition voting network and find the groups of different users who vote in the coalition. Louvain algorithm can find the community structure in a short time based on the modularity optimization algorithm, without missing the small community structure, and the divided communities are non-overlapping. Therefore, this study selects the Louvain algorithm to divide the community of the coalition voting network.

Finally, the number of participants of a coalition in each voting is counted, and the user voting behavior is divided into coalitional voting and non-coalitional voting based on the Otsu threshold method. That is to say, for each group G, the threshold of judging coalition voting behavior and non-coalition voting behavior is th, and the proportion of coalition voting behavior is w_0 and its average number is u_0. The proportion of non-coalition voting behavior was w_1 and its average number is u_1, the average number of users voting is $u = w_0u_0 + w_1u_1$. Establish the objective function for coalitions k:

$$g_k(th) = w_0(u_0 - u)^2 + w_1(u_1 - u)^2 \tag{2}$$

This study constructs a voting behavior network based on the co-voting relationships of users. However, users with similar interests may vote similarly, which, if not accounted for, could skew empirical tests. We use the Latent Dirichlet Allocation (LDA) model to capture users' knowledge topic distributions and adjust for voting similarities due to shared interests. Coalitional voting occurs when multiple users adopt the same voting strategy over a time period. This study uses the Louvain algorithm to identify groups participating in coalitional voting. We measure the number of coalition votes using the Otsu threshold method, which classifies voting behavior into coalition and non-coalition voting based on differences in the mean and variance of voting participants. This paper measures an author's coalitional voting by the proportion of coalition votes in total votes. The specific algorithm first counts the total number of $Voted_i^n$ of the user n's article i, and then judge whether the coalition k has coalition voting behavior in the article through the previous Otsu threshold. If there is coalition voting behavior, the number of the coalition k appearing in the article will be counted as $Voted\_k_i^n$, assuming that the total number of coalitions is m, through $\sum_{k=1}^{m} Voted\_k_i^n$ the total number of coalition votes for article i can be calculated, and use $\sum_{k=1}^{m} Voted\_k_i^n/Voted_i^n$ is used to measure the voting of knowledge producers coalition.

Suppose that the internal knowledge quality of user i is IKQ_i and the accessible knowledge quality is AKQ_i, and the knowledge quality KQ_i of user i is obtained by weighting the two kinds of knowledge quality, σ_1 and σ_2 weights, which are calculated by the standard deviation of coefficient.

$$KQ_i = \sigma_1 IKQ_i + \sigma_2 AKQ_i \tag{3}$$

In order to measure a user's contribution to the community's knowledge innovation, this study constructs a knowledge co-occurrence network based on a user's original articles over a certain period, selecting knowledge depth and breadth as the dimensions to describe knowledge innovation. The specific calculation process is as follows.

---

for user i's post $p \in P_t$ in a period of t **do**
    the topic set T_P for each topic by LDA topic model
    for each topic $t \in T_p$ **do**
        if the probability of topic t is greater than the threshold **then**
        topic t is added to the co-occurrence knowledge element set K
        end if
        for knowledge element $k_1 \in K$ **do**
            for knowledge element $k_2 \in K$ **do**
                add an undirected edge to K_1 and K_2
            end for
        end for
    end for
end for

4.2 The Panel-VAR Methodology

We use the Panel Vector Autoregression (PVAR) model to explore the dynamic relationship between voting behavior and knowledge quality. The process of establishing the PVAR model involves unit root testing and lag selection. We conduct the Augmented Dickey-Fuller (ADF) unit root test on all endogenous variables to ensure the stationarity of time series data, a prerequisite for the Granger causality test in the PVAR model. Only data that pass this stationarity test are used in subsequent analyses. The results confirm that our data meet the stationarity requirement.

5   Empirical Analysis with the PVAR Model

5.1 User Classification

This study classifies users into knowledge producers, knowledge discoverers, and browsers based on their posting and voting behaviors. This classification allows for more detailed analysis across different user groups. The K-means clustering method is used to segment user roles based on these behaviors (Table 1).

Table 1. User clustering results

	Class number		
	1	2	3
Number of posts	0.8576	0.2063	2.0529
Number of votes	0.3584	0.14	0.2918
Number of users	1695	3543	436
Percentage	29.87%	62.44%	7.68%

5.2 Granger Causality Analysis

The Granger causality test is a prediction methodology for stationary time series data. If including the past information of variables X and Y improves the prediction of Y beyond the use of past information of Y alone, or if variable X helps to explain the variation in variable Y, then variable X is considered to be the Granger cause of variable Y. Based on previous research, we conducted a Granger causality test, with the results presented in Fig. 1.

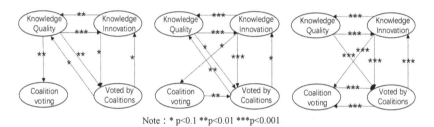

Note : * p<0.1 **p<0.01 ***p<0.001

Fig. 1. Results of Granger Causality Analysis.

5.3 Generalized Method of Moments

The Granger causality test verifies whether a Granger causality exists between variables, but it does not verify the degree of interaction between variables. To address this, the Generalized Method of Moments (GMM) was employed to describe the results of the PVAR model estimation, which are reported in Table 2.

Table 2. Generalized method of moments results

Independent variable	Dependent variable	Knowledge producer	Knowledge discoverer	Browser
KQ	KQ	+***	+***	+***
	KI	+***	+***	n.s
	CV	+**	n.s	n.s
	VC	+**	+^	+***
KI	KQ	+**	+***	+***
	KI	+**	+***	+***
	CV	n.s	+*	+***
	VC	+^	+***	+***
CV	KQ	n.s	n.s	n.s
	KI	n.s	n.s	n.s
	CV	+***	+***	+***
	VC	n.s	+**	+***
VC	KQ	−*	+**	−^
	KI	−*	+***	−***
	CV	n.s	n.s	+***
	VC	+***	+***	+***

Note: ^ $p < 0.1$, * $p < 0.05$, ** $p < 0.01$, *** $p < 0.001$, n.s not significant, + positive influence, − negative influence.

5.4 Impulse Response Functions

The impulse response function is used in this study to explore the dynamics of the mutual influences between voting behavior and user knowledge contribution.

The impulse response function for knowledge producers (Fig. 2) shows a positive correlation between knowledge quality and knowledge innovation, peaking in the first period and then rapidly decreasing. Knowledge quality also positively impacts coalition voting, with the effect peaking in the third period and then slowly tapering off. The relationship between knowledge quality and coalition voting follows a similar pattern.For knowledge discoverers (Fig. 3), the interplay between knowledge quality and innovation is positive initially but lessens over time. The impact of knowledge quality on the frequency of being voted by coalitions also follows a similar pattern. Knowledge innovation positively affects knowledge quality and coalition voting. Interestingly, the initial positive effect of being voted by coalitions on knowledge quality turns negative over time.For browsers (Fig. 4), knowledge quality negatively influences the incidence of being voted by coalitions. Their knowledge innovation positively impacts both knowledge quality and coalition voting. However, while coalition voting positively influences the frequency of being voted by coalitions, being voted by coalitions in turn hampers knowledge innovation.

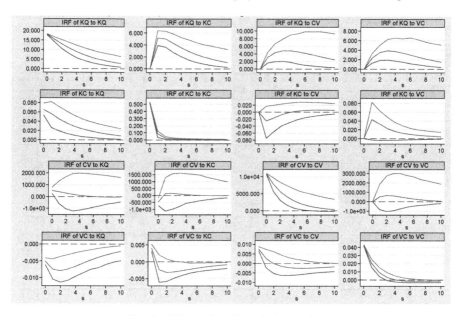

Fig. 2. IRFs results of knowledge producer.

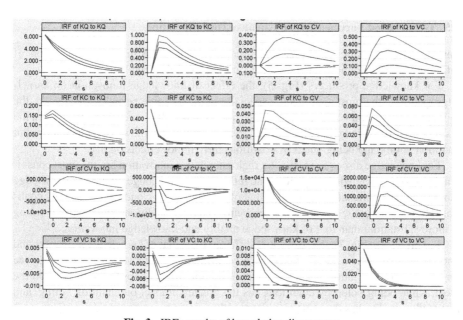

Fig. 3. IRFs results of knowledge discoverers.

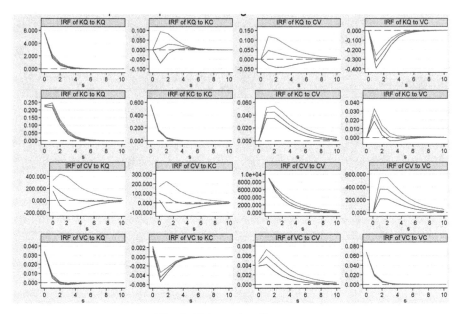

Fig. 4. IRFs results of browser.

5.5 Generalized Forecast Error Variance Decomposition

Variance decomposition analysis was used to evaluate the relative contribution of different variables to the prediction error variance of each variable in the system.

For knowledge producers (Fig. 5), the effect of knowledge innovation on knowledge quality is the most significant, especially in the first five periods. Meanwhile, the influence of coalition voting on knowledge quality gradually increases starting from the sixth period. Knowledge quality has the greatest impact on knowledge innovation, but coalition voting and being voted by coalitions have minor effects on it.

For knowledge discoverers (Fig. 6), their knowledge innovation has the most considerable impact on knowledge quality. The influence of knowledge quality on knowledge innovation is significantly higher than that of coalition voting and being voted by coalitions. Furthermore, the relation between being voted by coalitions and coalition voting is more substantial than that of knowledge quality and knowledge innovation.

For browsers (Fig. 7), the scenario is different from that of knowledge producers and knowledge discoverers. Being voted by a coalition has the most substantial influence on knowledge contribution. Knowledge contribution affects knowledge innovation the most, while being voted by coalitions has the greatest impact on coalition voting, and knowledge quality has the most significant influence on being voted by coalitions.

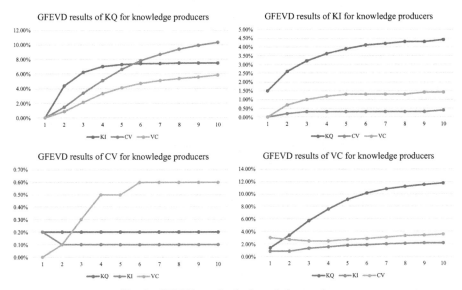

Fig. 5. GFEVD results for knowledge producer.

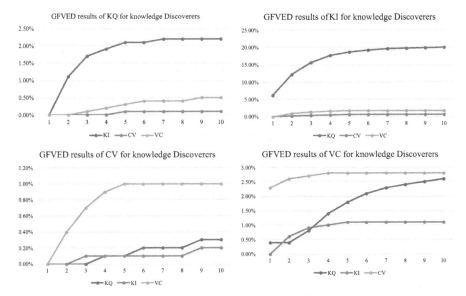

Fig. 6. GFEVD results for knowledge discoverers.

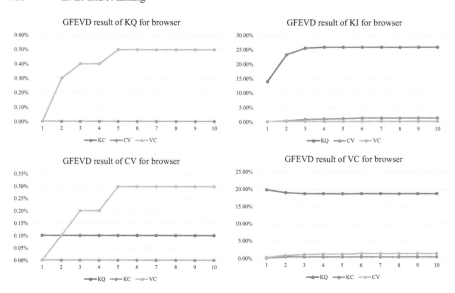

Fig. 7. GFEVD for browsers

6 Discussion

This study uses time series data to explore the dynamic relationship between user coalition voting and user knowledge contribution. Key findings include the limited impact of active participation in coalition voting on users' knowledge contribution. This is supported by the Granger causality test and GMM estimation. The effects of being voted by coalitions vary among users. For knowledge producers, their knowledge quality decreases when their articles are voted by coalitions. Knowledge discoverers also experience a negative impact on their knowledge quality and innovation due to coalition voting. However, the main factor influencing their knowledge contribution after the fifth period is their coalition voting behavior. For browsers, being voted by coalitions tends to inhibit their knowledge quality. Coalition voting behavior demonstrates a positive self-promotion effect. The primary reason articles from knowledge producers are voted by coalitions is not the quality and innovation of their content, but the coalition voting behavior of other users.

6.1 Implications

This study contributes to knowledge management research by examining the relationship between coalition voting and user knowledge contribution in communities under blockchain token incentives. It uses panel data to study this relationship over time, which is important for understanding the dynamic role of incentive mechanisms.

In practical terms, the coalition voting detection algorithm proposed here can be used for community governance in many blockchain-based platforms. The study also suggests that different user groups respond differently to token incentives, so different incentives or management measures might be needed for different users. Lastly, controlling coalition voting is crucial in blockchain-based knowledge communities, as unchecked coalition voting could demotivate contributing users and reduce the community's value.

6.2 Limitations and Future Studies

This study, while informative, has limitations that future research could address. First, we based coalition voting on long-term user voting behavior, but future studies should distinguish between different coalitions for a more nuanced understanding. Second, we focused on the impact of coalition voting on knowledge contribution, leaving room for future research on other aspects like knowledge interaction and searching. Lastly, while we used the PVAR model, future research could use system dynamics simulation to study the dynamic model of a knowledge community ecosystem with complex blockchain token incentives.

Overall, this research explores the relationship between user coalition voting and knowledge contribution in blockchain communities, laying an empirical groundwork for the governance of blockchain platforms based on token-weighted voting and inspiring future studies on incentive mechanisms and user contributions.

Acknowledgment. This work was supported by grants No. 72171089 and 72071083 from the National Natural Science Foundation of China, grant No. 2021A1515012003 from the Guangdong Natural Science Foundation of China, and grant No. 2021GZQN09 from the Project of Philosophy and Social Science Planning of Guangzhou in 2021.

References

1. Wasko, M.M., Faraj, S.: Why should i share? Examining social capital and knowledge contribution in electronic networks of practice. MIS Q. **29**(1), 35–57 (2005)
2. Bock, G.-W., et al.: Behavioral intention formation in knowledge sharing: examining the roles of extrinsic motivators, social-psychological forces, and organizational climate. MIS Q. **29**(1), 87–111 (2005)
3. Baym, N.K.: Tune in, Log on: Soaps, Fandom, and Online Community, vol. 3. Sage, Thousand Oaks (2000)
4. Hardin, G.: The tragedy of the commons. J. Nat. Res. Policy Res. **1**(3), 243–253 (2009)
5. Zhihong, Li., Jie, Z.: Online knowledge community governance based on blockchain token incentives. In: Chen, J., Huynh, V.N., Nguyen, G.-N., Tang, X. (eds.) KSS 2019. CCIS, vol. 1103, pp. 64–72. Springer, Singapore (2019). https://doi.org/10.1007/978-981-15-1209-4_5
6. Wang, Y., Fesenmaier, D.R.: Towards understanding members' general participation in and active contribution to an online travel community. Tour. Manag. **25**(6), 709–722 (2004)
7. Liu, Z., et al.: User incentive mechanism in blockchain-based online community: an empirical study of steemit. Inf. Manag. **59**(7), 103596 (2022)
8. Tsoukalas, G., Falk, B.H.: Token-weighted crowdsourcing. Manag. Sci. **66**(9), 3843–3859 (2020)
9. Li, C., Palanisamy, B.: Incentivized blockchain-based social media platforms: a case study of steemit. In: Proceedings of the 10th ACM Conference on Web Science (2019)
10. Tang, H., Ni, J., Zhang, Y.: Identification and evolutionary analysis of user collusion behavior in blockchain online social medias. IEEE Trans. Comput. Soc. Syst. (2022)

11. Ryan, R.M., Deci, E.L.: Self-determination theory and the facilitation of intrinsic motivation, social development, and well-being. Am. Psychol. **55**(1), 68 (2000)
12. Gneezy, U., List, J.A., Wu, G.: The uncertainty effect: when a risky prospect is valued less than its worst possible outcome. Q. J. Econ. **121**(4), 1283–1309 (2006)
13. Hsieh, G., Kraut, R.E., Hudson, S.E.: Why pay? Exploring how financial incentives are used for question & answer. In: Proceedings of the SIGCHI Conference on Human Factors in Computing Systems (2010)
14. Goldsmith, K., Amir, O.: Can uncertainty improve promotions? J. Mark. Res. **47**(6), 1070–1077 (2010)
15. Nahapiet, J., Gratton, L., Rocha, H.O.: Knowledge and relationships: when cooperation is the norm. Eur. Manag. Rev. **2**(1), 3–14 (2005)
16. Chai, S., Kim, M.: A socio-technical approach to knowledge contribution behavior: an empirical investigation of social networking sites users. Int. J. Inf. Manag. **32**(2), 118–126 (2012)
17. Scott, N., Larimer, D.: Steem. An incentivized, blockchain-based, public content platform (2017)
18. Kiayias, A., et al.: A puff of steem: security analysis of decentralized content curation. arXiv preprint arXiv:1810.01719 (2018)
19. Wang, Z., et al.: ColluEagle: collusive review spammer detection using Markov random fields. Data Min. Knowl. Disc. **34**, 1621–1641 (2020). https://doi.org/10.1007/s10618-020-00693-w
20. Zhang, F., et al.: Label propagation-based approach for detecting review spammer groups on e-commerce websites. Knowl.-Based Syst. **193**, 105520 (2020)
21. Killeen, P.R.: Incentive theory. In: Nebraska Symposium on Motivation. University of Nebraska Press (1981)
22. Deci, E.L., Ryan, R.M.: Self-determination theory: a macrotheory of human motivation, development, and health. Can. Psychol. **49**(3), 182 (2008)
23. Cook, K.S., Emerson, R.M.: Social exchange theory (1987)
24. Otsu, N.: A threshold selection method from gray-level histograms. IEEE Trans. Syst. Man Cybern. **9**(1), 62–66 (1979)

Exploring the Evolutionary Dynamics of Wikipedia: The Interplay of Knowledge Community Formation and Editorial Behavior

Hongyu Dong⬨ and Haoxiang Xia⁽⊠⁾⬨

Dalian University of Technology, Dalian 116024, Liaoning, China
hxxia@dlut.edu.cn

Abstract. Understanding the development patterns of large-scale knowledge networks is of significant importance for knowledge integration. Wikipedia, as one of the world's largest online knowledge platforms, provides important references for group knowledge construction and integration in terms of its knowledge construction patterns and link structures. The development pattern of Wikipedia can be derived by analysing the changes of entries and links. We analysed the evolutionary characteristics of the Wikilink network from the perspective of knowledge community formation and editing behaviour. It is found that the connections of new entries introduced in the network have significantly different tendencies, including two types, linking to popular entries and linking to entries within related knowledge communities. This tendency is affected by changes in editor behaviour and the focus of website operation, and is an important factor in the change of structural features during the evolution of the Wikilink network, which makes the evolution of the Wikilink network shows a unique pattern that is different from that of existing models. We quantitatively portrayed this tendency and reveals important laws in the evolution of Wikipedia, which is instructive for a deeper understanding of the development and evolution of Wikipedia and other online knowledge communities.

Keywords: Wikipedia · Link Structure · Knowledge Communities · Network Evolution · Editorial Behaviour

1 Introduction

The rapid development of Internet technology has led to an explosive growth in the amount of knowledge, and traditional forms of knowledge organisation are no longer able to cope with the requirements of knowledge production and dissemination in the era of knowledge explosion, so a group collaborative approach to knowledge construction has emerged [1]. The user-led knowledge creation and construction process not only makes it possible to build large-scale knowledge networks, but also makes knowledge more accessible and widespread, represented by Wikipedia. Since its launch in 2001, Wikipedia has grown rapidly to become one of the world's largest online knowledge exchange platforms, a testament to the success of Wikipedia's own open and self-organising structure.

J. Chen et al. (Eds.): KSS 2023, CCIS 1927, pp. 165–177, 2023.
https://doi.org/10.1007/978-981-99-8318-6_12

There have been extensive researches on Wikipedia's knowledge structure. For example, Capocci et al. found that the degree distribution in the link network is scale-free and consists of a few major components in the macro structure [2]. Zesch et al. constructed and analysed two networks based on lexical links and found them to have scale-free and small-world properties [3]. Some scholars also focused on the differences between various language versions, or studied sub-networks in different do-mains and the similarities between them [4, 5]. These studies speak volumes about the value of linking networks for research.

Wikilink network removes the non-word page elements from Wikipedia and retains only the core inter-word link structure, which itself constitutes a huge mind map and its own dynamic changes reflect the evolution of the network. Therefore, Wikilink is an ideal source to explore the link structure of Wikipedia. Previous researches on the link structure of Wikipedia have been focusing on static structures, and there is a lack of research related to the evolutionary development of Wikilink network. To remedy this, we used the English Wikilink network from 2002 to 2018 as a data source, portrays the development process of knowledge topics and communities, analyses the evolution characteristics using network dynamics, and constructs an evolutionary model of Wikipedia and validates it with simulation experiments.

2 Methods

2.1 The Structure of Network Communities

A community is a sub-structure that exists within a complex network, where the nodes within these structures are tightly interconnected, while the connections between the structures are looser [6]. The presence of community structure reflects the heterogeneity within the network and can be measured using modularity [7]. The higher the value, the clearer the community structure in the network. When all the nodes in the network belong to one community, the modularity of the network is zero, and there is no community structure in the network. It is calculated as follows

$$Q = \frac{1}{2m} \sum_{i,j} \left(A_{ij} - \gamma \frac{k_i k_j}{2m} \right) \delta(i, j) \tag{1}$$

where m represents the total number of edges in the network, A_{ij} is the adjacency matrix of the network, each element represents whether the ith node and the jth node are connected, if connected then $A_{ij} = 1$, otherwise it is equal to 0. k_i and k_j are the degrees of the i-th and j-th nodes in the network. γ is the resolution [8]. $\delta(i, j)$ is the or function, if $i = j$ then $\delta(i, j) = 1$, otherwise it is equal to 0.

2.2 Community Partitioning Methods

Many algorithms have been proposed for network community partitioning, and the main ideas can be divided into two types, based on topological analysis of the network and flow analysis. The former is based on the idea that the density of connected edges varies within and between communities, while the latter is mainly based on a network-based flow analysis approach.

One of the more widely used division methods is the Louvain algorithm [11]. The basic algorithm is to initially divide all nodes into separate communities and to continuously try to merge communities to obtain the fastest increase in the modularity of the network. The algorithm terminates when merging communities does not lead to further increases in modularity. The algorithm usually obtains a multilevel structure for the community division, and is therefore also known as the multilevel algorithm in some network science computing packages. This algorithm is also used in this study mainly for community identification and detection in Wikilink network.

3   Network Construction and Analysis

Cristian et al. crawled and processed the complete Wikipedia data in 2019 to obtain highly formatted Wikilink network data by parsing xml files and using regular expressions to match node and link data [12]. This dataset was used in this study and includes Wikipedia link data from 2001 to 2018. The dataset contains a complete snapshot of Wikilink network as of March 1 of each year. The data are cumulative and incremental, with each year of data being complete from the beginning of Wikipedia's existence to that year, and a comparison of the word catalogues also gives a record of deletions between the two years.

After pre-processing the data and removing redirects and duplicate edges, we constructed unweighted link networks, of which the trends in network size for several languages are shown in Fig. 1. We found that Wikilink networks for dif-ferent languages show an approximately linear growth.

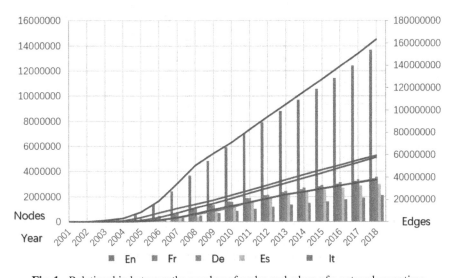

Fig. 1. Relationship between the number of nodes and edges of a network over time

This type of growth is rare, probably because during the development of the linking network, the core entries were created earlier and later nodes were added at the more peripheral locations, with subsequent additions only needing to connect to a few core nodes to make the network connected. Whether the way in which Wikipedia has developed has profoundly affected the ability of the linked network to exchange and transmit information and other characteristics is a question that will be explored later, as the new edges of the network are unable to compensate for the reduced level of linking caused by the addition of new nodes.

The analysis that follows focuses on the English language wiki. We calculate trends in the number and modularity of online communities, as shown in Fig. 2. It can be seen that the Wikipedia Connected Network is a knowledge network with a highly modular structure, with a much higher degree of modularity than the random network with the same degree of distribution, and that both the degree of modularity and the number of communities are steadily increasing. The degree of modularity and the number of communities are steadily increasing, while the degree of modularity of the random network is slowly decreasing due to the decreasing density of the network, which reflects the very different characteristics of Wikipedia. The increase in the number of communities in the later stages of the network's development causes an increase in the average distance of the network, which has a detrimental effect on the efficiency of global information transfer and interaction on the network.

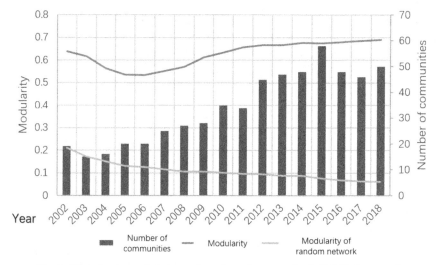

Fig. 2. Variation in modularity and number of community in Wikilink network

4   Knowledge Communities and the Evolution of Topics

There is a clear modular structure in the Wikilink network, which corresponds to a relatively independent knowledge community. In turn, the highly connected nodes in the network are not tightly connected, while the strength of the connections between nodes with high participation coefficients is unusually high, and the knowledge communities in the network are tightly connected to each other through a plurality of club nodes [6]. Knowledge is not static for human beings, but is always in a dynamic evolutionary process, and the structure of knowledge networks should also adapt to changes in the knowledge system.

In order to further demonstrate the formation process of the knowledge communities of the network, we divided Wikilink network in 2002 and obtains a total of 23 communities, among which there are 19 communities with the number of nodes over 100, and plots their distribution in the network with the global efficiency and local efficiency of each community respectively, as shown in the following figure (a). The nodes with a degree less than 20 in the 2002 link network are hidden and coloured according to the community they belong to, and the node sizes in the figure represent the magnitude of the participation factor. The global and local efficiencies of each community component sub-graph were also calculated and are shown in figure (b) below (Fig. 3).

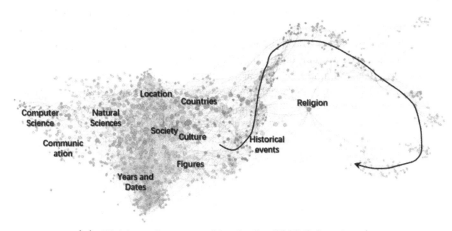

(a) Division of communities in the 2002 link network

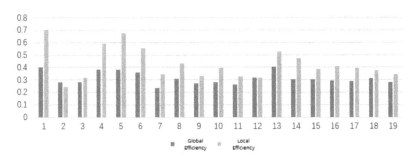

(b) Efficiency of communities in 2002 network

Fig. 3. Community segmentation and sub-network efficiency in 2002 Wikilink network

The black curve marked in figure (a) represents the timeline consisting of historical years and historical events, and the nodes within this timeline have very high clustering coefficients, corresponding to communities 1, 4, 5 and 6 in the figure below, flanked on either side of the timeline by historical events that are closely linked to time. Historical events and religious culture correspond to communities 2, 3 and 7–10, which are less globally efficient compared to the network average, and it is possible that their reliance on historical chronological indexing is necessary to achieve high knowledge connection efficiency. On the far left of the graph below are entries related to computers and communications, corresponding to communities 14–16. The size of these communities is relatively large for the network, suggesting that the early editors of Wikipedia preferred to introduce entries related to computer technology into the network, and that these entries were closely linked to the natural sciences. Communities 17–19 correspond to the natural science communities, which have an average network efficiency compared to the global picture.

As the size of the network continues to grow and the sub-networks made up of highly nested nodes in the network become denser, the boundaries between communities are further blurred and the modularity of the network decreases significantly. Whereas in 2004 the presence of river-shaped historical communities could still be observed in the network, in 2005 the pre-medieval years no longer appear in the network, and historical communities have become increasingly highly integrated with other communities. Neutral knowledge communities such as countries, people and geographic locations gradually diffuse, becoming more and more dispersed in the middle of the network and more closely communicating with other knowledge communities. This phase saw the emergence of a large number of list entries, which existed widely within community boundaries and had a high participation factor, playing an important linking role in knowledge integration (Fig. 4).

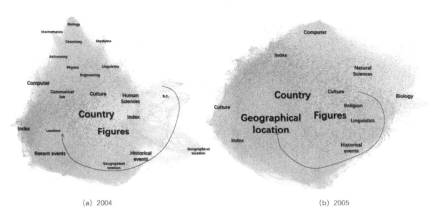

(a) 2004   (b) 2005

Fig. 4. Community segmentation in 2004 and 2005 Wikilink network

The evolution of Wikipedia shows that the whole network started with a skeleton of highly valued core nodes, which gradually developed over the years into a giant knowledge network containing several knowledge communities. Time-indexed entries played an important role in the development, with other knowledge communities surfacing in the vicinity of historical chronological communities. The different knowledge communities have developed differently depending on their disciplinary characteristics and the type of lexical links. Communities based on word content links have a higher modular structure and clustering coefficient, while communities based on inter-word citation relations have higher inter-community connectivity and blurred community boundaries. Index nodes play a positive role in knowledge integration, but generally only connect nodes within their communities and can disrupt the small-world nature of the network. Once index nodes appear on a large scale in the network and continue to aggregate different knowledge communities, the global and local information transfer efficiency of the network will decrease simultaneously. From the perspective of knowledge community development, if the boundaries of individual knowledge topics are not clear enough, and knowledge from multiple communities is mixed and connected, it will not only affect the classification structure within the knowledge clusters, but also increase the cost of understanding and using knowledge.

5 Evolution of Editorial Behaviour

The evolution of the network can be described at different levels, such as micro and macro, depending on the perspective from which it is observed. Wikipedia is a constantly updated knowledge system, and there are three phenomena in the network: additions, deletions and updates of nodes, and additions and deletions of edges. We count the number of new and modified entries and edges in the linked network for each year, as shown in the figure below. The left side shows the number of additions, updates and deletions of nodes and the proportion over time, while the right side shows the number of additions and deletions of edges over time. The horizontal axis represents time and the vertical axis represents the number and proportion. It can be seen that in the early stages of Wikipedia's development, the proportion of nodes and edges added was high. And once the thematic structure of the network was established, the number of word content updates grew steadily, but the proportion declined (Fig. 5).

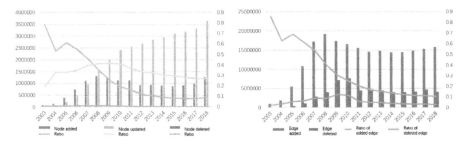

Fig. 5. Evolution of nodes and edges in Wikilink networks

Users are the producers and providers of content in Wikipedia and have been the main driving force behind its evolution [9]. stats.wikimedia.org provides statistics on the number of Wikipedia views, users and pages. We obtained monthly user and editor count data from this website from 2002 to the present, as shown in the graph below. Figure a shows the number of Wikipedia editors and active editors, where editors are the number of users who have made a valid edit (in the month) and active editors are those who have made at least five edits. It can be seen that active editors make up only a small proportion of editors, and after an initial rapid decline the proportion has remained at around 10%. The graph on the right shows the trend of new Wikipedia users and total Wikipedia users, with active editors not only contributing more to content updates, but also being more sticky users (Fig. 6).

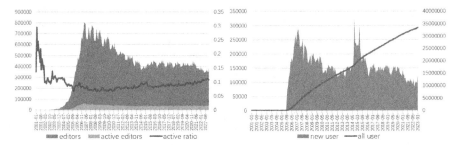

Fig. 6. Monthly data on the number of Wikipedia editors and users

By examining the phenomena of evolution at the node and user levels and the phenomena of associative change within each level, we proposed a hypothesis that the editing behaviour of users causes changes at the edge and node levels in the knowledge network, and that the evolution at the micro level causes the evolution of the global structure of the network. This mechanism of influence can be represented in the following diagram, with the specific implication that editors in the user community edit words, and that different types of editors have different editing tendencies and preferences. The act of editing causes the updating of words and links, which in turn indirectly affects the structure and macro features of the network (Fig. 7).

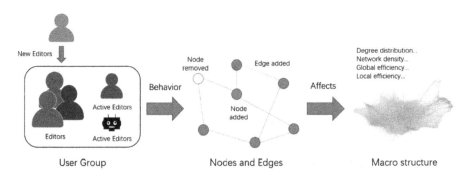

Fig. 7. A dynamic model of Wikipedia evolution and editorial behaviour

In order to measure the impact of user behaviour on the node and edge level of the network, we analysed the correlation between the number of users and editors and the evolutionary phenomena in the network. Editors were more inclined to maintain entries and links between entries, and core editors were more highly correlated with these metrics, suggesting that updates to existing knowledge in the network are updated and maintained by core editors, a result that confirms that editors play a major role in driving the evolution of Wikipedia within the user community, and that their editing behaviour profoundly influences the development of Wikipedia (Table 1).

Table 1. Correlation coefficient matrix between users number and network micro-evolution

	Node added	Node updated	Node removed	Edge added	Edge Removed	Editors	Active Editors
Node added	1.000						
Node updated	0.680	1.000					
Node removed	0.693	0.248	1.000				
Edge added	0.980	0.701	0.742	1.000			
Edge Removed	0.815	0.734	0.702	0.824	1.000		
Editors	0.913	0.710	0.843	0.830	0.718	1.000	
Active Editors	0.907	0.984	0.952	0.942	0.878	0.980	1.000

The nodes that existed in the early days of Wikipedia later developed into core nodes in the network; some became high Z-value nodes in the community, closely connected to nodes in the community; others became bridge nodes between communities in the network; these nodes had a high participation factor and maintained the strength of the connections between communities [10]. In the early days of the network the number of open registrations in Wikipedia was small, and at this time the addition of new entries and links to the network was mainly the responsibility of the editors. And as time progressed, later additions to the network were dominated by edge nodes, when the density of the network was still low and therefore the average distance grew slowly. After a certain stage of development, the links between the core nodes in the network stabilised and the focus of Wikipedia's operations shifted to the updating and maintenance of existing content, which was mainly influenced by the behaviour of the editors.

6   Evolutionary Mechanisms and Simulation

From the previous analysis we found that Wikipedia shows a sparse growth pattern and that the modularity of the network increases while the efficiency of information transfer decreases. It can be speculated that this is related to the selective connection of late joining nodes in the network, i.e. new entries are usually connected to nodes of the same knowledge community in the network and less likely to be connected to nodes of other communities in the network, which violates the assumption of degree-first connection in the BA scale-free model [13]. According to the analysis in the previous section, the growth of nodes is mainly caused by editors creating new entries. When editors add or edit nodes they usually tend to add links to common or authoritative entries in the network. This leads to a tendency for new entries to link to nodes with high degree values (popular entries), but this tendency is not global, but limited to a certain area around the knowledge community where the entry is located.

Based on the evolutionary mechanism we proposed a random network linking method. The basic idea of the algorithm is that a small-scale ER random graph is generated initially. A new node and a number of edges are cyclically added to the network, with the number of edges added each time conforming to a Poisson distribution with the expectation of the average degree of the network. For each newly added edge, a random number between 0 and 1 is generated. If it is greater than the reconnection probability, a node is randomly selected from the network with probability proportional to the degree value, otherwise it is connected to a neighbouring node. To determine the effect of the reconnection probability α on the network characteristics, we analyses the network structure parameters for different reconnection probabilities, as shown in the figure below. The trends of some network metrics are plotted for a random network of 1000 nodes with 5000 edges on the left and a random network of 1000 nodes with 25000 edges on the right (Fig. 8).

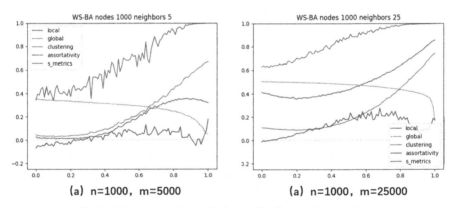

(a) n=1000, m=5000   (a) n=1000, m=25000

Fig. 8. Simulation of network degree distribution characteristics

It can be found that as the reordering probability increases, the local information transmission efficiency and the congruency of the network also show an upward trend, while the global efficiency decreases, which is in line with the trend of the network's transformation from a scale-free network to a small-world network. The network ensures both high global and local efficiency when α is around the range of 0.85 to 0.95. We then analysed the network properties of the constructed networks with α values of 0.85, 0.9 and 0.95 according to the previously proposed evolutionary model of Wikilink networks and compare them with the actual data to reveal the evolutionary mechanism of the networks from a deeper perspective. Due to the limitations of the model, it is not possible to simulate the properties of α over time, so we mainly focused on a random network of the same size as the 2002 link network (n = 27526, m = 195385).

The degree distribution can reflect the distribution probability of node degree values as well as the scale-free degree characteristics. In this case, the degree distribution curves of the original network and the random network generated according to the three parameters are plotted separately. It is shown that the degree of the simulated network obeys the power-law distribution within a certain range, and the phenomenon of "low head" and "trailing tail" appears, which generally conforms to the characteristics of the generalised degree distribution. However, since the expectation of the number of edges added to the network is the average degree of the network, a completely non-rearranged degree distribution should also be close to the Poisson distribution. The number of low degree nodes is small, but they are usually the edge nodes in the community, which only affects the size of the network but not the global structure of the network, and therefore does not affect the discussion of the alpha value (Fig. 9).

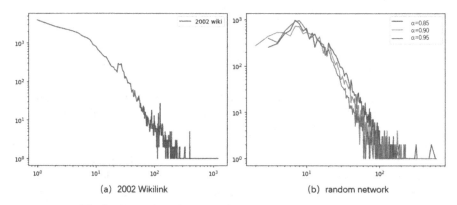

(a) 2002 Wikilink   (b) random network

Fig. 9. Simulation of network degree distribution characteristics

The structural parameters of the original network and the three simulated networks were also calculated, as shown in the table below. It can be found that for the same network size, the higher the value of α the higher the global efficiency and modularity level of the network, while the lower the local efficiency and clustering coefficient. The local efficiency and clustering coefficients are lower than those of the actual Wikiplink network because there are fewer low-value nodes in the network than in the actual network, but the results are still comparable for the simulated networks generated by the same method, i.e. the reconnection probability affects to a greater extent the global network and the connectivity and information transfer efficiency within the knowledge community (Table 2).

Table 2. Results of network specific parameter simulation

Network	2002 wiki	α=0.85	α=0.9	α=0.95
Number of Nodes			27526	
Number of Edges			195385	
Degree distribution parameter	-1.752	-1.598	-1.713	-1.786
Global efficiency	0.257	0.250	0.235	0.225
Local efficiency	0.325	0.096	0.111	0.123
Average path length	4.079	4.193	4.374	4.489
Cluster coefficient	0.211	0.131	0.134	0.141
Modularity	0.634	0.247	0.270	0.336

7   Conclusion

In this paper, we conducted a preliminary analysis and research on the evolution process of Wikipedia mainly from different perspectives, and found that the evolution of knowledge communities is related to the strength of community knowledge topics and word connections, and that Wikipedia early on used historical year entries as an index to string together various knowledge communities. The existence of index nodes facilitates the construction of knowledge and the understanding and use of knowledge by users, but this hinders the connectivity of knowledge in the network. Based on the way in which nodes are connected to edges, we divided the connectivity of nodes into nodes that are similar to popular (highly) nodes and knowledge topics within communities, and this is influenced by the behaviour of editors, whose editing behaviour affects the creation and updating of words and links, and subsequently the changes in the global characteristics of the network. Shifts in editorial behaviour affect differences in how nodes in the network are linked, with new knowledge areas becoming increasingly difficult to emerge and a shift in the focus of network operations to maintaining existing entries, resulting in a tendency for new nodes to link more centrally to their communities. We conducted simulation experiments to verify the impact of changes in node linking propensity on network properties. The experimental results show that the propensity to add new edges to the network is an important factor in the change of structural features during the evolution of Wikipedia, and that Wikipedia does not conform to the scale-free growth model of the BA model, but is profoundly influenced by editorial behaviour.

The significance of our work is that it explores the evolutionary process of Wikipedia and provides a preliminary analysis of the evolutionary features of the network from the perspective of knowledge community formation and editing behaviour, which is important for a deeper understanding of the knowledge community and network structure evolution patterns. It also provides insights for further exploration of the knowledge evolution of Wikipedia. Some of the methods and technical tools adopted in this paper are also innovative, but there are still limitations and room for improvement.

References

1. Engstrom, M.E., Jewett, D.: Collaborative learning the wiki way. TechTrends **49**(6), 12 (2005)
2. Capocci, A., et al.: Preferential attachment in the growth of social networks: the internet encyclopedia Wikipedia. Phys. Rev. E **74**(3), 036116 (2006)
3. Zesch, T., Gurevych, I.: Analysis of the Wikipedia category graph for NLP applications. In: Proceedings of the Second Workshop on TextGraphs: Graph-Based Algorithms for Natural Language Processing (2007)
4. Zlatić, V., et al.: Wikipedias: collaborative web-based encyclopedias as complex networks. Phys. Rev. E **74**(1), 016115 (2006)
5. Silva, F.N., et al.: Investigating relationships within and between category networks in Wikipedia. J. Inform. **5**(3), 431–438 (2011)
6. Clauset, A., Newman, M.E.J., Moore, C.: Finding community structure in very large networks. Phys. Rev. E **70**(6), 066111 (2004)
7. Newman, M.E.J.: Modularity and community structure in networks. Proc. Natl. Acad. Sci. **103**(23), 8577–8582 (2006)
8. Newman, M.E.J.: Equivalence between modularity optimization and maximum likelihood methods for community detection. Phys. Rev. E **94**(5), 052315 (2016)
9. Blondel, V.D., et al.: Fast unfolding of communities in large networks. J. Stat. Mech. Theory Exp. **2008**(10), P10008 (2008)
10. Consonni, C., Laniado, D., Montresor, A.: WikiLinkGraphs: a complete, longitudinal and multi-language dataset of the Wikipedia link networks. In: Proceedings of the International AAAI Conference on Web and Social Media, vol. 13 (2019)
11. Salutari, F., et al.: Analyzing Wikipedia users' perceived quality of experience: a large-scale study. IEEE Trans. Netw. Serv. Manag. **17**(2), 1082–1095 (2020)
12. Dong, H., Xia, H.: Who connects Wikipedia? A deep analysis of node roles and connection patterns in Wikilink network. In: Meng, X., Chen, Y., Suo, L., Xuan, Q., Zhang, Z.-K. (eds.) BDSC 2023. CCIS, vol. 1846, pp. 107–118. Springer, Singapore (2023). https://doi.org/10.1007/978-981-99-3925-1_7
13. Barabási, A.-L., Albert, R.: Emergence of scaling in random networks. Science **286**(5439), 509–512 (1999)

Construction and Simulation of Major Infectious Disease Transmission Model Based on Individual-Place Interaction

Jingwen Zhang, Lili Rong[✉], and Yufan Gong

Dalian University of Technology, Dalian 116024, China
llrong@dlut.edu.cn

Abstract. In recent years, places have played an important role in both the spread of the virus and the implementation of prevention and control. Therefore, we start from the place elements that affect the development of infectious disease transmission, we comprehensively study the evolution process of infectious disease transmission in places through case analysis, abstract modeling, simulation experiments, and build a model of major infectious disease transmission that takes place into account. Based on the influence relationship between individuals and places in the process of infectious disease transmission and prevention and control, combined with the human individual activity law, the basic attributes of individual residents and places are designed considering the characteristics of human age and place function, and the behavioral rules of the two types of agents concerning the three key activities of travel, transmission and prevention and control are constructed, and the state transition process of the two types of agents is clarified in various activities Subsequently, a simulation analysis of typical transmission and prevention and control scenarios was conducted, focusing on the impact of initial transmission place characteristics on transmission outcomes and the impact of different prevention and control measures related to the place on the spread of the infectious disease. The results may be useful for the formulation of differentiated prevention and control measures at different periods of the development of epidemics.

Keywords: Major Infectious Diseases · Places · Transmission Model · Agent-Based Simulation

1 Introduction

As major outbreaks of infectious diseases sweep the world, in addition to their impact on human health, widespread infectious diseases will also have a lasting impact on economies and societies. In this context, the study of transmission mechanisms and prevention and control strategies for such major infectious diseases has become a focus of global attention. Reviewing key epidemic transmission events during the period, it can be found that some places, due to the characteristics of closed space, dense personnel and large mobility, lead to high risk of population infection and difficult investigation, and become epidemic "amplifiers" that need to be focused on. Such as Zhangjiajie scenic spot, Xiamen Shoe Factory, Jilin College of Agricultural Science and Technology in 2022, Liaoning Agricultural Vocational and Technical College, Shanghai Hotel, etc., are all important places that cause the spread of the epidemic, greatly increasing the complexity of the transmission chain and the impact of the epidemic.

2 Literature Review

Many scholars are committed to exploring its complex transmission mechanism through various models, and further searching for the most appropriate prevention and control mechanism in different scenarios through simulation experiments [1]. In the modern economic and social context, with the increasingly frequent movement of goods and people between regions, infectious diseases such as COVID-19, which have complex transmission modes, highly infectious and fast spread, will face a more complex and severe spread situation in human society. In order to control the spread of such major infectious diseases in the human world, the most important work is to fully understand the transmission law, epidemic characteristics and other mechanism knowledge of infectious diseases.

Infectious disease transmission model is an important means to help people understand the mechanism of infectious disease transmission deeply and evaluate and judge realistic intervention measures effectively [1]. The most commonly used models for infectious disease research are the classical compartmental model and its extensions, which classify populations according to their state as a whole and quantitatively fit and predict the spread of epidemics through model calculations. The most basic cell model includes SI, SIS and SIR [1, 2]. On this basis, according to the law of reality, some scholars have subdivided and expanded the cell types by considering the population classification factors such as latent [3], asymptomatic [4], death [5], isolation [6]. Infectious disease transmission models and simulations are important approaches to help one understand the mechanisms of infectious disease transmission. In the existing research on the transmission model of major infectious diseases, the model and simulation research were carried out at different levels in combination with mathematical model [7], population flow data [8] and micro-simulation methods [9]. However, most models mainly focus on individuals and rarely consider the factors associated with the key places that influence the spread of infectious diseases. Some studies deal with the interaction between human activities and different types of places, and focus on places as the initial spatial environment in which individual models operate, lacking a description of the transitions in intra-individual states and the further effects of these transitions on individual actions in subsequent evolution. Simulations are designed to obtain virtual data corresponding to the evolution of a real epidemic with the help of more abstract model design and parameter settings, enabling more flexible and detailed analysis of various transmission scenarios. The degree of correspondence between the model and reality will directly affect the validity of the analytical results. Therefore, important factors that affect the actual transmission process of infectious diseases should be fully considered when building the model, and realistic factors such as the basic characteristics of infectious diseases, the activity and contact laws of human individuals, and various specific prevention and control measures should be abstractly designed.

Based on the above background, this paper intends to study the evolutionary mechanism of the transmission of major infectious diseases based on the interaction between people and places by referring to recent actual transmission cases of typical major infectious diseases.

3 Construction of Major Infectious Disease Transmission Model Considering Place

3.1 Multi-agent Model Construction Ideas

Focusing on the influence relationship between individuals and places during transmission and prevention and control, we analyze and extract the attributes, changes and interaction processes of the two types of subjects and design a multi-agent model. The basis of infectious disease transmission is that individuals come into contact with each other in different places through various activities, the specific activities being related to the travel rules of the individuals and the functional properties of the places. Dedicated to model construction, individual age features and place functional features that may affect individual travel are abstracted to characterize agent attributes. Regions where individuals and places coexist are divided into uniformly distributed grids, and a grid is used to represent a place. At this point, the travel activity of individuals can be abstracted as migration between different grids. By designing individual travel rules to depict specific interaction processes, contact and transmission between different people in different places can be simulated from a microscopic perspective. The state transition process from susceptible, latent to infected, followed by recovery or death is described with the help of the setting of individual health status in traditional infectious disease dynamics models. The state transition process during the transmission is supplemented and analyzed. Based on this, control measures for individual residents and places were designed, along with prevention and control rules and state transition instructions under the rules.

In the agent model constructed based on the above analysis, agents are mainly divided into two categories: places in the region and individuals living in the region (i.e. residents). The spreading environment in a larger region and the prevention and control policy background in that region are the base environments of the model for both types of agents, and the other agents and the base environment together constitute the external environment for one type of agent. The behavioral rule base for both types of agents can be divided into three main parts: individual travel, transmission, and prevention and control. During the operation, both types of agents perceive changes in the environment and the properties of other agents, and feed back to the rule base. The rule base will respond to this and trigger actions that change its own properties and affect the external environment, thus directly or indirectly influencing the perception of other agents. The basic idea of the whole model construction is shown in Fig. 1.

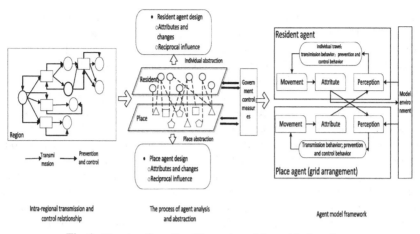

Fig. 1. Construction of multi-agent model considering places

3.2 Design of Agent Attributes

3.2.1 Attributes of Resident Agents

In order to simulate the various behaviors and states of individual residents in the actual infectious disease development process more comprehensively, in addition to marking the unique ID of resident agents, the attributes of resident agents in the area are designed from the aspects of age, basic address information, health, travel, etc. Taking into account the different travel patterns of people of different ages, urban residents were classified into three categories: students, office workers, and retirees, by re-referencing demographic structure data from census results and related literature settings. At the same time, given the actual travel situation and the characteristics of people associated with different types of places in the transmission, each resident agent is randomly assigned and locked to a residential place, which is set as the address information. Randomly assign and lock a unique office learning place for each student. Considering the different places of work brought by different specific occupations, each office worker is randomly assigned and locked an office learning or other service place and set it as the study/work address. These two types of places are fixed places in the travel rules of resident agents, and will not change during the operation of the model, and residents' activities are carried out on this basis.

According to epidemiological studies, infected individuals always have an infectious incubation period with a low likelihood of recurring infections in the short term. Therefore, with reference to the setting of SEIR model and considering the possibility of death, the health status of residents is mainly divided into five categories: susceptible state (S), latent state (E), infected state (I), recovered state (R) and dead state (D). At the same time, considering that contact with individuals with infectious sources may lead to control of close contacts and other prevention and control behaviors, the information of individuals with infectious sources contacted by each individual is recorded by maintaining the attribute of exposure records.

In addition to the basic travel patterns of different age groups, the travel of residents is also restricted by two properties: travel restrictions and travel range. Travel restrictions determine whether individuals are able to travel: regardless of any controls, residents are in a state of unrestricted travel; Travel restrictions are mainly considered in three cases: First, after the onset of the disease is admitted to the hospital (in order to distinguish the isolation from other individuals, specifically referring to the isolation treatment of infected people); Second, infected residents may also change from a latent state to an infected state during quarantine and then be transferred to the hospital due to the epidemiological quarantine of individuals; Third, when there is an infected person in the residence or when there is a high risk of transmission within the area, the resident is at home and cannot move while the residence is closed. The range of travel represents the furthest distance that residents can travel. In the absence of an epidemic, residents are allowed to travel without any restrictions, and the travel scope covers all places in the region. When epidemics occur in the region, from a conservation point of view, residents may reduce their long-distance travel and narrow their travel scope under the government's restriction orders and residents' self-protection and control awareness. Based on the above analysis process, the basic attributes of resident agents are shown in Table 1.

Table 1. The attribute table of resident agents.

Agent attributes	Attribute illustration
ID	Resident agent identity, each individual has a unique value
Age	According to different age, individuals are divided into students, office workers and retired elderly people, whose activities characteristics are different
Address	The coordinates of the resident's unique residence place
Study/work address	The coordinates of each student or office worker's unique place for study or work
Health State	Classified as susceptible (S), latent (E), infected (I), recovered (R), and dead (D)
Exposure record	Record which sources of infection the individual has been in contact with, mainly for close screening
Travel restriction	It can be divided into four states: unlimited, admitted to hospital, isolation and at home
Travel range	The maximum travel distance reflects the travel range of residents in different epidemic situations

3.2.2 Attributes of Place Agents

In order to fully model the influence of place in the process of individual travel, transmission of infectious diseases, and prevention and control, the basic properties of place agents are mainly designed in terms of place function, transmission-related records, and prevention and control-related states. The most fundamental feature of each type of place is its functional character, which will determine the probability of its being visited by different types of resident agents, and thus the amount and scope of its influence in a particular transmission. In this paper, place agents are divided into six categories: residential, office learning, dining and shopping, leisure and entertainment, life services and medical treatment. In the specific transmission process, from the perspective of affecting transmission, factors such as contact conditions and impermeability in the place can be unified into the risk coefficient of the place, which will be affected by factors such as the prevention and control status of the place. From the point of view of transmission occurrence, there are two main types of attributes, exposure and whether transmission has occurred, which records whether the place has been exposed to the virus and whether anyone in the place has been infected. At the same time, information about the source of infection visited in connection with the exposure will be recorded, known as the exposure record. If an individual is indeed infected at the location at this time, the transmission is recorded.

From the point of view of prevention and control, the place may be in a closed state due to control, and the closed place will not be visited by foreign individuals and there will be no transmission within the place. In order to meet the cycle requirements of different control measures, the place will record the time when it starts to be controlled, that is, the start time of closure. At the same time, depending on the classification of the control requirements, the specific time requirements for the management and control, namely the closing period, will be different. Depending on the start time of the closure and the closing period, the attribute of whether or not it is open will change again. Related properties are mainly affected by resident agents and government policies and are in a state of dynamic change.

Based on the above analysis, the basic attributes of place agents are shown in Table 2.

Table 2. The attribute table of the place agents.

Agent attributes	Attribute illustration
Functional characteristics	It is mainly divided into residence, office learning, catering and shopping, leisure and entertainment, life service and medical treatment
Risk coefficient	The transmission probability of the population in the place is affected, and the prevention and control measures will affect the actual value
Exposure or not	According to whether an infected person has stayed there, it is divided into two states: exposed and unexposed

(continued)

Table 2. (*continued*)

Agent attributes	Attribute illustration
Exposure record	Record which source of infection the place has been affected and it is used for epidemiological investigation
Open or not	It is divided into open and closed states, which are related to the control measures of specific places
Exposure record	Record which sources of infection the individual has been in contact with, mainly for close screening
Closing start time	Record the time when the closure began
Closed period	According to the control requirements, the closing cycle requirements of different types of places are different

3.3 Agent Behavior Rules

3.3.1 Travel Rules of Resident Agents

Some studies of temporal and spatial rules of human activity have shown that human activities have distinct periodic and predictable characteristics, and that the range of activities is mostly related to home and work places. In combination with the influence range properties of the transmission places, the travel behavior of the resident agents is classified into two categories: fixed and random. Fixed travel is mainly home, work/school, and its destination address does not change in a short time, which is set as a unique value in the model. In addition to this kind of fixed travel, human travel is also full of randomness, which is mainly reflected in the visit of service places. This travel choice has a large individual variation depending on the demand for services and the distribution of places.

Therefore, from the point of view of micro-simulation, the periodic and stochastic features of reality are fully taken into account, and different action rules are assigned to the various agents according to their activity characteristics at different ages and the settings of the relevant studies, as shown in Table 3.

Table 3. Travel choice of resident agents.

Destination place	Working day			Weekends
	Student	Office worker	Retired elderly	
Place 1	School	Work place	Other place	Other place
Place 2	Home	Other place	Other place	Other place
Place 3	Home	Home	Home	Home

In the operation of the model, when the destination is home or school/company, the agent migrates to the only place, and when the destination is other places, it is set to any other type of place including other living places and service places, taking into account the need to visit relatives and friends. Studies on the activities of older people have found that due to restrictions on travel patterns, the high-frequency activities of older people in their daily lives are concentrated in the vicinity of their residence, with 80% of activities concentrated in residential areas of about 1 km. Therefore, in the design, a circle of daily activities centered around the home with a radius of 2 units is set up for the elderly. In the selection of places for travel purposes for the elderly, there is an 80% likelihood of choosing places within the daily activity circle and a 20% likelihood of visiting other places outside the circle within the travel scope. Based on the above rules, the travel diagrams of various types of intelligent agents are roughly shown in Fig. 2(a) individual travel diagram. Weekday and weekend activities are quite different for students and office workers, while other activities chosen by older people are mostly near their homes.

The above rules are basic activity descriptions of resident agents that are not restricted to travel and are not restricted to destination places. However, when prevention and control is taken into account, only travel agents without restrictions can travel, and their travel will be affected by the status of the destination. If the purpose of travel is to work and study, there is only one destination place, and when that place is closed, it will not be traveled. When the purpose of travel is to other places, randomly selects a type of place as the destination place, selects an open place that meets the travel requirements of the type of place, and if no place is available, the individual does not travel. The specific process is shown in the flow Fig. 2(b) judgment process of individual travel.

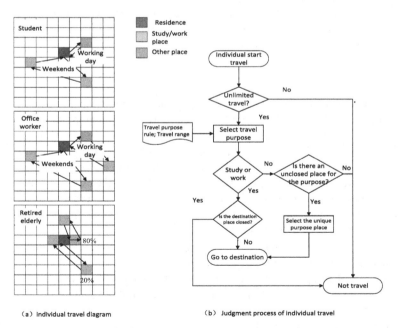

(a) Individual travel diagram   (b) Judgment process of individual travel

Fig. 2. Travel rules of resident agents.

3.3.2 Transmission Rules Considering the Place

Based on the theory of disaster systematics, the dynamic process of disasters can be described as the interaction and mutual influence of the stability of disaster-bearing environment, the risk of disaster-causing factors and the vulnerability of disaster-bearing bodies. The occurrence of infectious disease transmission events in the place can also be analyzed from these three aspects. The risk characteristics of the disaster-causing factors are mainly related to the infectivity of the infectious disease (basic regeneration number R_0), the number of infectious sources ($N_{E,I}$, , the number of latent and infected).The stability of the disaster environment is mainly related to the transmission conditions in the place (including main contact mode, impermeability, C_{place}). Without considering individual differences, the vulnerability of disaster bearing body is mainly reflected in the number of susceptible persons in the place (N_S). The transmission behavior of infectious diseases among individuals in the place (D_f) is the result of the joint action of several factors, as shown in Formula 1.

$$D_f = f(R_0, N_{E,I}, C_{place}, N_S) \tag{1}$$

Based on the above factors, only the possibility of transmission of co-existing infectious sources to susceptible individuals under the influence of infectious disease infectious factor R_0 is considered. The basic transmission probability in the co-place is defined as P_0, and the transmission conditions in the place can be understood as the place risk coefficient r. The actual transmission probability P_t of a source of infection to a susceptible person in the place can be defined as the product of P_0, and r. At this time, the result of individual transmission in the site is expressed as the result shown in Formula 2, which is determined by the probability P_t and the number of infectious sources in the place $N_{E,I}$ and the number of susceptible persons N_S.

$$D_f = f(P_0, N_{E,I}, r, N_S) = f(P_t, N_{E,I}, N_S) \tag{2}$$

The source of infection in the place may be individuals in latent state or infected state. For the transmission of a single infectious source individual in one of the places, the process is shown in Fig. 3. According to the actual situation, after the infected person is diagnosed, through traceability investigation, the relevant trajectory places will be investigated during the infection period. Therefore, as long as the infectious source has visited the place, the place will be affected to enter the exposure state, and the exposure history will be recorded. In a place, for each individual source of infection and for all susceptible individuals in the same place at the same time, the process of virus transmission is simulated by generating random values: Each susceptible individual generates a random value. If it is less than or equal to P_t, it will be infected and enter the latent state. At the same time, since all individuals with co-existing relationships have the possibility of being infected, all individuals will record that they had co-existing relationships with the source of infection. If there are multiple sources of infection in the place, the process is repeated for each source of infection, and for each source of infection that is not restricted to travel, the process is repeated for each place arrived.

Under this transmission rule, combined with SEIR model's setting of infectious disease transmission mode, the health state transformation process of the agent is shown

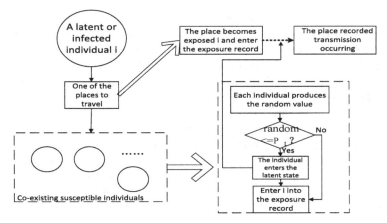

Fig. 3. Schematic diagram of single transmission behavior.

in Fig. 4: Susceptible individuals (*S*) coexist with latent or infected individuals during daily travel, and enter the latent state if infected. Individuals in latent state (*E*) change into infected state (*I*) with a probability of μ every day, and μ is the reciprocal of the average incubation period. An infected individual returns to a restored state with a probability of γ every day (*R*), the value of which is the reciprocal of the average recovery period, and ω is the probability of treatment failure and death (*D*).

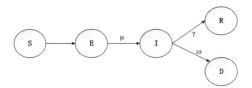

Fig. 4. Transformation of health state of resident agents.

3.3.3 Prevention and Control Rules at Different Prevention and Control Stages

The prevention and control of infectious diseases can be divided into three stages: prevention pre-event, emergency control in-event and recovery post-event. In the prevention phase, the risk of transmission among local people is mainly reduced through various preventive measures in advance, which is reflected in the model comprehensively as the probability of transmission among local individuals is reduced. The recovery phase refers to the recovery of infected individuals and life in the recovered region after the transmission chain has been cut, and to the consideration of prevention against the next possible risk of infectious disease. It is reflected in the model as the recovery of the two types of agents, individuals and places, and the analysis and summarization of the results of the operation.

During the operation of the model, it is the various measures in the emergency management and control phase that have a large impact on the subsequent development of the epidemic situation and bring about complex changes. Therefore, the prevention and control measures at this stage are mainly analyzed in detail in the foliation section. The main prevention and control measures at this stage include four parts: risk source monitoring, epidemic risk management and control, epidemic investigation and potential risk control. In the model, risk source monitoring mainly improves the admission rate of infected persons by enhancing the investigation of infected persons, while epidemiological investigation mainly provides the basis for the control of the associated risk. Epidemic risk management and control, that is, the management and control process of at-risk individuals and places involved in the epidemic is roughly shown in Fig. 5: Infected individual i is found and admitted through voluntary medical treatment or large-scale testing, and becomes an admitted individual; For the admitted individuals, through the screening of exposure records, the individuals and places they had contact with during the infection period were controlled, which were mainly divided into two categories: close contact control and trajectory-related place control.

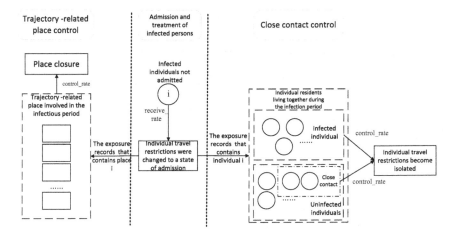

Fig. 5. Control process of epidemic-related individuals and places.

In the Figure, the receive_rate of infected persons refers to the timely admission rate, that is, the probability of each infected individual being found and admitted to the hospital every day from the first day of entering the infected state. Control_rate) refers to the proportion of close contacts and trajectory places related to infected persons under control after they are admitted to the hospital.

Trajectory-related places control aiming at all place agents whose exposure records contain i, is classified into three categories of enclosed control, general control and temporary control by referring to the classification standard of prevention and control levels. The calculation of the closure period takes the admission of the last relevant infected person as the starting point, corresponding to the unblocking judgment process in which there are no new confirmed cases can be unblocked within a certain period after the emergence of the last confirmed case in reality.

In addition to the above control of trajectory-related places, when the risk of infection in the region further increases, that is, when the number of confirmed patients in the region is large, upgraded control measures may be taken for some potential risk individuals and places in the region.

4 Simulation Analysis of Transmission and Prevention of Major Infectious Diseases

4.1 Model Initialization

The simulation experiments in this paper are implemented based on NetLogo software, which is widely used in multi-agent modeling and simulation experiments and is well studied and analyzed for realistic complex problems. The main parameters involved in the running of the model, their associated descriptions and initial values are shown in Table 4. The total number of places in the abstract area is preset to be 400 (20*20). According to the age structure of the national population in the data of the seventh Census in China, the proportion of different types of population is roughly set as students 18%, working people 63%, and retired people 19%. In the specific setting of places in the region, POI (Point of Interest) data is mainly used for setting. Select Beijing, Shanghai, Nanjing, Shenyang, Changchun, Suzhou, Yang-zhou, Jilin, Yingkou and Heihe. Most cities were found to have a ratio of resident population to number of POI points of about 35:1. Therefore, in this paper, we set the ratio of the population to the number of places in the idealized region of the model to be 35:1, and the number of resident agents can be set to be 14, 000. In setting the model, the average value of each city was taken as the initial value, and the final proportions were 6%, 16%, 48%, 4%, 21% and 5% for residential, office and study, dining and shopping, leisure and entertainment, life services and medical places, respectively.

Table 4. Control rules of potential risks in the region.

Parameter type	Name	Parameter description	Initial value
Agent parameter	Number of place agents	Experimental hypothesis	400
	Number of resident agents	It is calculated according to the ratio of real residents to places	14000
	Proportion of various places	The proportion of various places, refer to the distribution and setting of real places in each city	6/16/48/4/21/5

(continued)

Table 4. (*continued*)

Parameter type	Name	Parameter description	Initial value
	Proportion of population by age	The proportion of population of different ages is set with reference to the real population situation	18/63/19
Transmission parameter	Initial number of infectious sources n_0	Number of incubation period at the beginning of the model, experimental hypothesis	1
	Regeneration number R_0	The average number of people infected by an infectious source reflects the infectivity of the infectious disease	2.5[10]
	Mean period of infection τ	The number of days that an individual can spread freely	10[11]
	Fundamental transmission probability P_0	It is calculated by R_0 and τ	0.067%[11]
	Actual transmission probability P_t	It is the product of P_0 and r	0.067% *r
	Daily incidence probability μ	Generally equal to the reciprocal of the average incubation period	1/5[10]
	Daily recovery rate γ	The rate of individual recovery is generally taken as the reciprocal of the average recovery period	1/16[12]
	Daily mortality rate ω	The daily death rate of infected person	0.002[12]
Prevention and control parameters	Place risk factor r	Reflect the overall protection status of individual and place	1
	Receive_rate	The probability of an infected individual being admitted per day	0
	Control_rate	Control ratio of individuals (and places) related to the infected person in the treatment	0

4.2 Simulation Analysis

From the perspective of the transmission of infectious diseases, the experiment in this section studied the infectivity of infectious diseases, the number of individuals at the initial source of infection, and the direct impact of the initial transmission place on the transmission process of infectious diseases through simulation experiments based on the model established considering the differentiated interaction process between individuals and places.

Figure 6 shows that with the increase of infectivity of infectious diseases, the outbreak time will be advanced and the outbreak speed will be greatly increased. Figure 7 shows that when there is no epidemic in the region, the risk of external import should be strictly controlled. When there is an epidemic in the region, it can be regarded as multiple sources

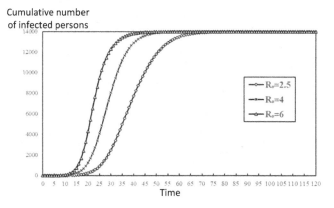

Fig. 6. Control process of epidemic-related individuals and places.

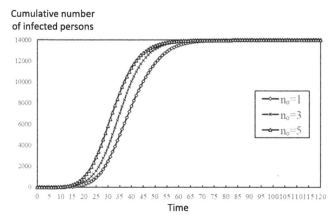

Fig. 7. Cumulative number of infected people with different initial sources of infection.

of infection spreading in the region when it is found late, and the speed of transmission and spread is greatly increased. Figure 8 shows that in the prevention and control of infectious diseases, various types of service places should be paid attention to. Figure 9 illustrates the significance of proactive preventive measures taken before the onset of the disease for the overall prevention and control of the epidemic is to delay the outbreak. Figure 10 shows that timely treatment of infected individuals can directly reduce the final infection scale of the epidemic and delay the spread of the epidemic in time. Figure 11 shows that under the same control conditions, the earlier the intervention time, the smaller the scale of infection, and the earlier the infected person is found in reality, the more timely the epidemic can be controlled within a smaller scale. Figure 12 shows that the effect of control on the trajectory-related places of infected persons is very significant. Figure 13 shows the overall change trend of the results is consistent with the results of increasing place control.

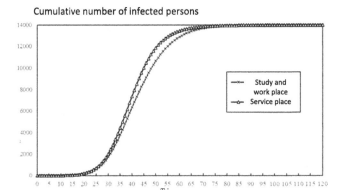

Fig. 8. Cumulative number of infected people in different initial transmission places.

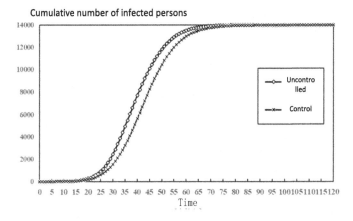

Fig. 9. Cumulative number of infected people with and without preventive measures.

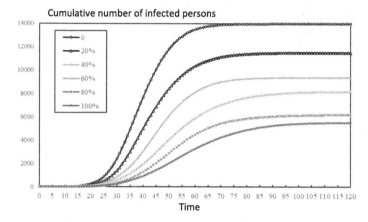

Fig. 10. Cumulative number of infected people under different treatment rates.

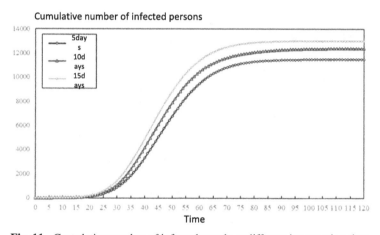

Fig. 11. Cumulative number of infected people at different intervention time.

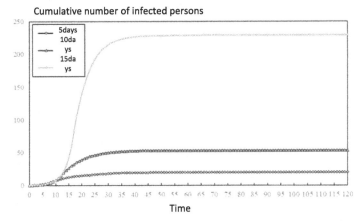

Fig. 12. Cumulative number of infected people considering the control of trajectory-related places.

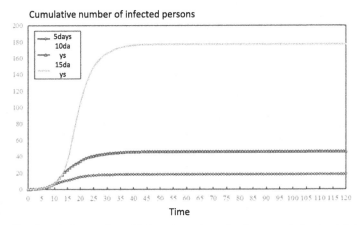

Fig. 13. Cumulative number of infected people under the control of potential risks.

5   Conclusion

Based on realistic analysis and classical infectious disease modeling methods, this paper constructs a multi-agent transmission model considering place. By clarifying the two main lines of transmission and prevention and control in the process of infectious disease transmission and evolution, the key two types of subjects -- individual and place -- are extracted. The behavioral rule base is established, and the construction of multi-agent transmission model considering place is completed. That the model is reasonable and realistic.

Acknowledgement. This work was supported by the National Natural Science Foundation of China (72271041; 7171039).

References

1. Fa Zhang, L., Li, H.X.: Review of infectious disease transmission models. Syst. Eng. Theory Pract. **31**(09), 1736–1744 (2011)
2. Allen, L.J.S.: Some discrete-time SI, SIR, and SIS epidemic models. Math. Biosci. **124**(1), 83–105 (1994)
3. Fan, R., et al.: Transmission model and inflection point prediction of COVID-19 based on SEIR. J. Univ. Electron. Sci. Technol. China **49**(03), 369–374 (2019)
4. Sang, M., et al.: Transmission dynamics model based on characteristics and prevention and control measures of novel coronavirus. Syst. Eng. Theory Pract. **41**(01), 124–133 (2019)
5. Viguerie, A., et al.: Simulating the spread of COVID-19 via a spatially-resolved susceptible-exposed-infected-recovered-deceased (SEIRD) model with heterogeneous diffusion. Appl. Math. Lett. **111** (2021)
6. Berger, D.W., Herkenhoff, K.F., Mongey, S.: An SEIR infectious disease model with testing and conditional quarantine. Political Economy - Development: Public Service Delivery eJournal (2020)
7. He, S., Peng, Y., Sun, K.: SEIR modeling of the COVID-19 and its dynamics. Nonlinear Dyn. **101**, 1667–1680 (2020)
8. Jia, J., et al.: Population flow drives spatio-temporal distribution of COVID-19 in China. Nature **582**(7812), 389 (2020)
9. Koo, J.R., et al.: Interventions to mitigate early spread of SARS-CoV-2 in Singapore: a modelling study. Lancet. Infect. Dis **20**(6), 678–688 (2020)
10. Wu, M., Liu, M.: Study on the incubation period of COVID-19 infected by different novel coronavirus variants. Chin. J. General Med. **25**(11), 1309–1313 (2022)
11. Chen, T., et al.: Research and analysis of infectious disease model based on individual behavior. Dis. Surveill. **37**(06), 813–820 (2019)
12. Zhang, Y., et al.: Analysis of length of stay of 365 patients with Novel coronavirus pneumonia (COVID-19) based on multilevel model. Mod. Prev. Med. **47**(22), 4210–4213 (2019)

Research on the Blockchain Technology Diffusion in China's Supply-Chain Finance Industry

Zhen Chen[1(✉)] and Wenjie Yang[2]

[1] Guangdong Coastal Economic Belt Developing Centre, Lingnan Normal University, Zhanjiang, China
chenzh@lingnan.edu.cn
[2] Business School, Lingnan Normal University, Zhanjiang, China

Abstract. The high level of risk in supply-chain finance can be overcome by the technical features of blockchain. To smooth the promotion and application of blockchain technology in China's supply-chain finance industry, this research constructed a system dynamics model based on a technology–organization–environment (TOE) framework to simulate blockchain diffusion and adoption. The results indicated that in the initial stage, the motivation and obstacles of blockchain adoption were mostly related to external factors such as government policies, organizations' technology levels, industry standards, and managers' willingness. It is necessary to use government policies to build a good ecology for the growth of professional blockchain enterprises. However, in the development and maturation stages of blockchain diffusion, the influence of environmental and organizational factors was found to diminish. It is necessary to deepen scientific research, consolidate the comparative advantages of blockchain itself, and maintain the sustainable diffusion of this technology.

Keywords: Supply-chain finance · Blockchain · Technology diffusion

1 Introduction

In the era after the COVID-19 epidemic, the world economy is slowly recovering. As the driving force for economic development, small and medium-sized enterprises (SMEs) are facing unprecedented financing difficulties. In response, governments and enterprises in various countries have explored using innovative financial tools to improve the efficiency of economic operation and reduce the costs of social operation. Among them, supply-chain finance is a cost-effective approach for SMEs.

Supply-chain finance serves SMEs by providing credit to upstream and downstream enterprises that do not meet traditional financing thresholds through a credible financial platform and by using real transaction information as a reference. Randall et al. (2009) argued that a supply-chain finance ecosystem could monitor and optimize cash flows in real time and achieve capital transparency across the supply chain. The information gap between suppliers, demanders, and financial institutions would be gradually leveled, and the overall financial cost would be significantly reduced. However, in the actual operation of supply-chain finance, information could not be transparent in real time, the credit of SMEs could not be truly guaranteed, improper management would result in frequent defaults (Du et al. 2020).

Blockchain is a distributed bookkeeping technology. It allows for the establishment of social trust relationships that do not require third-party intermediaries, and improves supply chain transparency. That has reinforced the advantages of trust itself (Xu et al., 2021). Owing to the characteristics of decentralization, information transparency, and tamper-proof, it is believed that applying blockchain to supply-chain finance could increase the credit of long-tail SMEs, avoid loan fraud, reduce operational costs, and improve performance efficiency (Chod et al., 2020; Du et al., 2020).

China has a good supply chain industry foundation, so the Chinese government and enterprises have always been encouraging financing through the supply chain, and also supporting the use of cutting-edge digital technology to improve financing efficiency and security. After 2019, the Chinese government has successively introduced a series of technology incentive policies, including China's 14th Five Year Development Plan, Guiding Opinions on Accelerating the Application of Blockchain Technology and Industrial Development, and etc. Thanks to support from the government, companies were increasingly trying to integrate blockchain into their supply-chain finance businesses to expand the application scenarios and overcome drawbacks (Ning et al., 2023). In order to promote the application of blockchain in supply-chain finance, this research constructed a system dynamics model based on a "technology–organization–environment" (TOE) theoretical framework that included blockchain adoption factors. In this way, this research examined the barriers to the diffusion of blockchain and explored possible measures for promoting its adoption. Some successful cases in China could be used for reference by other countries to jointly overcome the financing difficulties of global SMEs.

2   Literature Overview

2.1   Mechanism of Blockchain Optimizing Supply-Chain Finance

Wang et al. (2022) proposed that the blockchain-driven supply-chain finance project provided services to its customers by applying key resources and conducting corresponding practices, which create value for the participants through meeting their motives. Li et al. (2020) suggested that by applying blockchain technology, supply-chain finance could effectively solve the triangular debt dilemma of Chinese SMEs by increasing their financing proportion. Complemented by other related technologies, such as big data and cloud computing, blockchain could provide a highly trusted data flow environment for business scenarios, including finance and logistics (Dutta et al., 2020; Winkelhaus and Grosse, 2020).

2.2   Typical Case of Blockchain Applied in Supply-Chain Finance

Arguing for the use of blockchain and Internet of things in supply-chain finance simultaneously, Liu et al. (2021) suggested that the tamper-proof nature of blockchain could match the characteristics of internet of things, real-time input of goods information, which could reduce risks and optimize internet financial services. Rijanto (2022) and Song et al. (2023) studied the typical fintech cases worldwide, respectively. They demonstrated that the ways blockchain improved performance with smart contracts, transparency and security of distributed ledger data feature, as well as promoting the transition to new business models. Blockchain advantages provided automation solutions in global supply-chain finance practices.

2.3 Summary

In summary, most previous studies were qualitative and considered blockchain as a suitable technology for overcoming the drawbacks of supply-chain finance. Few studies, however, have investigated the barriers to the diffusion of blockchain technology and countermeasures to address them. Whether blockchain can be accepted and successfully implemented by supply-chain finance decision-makers depends on three factors: technology, organization, and environment. This research, therefore, used the TOE framework to analyze the diffusion and adoption process of blockchain technology, aiming to more accurately depict its evolutionary trajectory. This would not only enrich existing theory but also provide strong evidence for policy formulation.

3   Model Construction

3.1 Modeling Framework

This research used the TOE framework to analyze the key factors affecting the diffusion of blockchain technology in China's supply-chain finance industry. The TOE framework is the main analytical framework for studying technology diffusion (Tornatzky et al., 1990). TOE assumes that when a new technology is put into practical use, user acceptance is not only influenced by technical factors, but also by organizational characteristics and their social environment. In recent years, many scholars have used the TOE framework to conduct research in the field of cutting-edge technology diffusion, including e-maintenance (Aboelmaged, 2014), social media (Pateli et al., 2020), artificial intelligence (Na et al., 2022), and so on. Chen et al. (2022) believed that the key factors affecting the performance of supply-chain finance platforms could be classified according to organization, technology, and cognition. This is consistent with the concept of the TOE framework. Therefore, the use of the TOE framework in this research is reasonable.

3.2 Modeling Method

This research used system dynamics to construct a technology diffusion model. System dynamics has the advantage of being able to analyze complex systems by examining system causality and determining the evolutionary trends of factors within the system. When relying solely on data for modeling, many practical problems arise—namely, the lack of sufficient data sources greatly hinders the search for real answers. By contrast, system dynamics supports the user in deducing reasonable conclusions through a feedback loop structure, even when sufficient panel data are not available. This method is suitable for studying long-term dynamic problems with insufficient data sources and nonlinear interactions between variables.

3.3 Basic Settings

As a cutting-edge digital technology, blockchain could significantly change the established operational processes and rules of the supply-chain finance industry. Given the commonality and specificity of this technology, before constructing the model, the following settings were conducted.

(1) Spatial scope of blockchain technology diffusion. The model was used to investigate the diffusion of blockchain technology in the supply-chain finance industry in mainland China rather than the whole social system. The variables constituting the model came from different indicators related to supply-chain finance and blockchain in mainland China.

(2) Temporal starting point of blockchain technology diffusion. According to available statistics, blockchain technology first started to be diffused and applied in China in 2016, and the market expenditure in that year was only about 100 million RMB. For this reason, 2016 was selected as the starting point for the model to better reflect the diffusion of blockchain technology from its initial adoption to the present.

(3) Intention to adopt blockchain technology is an important indicator for measuring the likelihood of effective diffusion. The TOE framework divides the composition of adoption into technical, organizational, and environmental factors. Because the adoption rate of blockchain in China was still very low in 2016, the initial values of these three factors were set to 0 for the convenience of calculation. Then, the loading coefficients of these three factors in intention to adopt in 2022 were obtained through questionnaire analysis.

The survey, conducted by this research by the end of 2022, included 312 middle and senior managers of various Chinese financial institutions, supply-chain companies and some well-known blockchain researchers distributed in the Pearl River Delta, Yangtze River Delta, and Bohai Rim, the most developed in China's industries. Based on the analysis of these 312 questionnaires, the loading coefficients of the technology, organizational, and environmental factors in willingness to adopt were 1.789, 2.002, and 2.470, respectively. With the initial value of 0 in 2016 as the starting point, and the loading coefficient in 2022 as the end point, this research used the historical data of the three factors from 2016 to 2022 to predict its future evolutionary trend. The model parameters were set on this basis.

(4) In the diffusion of blockchain technology, the industry size of supply-chain finance also changed year by year. To determine whether the market spending of blockchain technology comes from the increase in willingness to adopt or the expansion of the scale of the whole supply-chain finance industry, this model set the annual growth rate of China's supply-chain finance industry to 6% according to reality. If the average annual growth of market spending on blockchain technology exceeds this value, it proves that the increase in willingness to adopt is more pronounced, and vice versa.

3.4 Main Feedback Loops in the Model

In order to measure the diffusion characteristics of blockchain technology in China's supply-chain finance industry, four state variables were set in the system dynamics model. These variables include the size of the supply-chain finance industry, technical factors, organizational factors and environmental factors that affect the willingness to adopt blockchain. And the main feedback loops that drives the evolution of the above state variables in this system are as follows.

(1) Supply-chain finance industry scale $\xrightarrow{+}$blockchain market expenditure scale $\xrightarrow{+}$number of national and ministry macro policies $\xrightarrow{+}$environmental evolution $\xrightarrow{+}$willingness to adopt.

Increases in the industry scale of supply-chain finance will drive the development of blockchain business (Hou, et al., 2019). To sustain the trend of blockchain technology adoption and discover new economic growth points, the central government and its ministries have introduced several macro policies, which has gradually corrected the impression of blockchain as a speculative tool in China. More people have come to recognize blockchain as a means of realizing the economics of trust and as a technical guarantee for economic growth and industrial development. A more tolerant environment for blockchain technology will increase the willingness to adopt it.

(2) Blockchain market expenditure scale $\xrightarrow{+}$number of national and ministry macro policies $\xrightarrow{+}$number of blockchain industrial parks $\xrightarrow{+}$number of real operating blockchain enterprises $\xrightarrow{+}$environmental evolution $\xrightarrow{+}$willingness to adopt.

The stable development of the blockchain technology market will prompt the introduction of more macro policies. The most direct way for these macro policies to penetrate the local level is to build specialized technology industrial parks (Zhu, 2020). When blockchain industrial parks are successively built and operated, more blockchain enterprises will move into them, and advanced production factors related to blockchain will gather and interact there, which will lead to the development of blockchain thinking in the innovation environment and enhance the willingness to adopt blockchain.

(3) Blockchain market expenditure scale $\xrightarrow{+}$number of blockchain enterprises in real operation $\xrightarrow{+}$number of blockchain patent applications $\xrightarrow{+}$number of patent applications on the theme of "supply-chain finance + blockchain" $\xrightarrow{+}$technological progress $\xrightarrow{+}$willingness to adopt.

When the application market of blockchain technology expands further, more specialized blockchain enterprises will be engaged in it, which will inevitably generate more application scenarios and more corresponding patents (Lu et al., 2020). According to data on blockchain enterprises published by the State Internet Information Office, supply-chain finance is the key area of blockchain technology application. Therefore, the number of patent applications related to the theme of "supply-chain finance + blockchain" will rise accordingly, which will improve the technical field in this direction and thus enhance the willingness to adopt.

(4) Number of national and ministry macro policies $\xrightarrow{+}$ number of blockchain papers published $\xrightarrow{+}$ number of papers published on the theme of "supply-chain finance + blockchain" $\xrightarrow{+}$ technological progress $\xrightarrow{+}$ willingness to adopt.

To comply with macro policies and get fund from the supporting projects issued by ministries and commissions, the researchers actively produce papers related to blockchain that have theoretical value (Mann et al., 2015). These policies generally have clear guidelines, telling researchers what fields blockchain should be applied to. The researchers nationwide can draw the inspiration from the guidelines and give the papers more application value. So reading these papers can promote the progress of blockchain and thus enhance the willingness to adopt it.

(5) Supply-chain finance industry scale $\xrightarrow{+}$ blockchain market expenditure scale $\xrightarrow{+}$ number of blockchain industry standards $\xrightarrow{+}$ organizational change $\xrightarrow{+}$ willingness to adopt.

China's supply-chain finance industry is growing at a steady pace. According to the previous analysis, the technical characteristics of blockchain are quite suitable for the regulatory requirements of supply-chain finance. While the supply-chain finance industry has ramped up the scale of blockchain market spending, this has also given rise to regulatory management requirements for the industry (Silva et al, 2019). China has set more industry standards for this purpose, putting more pressure on participating organizations and prompting them to enhance their willingness to adopt blockchain.

(6) Number of national and ministries' macro policies $\xrightarrow{+}$ number of blockchain industrial parks $\xrightarrow{+}$ number of real operating blockchain enterprises $\xrightarrow{+}$ organizational changes $\xrightarrow{+}$ willingness to adopt.

The introduction of macro policies to promote blockchain development has resulted in the completion of more local blockchain industrial parks. The blockchain talent-attraction policies, preferential rent, and special loans offered by these industrial parks will encourage more enterprises to engage in blockchain business innovation (Vlck-ova, 2020). More blockchain enterprises are emerging, which improves the level of blockchain application in supply-chain finance and will prompt more supply-chain finance enterprises to adopt the technology.

3.5 Flow Diagram Creation

Based on the analysis of the causal feedback loops, a system dynamics flow diagram has been created, as shown in Fig. 1.

The model contains four state variables and several auxiliary variables. According to the previous setting, the size of the supply-chain finance industry in this model maintains an annual growth rate of 6%. Under this premise, the growth of the supply-chain finance industry and the change in willingness to adopt blockchain lead to a significant increase in the expenditure of the blockchain market. The various auxiliary variables in the model interact with each other, bringing different effects to the factors that fall under the technical, organizational, and environmental modules. Meanwhile, the overall

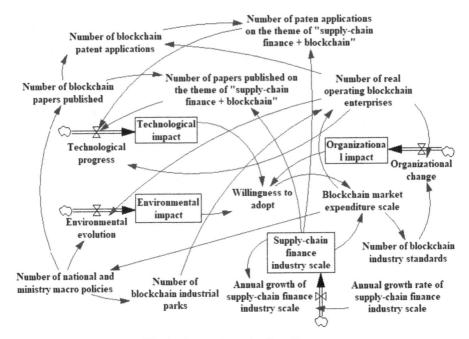

Fig. 1. System dynamics flow diagram

change in willingness to adopt changes blockchain market spending in the following year, which in turn affects the future rate of blockchain proliferation.

Vensim simulation software was used for system dynamics model construction and testing. The historical data from 2016 to 2022 were used for the validity check of the model. The simulation data of the variables in this time period were compared with real statistics, and the model parameters were modified to reduce the deviation rate of each feedback loop, thus improving the goodness-of-fit of the whole model. After the model passed the three tests of boundary appropriateness, structural validity, and behavioral compatibility, it could simulate the diffusion trend and support studying the intrinsic correlation mechanism between the whole supply-chain finance system and the adoption of blockchain technology.

4   Simulation Prediction

4.1   Prediction of Diffusion Effects

Model simulation is a key step in the analysis of system dynamics models. When we kept all the model parameters not changed, the trajectory of the key variables within that time period was analyzed. The period was set to 2016–2025. Among them, the data from 2016 to 2022 was real historical data, while the data from 2023 to 2025 was simulated data. Figures 2–4 show the evolutionary trends of the key variables in this system model.

Compared the above three figures, under the average annual growth rate of 6% in the entire supply chain finance industry (Fig. 2), the willingness to adopt blockchain

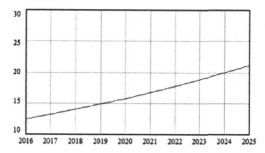

Fig. 2. Growth curve of supply-chain industry scale

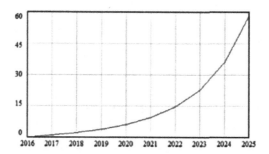

Fig. 3. Growth curve of willingness to adopt

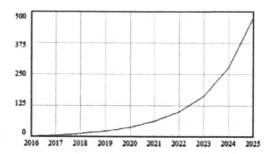

Fig. 4. Growth curve of blockchain application expenditure

(Fig. 3) and the expenditure on blockchain applications (Fig. 4) are significantly higher than the former. This trend is similar to the prediffusion stage of the S-curve. Blockchain technology is increasingly being applied in supply-chain finance, which is in the high growth stage of the curve and has not yet entered the stable stage.

4.2 Analysis of Changes in Willingness to Adopt

In general, the intention to adopt blockchain has shown a high growth rate from 2016 to 2025. Are there any new changes in the various factors that constitute the intention to adopt? To measure this change, this research constructed a map of willingness to adopt that includes technical, organizational, and environmental factors. Figure 5 shows the proportional changes of each factor over the years.

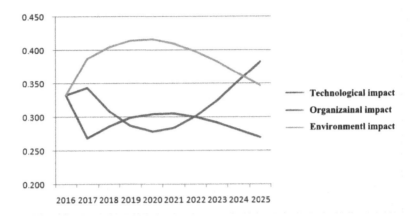

Fig. 5. Proportional change of constituent factors of willingness to adopt

In terms of technology factors, organizational factors, and environmental factors, according to the setting, the proportion of each factor was equal in 2016, and all were one-third. In Fig. 3, it can be seen that the proportion of environmental factors accounts for the highest proportion in the early and middle periods of the simulation period. The proportion of organizational factors is the lowest at the beginning, starts to become higher than the technical factors in the middle period, and falls back to being the lowest at the end. The proportion of technology factors maintained the middle position in the early stage and fell to the lowest in the middle stage but became the most dominant factor in willingness to adopt in the late stage.

This trend change reveals that the early diffusion of blockchain technology is mainly driven by the policy guidance of governments at all levels, creating a favorable environment for accepting blockchain applications in the real economy. It is because of government efforts to promote the technology that blockchain gradually transforms from a speculative tool in the eyes of the public into a tool for realizing trust economics. Although blockchain has technical advantages in terms of traceability and being tamper-proof, its distributed and decentralized characteristics require a certain scale of adopters to manifest such advantages. Without the influence of environmental factors, the willingness to adopt will not be as significant.

Technological advances are essential for maintaining the overall level of willingness to adopt in the long run. Many fintechs have functions that are similar to blockchain, and their relationship with blockchain can be either substitutional or complementary. As the government creates an environment increasingly conducive to the diffusion of blockchain, the technology becomes increasingly familiar, and the diffusion momentum gradually shifts to the comparative advantages of the technology itself. In such a situation, the technical potential of blockchain needs to be continuously explored through scientific and technological vehicles such as papers and patents. More feasible application scenarios need to be discovered through market-oriented business behaviors to maintain its advantage over other technologies in supply-chain finance.

The effect of organizational factors on willingness to adopt has been changing at a low level. According to this trend direction, there is no sign of rebound in the near future, and it is becoming lower and lower. This indicates that the factors of organizations related to supply-chain finance are the weakest in influencing the industry's willingness to adopt blockchain technology. The motivation to adopt blockchain comes more from direct or indirect support from the government in the early stage and from the advantage of technological progress in the later stage. Even if the technical standards of blockchain and industrial application standards are improved, the enterprises' own blockchain technology reserves alone are not sufficient to fully utilize the technology, and they need specialized blockchain enterprises to assist.

In summary, the composition of the factors of willingness to adopt is not constant. Incentives from government policies are not sustainable in the long run. Because enterprises have many alternative technologies of their own, their demand for blockchain decreases when the effect of policy incentives diminishes. At that point, the technological advantage of blockchain needs to be enhanced through greater investment in research to keep increasing the willingness to adopt. When the scale of blockchain adoption rises to a certain level, its technical advantages will be better reflected.

5   Optimizing Experiments

In addition to predicting the future development of system variables, the system dynamics model can also simulate the implementation of different policies to explore the optimal choice of system operation and provide a basis for formulating scientifically sound measures. This research has simulated the effect of different policies on organizations' willingness to adopt in a future period. By regulating the proportion of changes in specific model parameters, the optimal scenario for adoption in supply-chain finance are proposed.

5.1 Scenario Design Ideas

This research has used the following coefficients as control variables to develop seven policy scenarios: paper publication incentive, patent writing inspiration, macro policy formulation, industrial park construction, market stimulation operation, and industry standard regulation (Table. 1). Scenario 1 is the original scenario, and all the variables are kept constant, while scenarios 2–7 increase the value of one of the control variables by 50%. The purpose of this higher magnitude is to make the effect more obvious so that a better optimization solution could be determined within a shorter simulation interval.

Table. 1. Scenario design list

Scenario	Design idea	Paper publication incentive coefficient	Patent application incentive coefficient	Policy formulation incentive coefficient	Industrial park incentive coefficient	Enterprise operation incentive coefficient	Standards development incentive coefficient
1	Original	49.727	33.2867	1.15174	0.384236	37.2656	0.568307
2	Encourage the paper publication	74.591	33.2867	1.15174	0.384236	37.2656	0.568307
3	Encourage patent application	49.727	49.93005	1.15174	0.384236	37.2656	0.568307
4	Formulate more macro policies	49.727	33.2867	1.72761	0.384236	37.2656	0.568307
5	Build more industrial parks	49.727	33.2867	1.15174	0.576354	37.2656	0.568307
6	Cultivate more real blockchain enterprises	49.727	33.2867	1.15174	0.384236	55.8984	0.568307
7	Develop more industry standards	49.727	33.2867	1.15174	0.384236	37.2656	0.852460

5.2 Analysis and Discussion

The simulation period is from 2023 to 2025. The coefficients of the variables have been modified separately according to the above scenarios. Changes in the willingness to adopt blockchain under different scenarios are compared to determine the effect of each control variable on blockchain diffusion under the premise that the supply-chain finance industry grows at an average annual rate of 6%. Figure 6 shows the results and reveals that all seven scenarios show steady growth in willingness to adopt blockchain, but the different scenarios have led to differences in the magnitude of growth.

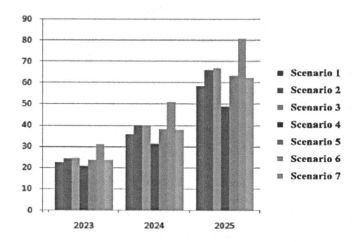

Fig. 6. The willingness of supply-chain finance industry to adopt blockchain

(1) Assessment of regulation on technical factors

According to the model, the variables that affect the technology factor are the number of papers, number of patents, and number of blockchain enterprises with real operating behaviors, corresponding to scenarios 2, 3, and 6, respectively. Comparing these three scenarios with scenario 1, the effects of encouraging the publication of papers and patent applications are similar, and both can improve willingness to adopt better than scenario 1 can. The effect of patents is slightly higher than that of papers. This indicates that patents are more applied, more targeted, and more conducive to solving problems in industrial development that can be addressed through technological innovation. Further, the effect is best if measures are taken to put more blockchain technology enterprises into real operation.

(2) Assessment of regulation on organizational factors

According to the model, the variables that affect organizational factors are the number of blockchain enterprises with real operating behaviors and the number of blockchain industry standards, corresponding to scenarios 6 and 7, respectively. The existence of blockchain enterprises cannot only improve the development of the technology itself but also enhance the ability and willingness of supply-chain finance organizations to apply blockchain. The effect of scenario 7, although better than that of scenario 1, is the lowest compared with other regulatory measures. This indicates that the formulation of industry standards should not be too detailed, and the development of the industry needs more market demand for promotion. If too many standards and rules are introduced, it will inhibit the innovative energy of the market.

(3) Assessment of regulation on environmental factors

According to the model, the variables that affect environmental factors are the number of macro policies and the number of industrial parks to be built, corresponding to scenarios 4 and 5. Scenario 4 does not have the same regulatory effect as the other variables, and the increase in willingness to adopt is lower than that of scenario 1. This indicates that after a certain stage of technology diffusion, the diffusion impetus also needs to originate more from the market, and too many macro policies from the government will instead reduce diffusion efficiency. The effect of scenario 5 is not as good as that of scenarios 2, 3, and 6, indicating that there should be an upper limit to the number of industrial parks to be built. There should be a reasonable fit between the number of parks and enterprises.

5.3 Suggestion

1) In the the preliminary stage

Because blockchain has been misunderstood as merely a speculative tool, there is heavy reliance on the government to introduce macro policies to promote it in the preliminary stage of adoption and diffusion. Based on the analysis of Scenario 4 and Scenario 5, after the central government issues macro policies and completes the top-level design of the industry, local governments translate the policies into the construction of industrial parks to prompt more professional blockchain enterprises to be established. In the development of professional blockchain enterprises, emphasis should be placed on the construction of financial infrastructure and underlying platforms to improve the blockchain ecology through cooperation between government, industry, academia, research institutions, and users.

2) In the the mid-to-late stage

When blockchain diffuses to a certain scale in supply-chain finance, the effects of policy gradually diminish, industry standards are gradually improved, industrial park construction reaches the upper limit, and the influence of organizational and environmental factors on willingness to adopt decreases. It then becomes necessary to increase investment in scientific research. Based on the analysis of Scenarios 2, 3, and 6, by exploring theory and application, encouraging paper and patent output, and exploring the integration of blockchain with other technologies, professional blockchain enterprises can sustainably provide suitable application scenarios for supply-chain finance enterprises instead of allowing other technologies to replace blockchain.

6 Conclusion

6.1 Contribution

Based on the TOE framework, this research simulated the spatiotemporal trajectory of blockchain technology diffusion in China's supply-chain finance industry. Most previous studies conducted analyses using panel data, which made it difficult to reveal the influencing factors that need to be analyzed in the spatiotemporal dimensions. This research used a system dynamics approach to model the adoption and diffusion of blockchain in China's supply-chain finance industry based on questionnaires and statistics from industry reports. In this way, this research confirms the key factors for the adoption of blockchain in the Chinese supply-chain finance industry. Improving the application of this technology through effective measures will help further expand the scale of the supply-chain finance industry.

6.2 Limitation and Future Research

The current blockchain technology is still in the introduction stage of its life-cycle, and its application scope is not too broad, which has led to the difficulty of collecting sufficient data in this research. On the other hand, due to the impact of the COVID-19 pandemic, the external economic and social environment has undergone major changes. Before and after the outbreak of the pandemic, there were different changes in the mindset and answers of the interviewees. In order to minimize the impact of external interference on the conclusions, objective statistical data was extracted on certain variables for reasonable substitution.

In response to the future demand of the supply-chain finance industry in the application of blockchain technology, the future research directions are as follows.

(1) Integrated application of blockchain and other cutting-edge digital technologies

The application of blockchain in the supply-chain finance industry has completed the concept validation and application pilot stage. But to meet the requirements of more application scenarios, blockchain needs to be integrated with other cutting-edge digital technologies to accelerate the sustainable development of supply-chain finance.

(2) Following the progress of integrating digital currency into supply-chain

Digital currency is built based on blockchain technology, and its application will open up new application scenarios for the supply-chain finance. We should closely pay attention to the progress of digital currency to build an effective management mechanism and seize the opportunities of the new financial model.

References

Aboelmaged, M.G.: Predicting e-readiness at firm-level: an analysis of technological, organizational and environmental (TOE) effects on e-maintenance readiness in manufacturing firms. Int. J. Inf. Manage. **34**, 639–651 (2014)

Chen, S.H., Du, J.Z., He, W., Siponen, M.: Supply chain finance platform evaluation based on acceptability analysis. Int. J. Prod. Econ. **243**, 108350 (2022)

Chod, J., Trichakis, N., Tsoukalas, G., Aspegren, H., Weber, M.: On the financing benefits of supply chain transparency and blockchain adoption. Manage. Sci. **66**, 4378–4396 (2020)

Du, M.X., Chen, Q.J., Xiao, J., Yang, H.H., Ma, X.F.: Supply chain finance innovation using blockchain. IEEE Trans. Eng. Manage. **67**, 1045–1058 (2020)

Dutta, P., Choi, T.M., Somani, S., Butala, R.: Blockchain technology in supply chain operations: applications, challenges and research opportunities. Transp. Res. Part E – Log. Transp. Rev. **142**, 102067 (2020)

Hou, B.J., Hong, J., Wang, H., Zhou, C.Y.: Academia-industry collaboration, government funding and innovation efficiency in Chinese industrial enterprises. Technol. Anal. Strateg. Manage. **31**, 692–706 (2018)

Li, M., Shao, S.J., Ye, Q.W., Xu, G.Y., Huang, G.Q.: Blockchain-enabled logistics finance execution platform for capital-constrained e-commerce retail. Robot. Comput.-Integr. Manuf. **65**, 101962 (2020)

Liu, L.X., Zhang, J.Z., He, W., Li, W.Z.: Mitigating information asymmetry in inventory pledge financing through the Internet of thins and blockchain. J. Enterp. Inf. Manage. **34**, 1429–1451 (2021)

Lu, Q., Liu, B.N., Song, H.: How can SMEs acquire supply chain financing: the capabilities and information perspective. Ind. Manage. Data Syst. **120**, 784–809 (2020)

Mann, A., Folch, D.C., Kauffman, R.J., Anselin, L.: Spatial and temporal trends in information technology outsourcing. Appl. Geogr. **63**, 192–203 (2015)

Na, S., Heo, S., Han, S., Shin, Y., Roh, Y.: Acceptance model of artificial intelligence (AI)-based technologies in construction firms: applying the technology acceptance model (TAM) in combination with the technology-organisation-environment (TOE) framework. Buildings **12**, 90 (2022)

Ning, L.J., Yuan, Y.Q.: How blockchain impacts the supply chain finance platform business model reconfiguration. Int. J. Log. Res. Appl. **26**, 1081–1101 (2023)

Pateli, A., Mylonas, N., Spyrou, A.: Organizational adoption of social media in the hospitality industry: an integrated approach based on DIT and TOE frameworks. Sustainability. **12**, 7132 (2020)

Randall, W.S., Theodore, F.M.: Supply chain financing: using cash-to-cash variables to strengthen the supply chain. Int. J. Phys. Distrib. Logist. Manag. **39**, 669–689 (2009)

Rijanto, A.: Blockchain technology adoption in supply chain finance. J. Theor. Appl. Electron. Commer. Res. **16**, 3078–3098 (2022)

Silva, A., Montoya, I.A., Valencia, J.A.: The attitude of managers toward telework, why is it so difficult to adopt it in organizations? Technol. Soc. **59**, 101133 (2019)

Song, H., Han, S.Q., Yu, K.K.: Blockchain-enabled supply chain operations and financing: the perspective of expectancy theory. Int. J. Oper. Prod. Manage. **3** (2023)

Tornatzky, L.G., Fleischer, M., Chakrabarti, A.K.: The Processes of Technological Innovation. Lexington Books, Washington (1990)

Vlckova, M.: Controlling in relation to division of enterprises and company size. Digitalized Econ. Soc. Inf. Manage. **49**, 369–376 (2020)

Wang, L., Luo, X., Lee, F., Benitez, J.: Value creation in blockchain-driven supply chain finance. Inf. Manage. **59**, 103510 (2022)

Winkelhaus, S., Grosse, E.H.: Logistics 4.0: a systematic review towards a new logistics system. Int. J. Prod. Res. **58**, 18–43 (2020)

Xu, P., Lee, J., Barth, J.R., Richey, R.G.: Blockchain as supply chain technology: considering transparency and security. Int. J. Phys. Distrib. Logist. Manage. **51**, 305–324 (2021)

Zhu, W.D., Tian, Y.F., Hu, X., Ku, Q., Dai, X.Y.: Research on relationship between government innovation funding and firms value creation using clustering-rough sets. Kybernetes **49**, 578–600 (2020)

Opinion Mining and Knowledge Technologies

Understanding of the Party's Construction and Governing Philosophy by an Analysis of the Reports of Successive CPC's Congresses

Xiaohui Huang[1,2] and Xijin Tang[1,2(✉)]

[1] Academy of Mathematics and Systems Science, Chinese Academy of Sciences,
Beijing 100190, China
huangxiaohui@amss.ac.cn
[2] University of Chinese Academy of Sciences, Beijing 100049, China
xjtang@iss.ac.cn

Abstract. Over the past century, the Communist Party of China (CPC) has led China and her people out of extreme poverty and towards prosperity, transforming into a great modernized nation. In order to understand the evolution of the construction and governing philosophy of the CPC along different historical phase, this paper studies the official documents of Party Congresses from 1921–2020 by descriptive statistics, Latent Dirichlet Allocation (LDA) topic model, and word co-occurrence networks. The analysis result reveals the characteristics in development and governance by CPC from the documents in different phases. For better understanding by visualization, a thematic evolution chart is constructed to show the evolution of the CPC's ideology and a word co-occurrence network is established to analyze the changes of words across different stages. Our study illustrates the CPC's governing philosophy, which is help to learn the China Road to modernisation.

Keywords: CPC · Party Congress · Text mining · LDA · Word co-occurrence network

1 Introduction

The Communist Party of China (CPC) established over a century ago, is leading China through significant transformations and turning the country from poverty to prosperity [1,2]. How does the CPC accomplish this? What is the CPC's governing philosophy? Has her philosophy evolved over the past century? The official reports of successive National Congress of the CPC have expounded the progress and challenges of the Party at different stages and set corresponding objectives and courses of action. They are thus be used to deepen our understanding of the CPC by using text mining methods to interpret the Party Congress documents.

Many researchers have studied the CPC's governing philosophy through quantitative text analysis toward the Party Congress documents in China. By

J. Chen et al. (Eds.): KSS 2023, CCIS 1927, pp. 215–229, 2023.
https://doi.org/10.1007/978-981-99-8318-6_15

using the descriptive statistics, Li [3] and Wang [4] respectively analyzed the reports of the previous Party Congresses of the CPC, discussed the transformation from the governing concept, governing basis, guiding ideology, and governing mode, to show the change from a revolutionary party to a governing party. Based on the reports of previous Party Congress and several important meeting communique, Guo and Wu used word frequency analysis and correlation methods to explore the journey of how the CPC implements her original mission [5]. Similarly, Wu summarized the development and experience of the Party's construction concept since the reform and opening up from three aspects: the overall layout of construction, the sub-layout of construction, and the construction path [6]. In addition, there are still many researchers who study the documents of the Party Congress utilizing quantitative text analysis [7,8].

The Latent Dirichlet Allocation (LDA) proposed by Blei et al. have been applied in various fields to reveal the evolution of topics across large collections of documents [9,10]. Yan and Tang used the LDA and the Kleinberg model to extract and filter the emergent topics of the "safety helmet event" [11]. They used Jaccard similarity to measure the evolutionary relationship among topics and constructed a thematic evolution Sankey diagram of "safety helmet event". Huang and Tang examined the development of societal events and social coping strategies while mining social public opinion and evolution of medical supplies during the COVID-19 outbreak using the LDA and complex network [12]. Rule et al. established the word semantic network based on all U.S. State of the Union addresses from 1790 to 2014 and constructed a Sankey diagram to capture the historical flow of American political terms [13]. Wei et al. employed the text analysis and word network to analyze the reports of The State Council and mined the topic and evolution of the government working reports [14]. Similarly, some researchers also used the LDA method to analyze political texts [15,16].

To summarize, text mining has great potential in analyzing official political documents. The CPC, as the governing party of the world's second-largest economy, has laid out its construction and governing philosophy in the reports of each Party Congress. On July 1, 2021, Xi delivered an important speech at a ceremony marking the 100th anniversary of the founding of the CPC in Beijing. Consequently, we try to study CPC's political philosophy over the past century by analysing the report documents of the 19 Party Congresses from 1921–2020.

The remainder of this paper is organized as follows. Section 2 gives the descriptive statistical analysis of the documents. In Sect. 3, LDA topic modeling and topic evolution chart are presented. Section 4 displays the word co-occurrence network. Finally, discussion and conclusions are given in Sect. 5.

2   Descriptive Statistical Analysis of Party Congress Documents

2.1   Research Framework and Data Preprocessing

Since her establishment, the Communist Party of China (CPC), official report documents for each Party Congress are released, which serve as comprehen-

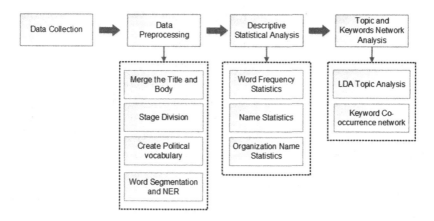

Fig. 1. Research framework.

sive records of the Party's core ideologies, strategic objectives, and tasks along her development. The official report documents of 19 Party Congresses of the CPC, from 1921 to 2020, were collected from the official database of the National Congress of the CPC[1]. This dataset contains 489 documents from each Congress Report, the Party Constitution, the Declaration, important resolutions, the Plenary Communique, etc. The report documents comprehensively summarize the key issues that capture the CPC's attention and the crucial guiding ideology at that stage.

Figure 1 is the research framework. After data collection, we merge the title and body of each document. For better understanding, as listed in Table 1, we divide the history of the Party Congress into six stages by combining the history and the leadership of the CPC. At different stages, the CPC has distinct objectives according to the prevailing challenges and circumstances within Chinese society. Table 1 lists the top leader and the objective of each stage. FoolNltk [2] is selected for word segmentation because of its high accuracy in natural language processing. We create a political vocabulary with 235 words for Party Congress documents based on the Sogou Pinyin [3] and related political word databases [5,6] to ensure word segmentation accuracy and prevent segmenting essential political words. Table 1 displays the document count and the word count per document (WPD) across each stage. For a total of 489 documents and 1,284,623 words, we filter out common stop words such as "我们", "然后","是", etc., and perform word segmentation and named entity recognition using FoolNltk. We analyse the documents of each stage to gain a better understanding of the CPC's governing philosophy.

---

[1] http://cpc.people.com.cn/GB/64162/64168/index.html.

[2] https://github.com/rockyzhengwu/FoolNLTK.

[3] https://pinyin.sogou.com/dict/.

Table 1. Stage division

St.	Time	Congress	Top Leader	Objective	#doc	#WPD
1	1921–1945	1st -6th	Chen & Mao	New-Democratic Mission	85	2908
2	1945–1977	7th - 10th	Mao	Socialist Revolution	79	3032
3	1977–1987	11th - 12th	Deng	Modernization Drive	76	1992
4	1987–2002	13th - 15th	Jiang	Chinese Socialism	119	2012
5	2002-2012	16th - 17th	Hu	Well-off Society	95	2631
6	2012-2020	18th - 19th	Xi	The Great Rejuvenation	35	4483
				Total	489	1284623

2.2   Word Frequency

The frequencies of those words in the documents are counted. We obtain a total of 20911 words and their corresponding frequencies after removing the common stop words. Figure 2 presents the top 20 words with the highest frequencies in the vocabulary of political terms. As shown in Fig. 2, the terms "Communist Party of China(中国共产党A)", "Central Committee(中央委员会B)", and "Proletariat(无产阶级C)" have the highest frequencies in the documents. Additionally, some important concepts, such as "Socialism with Chinese Characteristics(中国特色社会主义D)", "Reform and Opening-up(改革开放E)", "Modernization(现代化F)", "National Economy(国民经济G)", "Well-off Society(小康社会H)", "Mao's Thought(毛泽东思想I)", and "Deng's Theory(邓小平理论J)" also have high frequencies.

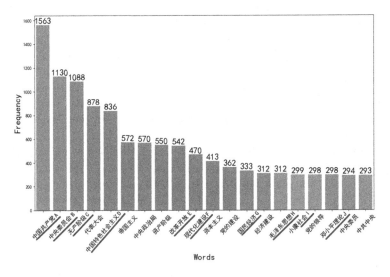

Fig. 2. The top 20 words with the highest frequencies in the vocabulary of political terms.

Figure 3 presents the word cloud, which are generated by Python tool word-cloud [4] after word segmentation. As shown in Fig. 3, during Stage 1, the words include "peasant(农民)", "labor union(工会)", "labour movement(工人运动)", "women's movement(妇女运动)", and "land revolution(土地革命)", which reflect that the CPC focus on the emancipation of the people at the bottom of society and the struggle against imperialism and feudalism at the early founding periods. In Stage 2, the primary words like "unity(团结)", "victory(胜利)", "country(国家)", "development(发展)", and "policy(政策)", indicate that with the end of the war, the CPC established the New China and committed to the national development. In Stage 3 and Stage 4, the words such as "reform and opening up(改革开放)", "societal development(社会发展)", and "institutional reform(体制改革)", etc.,suggest that during the reform and opening up, the CPC has emancipated the mind and productive forces, resulting in the development of the economy and the establishment of Socialism with Chinese characteristics. The words such as "deepening reform(深化改革)" and "great rejuvenation(伟大复兴)" in Stage 5 and Stage 6 infer that the main tasks during these stages are relevant with establishing a well-off society and enhancing ideological and cultural development, while maintaining economic development.

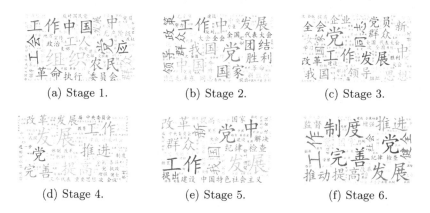

(a) Stage 1.	(b) Stage 2.	(c) Stage 3.
(d) Stage 4.	(e) Stage 5.	(f) Stage 6.

Fig. 3. Word clouds of Party Congress at various stages. The larger the font size, the more often the word appears.

2.3   Named Entity

The Party Congress reports contain many important entities such as personal names and organization names. Effective identification and statistics of these entities will help us understand the role and impact of leaders and organizations in different phase. We use the Python tool fool.analysis function to get the names of person and organizations in each stage.

---

[4] https://pypi.org/project/wordcloud/.

Table 2 lists top 10 names at each stage and the top 10 of the total documents based on frequency. The frequency of these names may represent these individuals' roles and impact during that particular stage. It is seen that top leader at each stage as listed in Table 1, all appear in respective stage. Mao's name occurs all through 6 stages and has the highest total frequency, indicating his role and great influence to the CPC.

Table 2. Statistics of names of Party Congresses at various stages

Stage 1		Stage 2		Stage 3		Stage 4		Stage 5		Stage 6		Total	
Name	Frq	Name	Frq	Name	Frq	Name	Frq	Name	Frq	Name	Frq	Name	Frq
陈独秀	43	毛泽东	601	毛泽东	181	江泽民	246	胡锦涛	179	习近平	206	毛泽东	1038
蒋介石	32	刘少奇	108	华国锋	123	邓小平	225	江泽民	102	毛泽东	51	邓小平	352
蔡和森	27	周恩来	80	邓小平	56	胡锦涛	122	毛泽东	65	胡锦涛	38	江泽民	348
张国焘	24	林 彪	60	胡耀邦	49	毛泽东	117	邓小平	33	李克强	17	胡锦涛	339
毛泽东	23	邓小平	38	叶剑英	44	赵紫阳	69	吴邦国	23	王岐山	14	习近平	214
周恩来	21	朱 德	29	陈 云	38	李 鹏	35	吴官正	13	刘云山	13	华国锋	134
瞿秋白	19	陈 云	28	刘少奇	34	乔 石	25	温家宝	13	王沪宁	11	刘少奇	110
李立三	15	陈伯达	22	李先念	32	吴邦国	25	李长春	12	赵乐际	11	周恩来	100
项 英	12	李富春	22	林 彪	29	温家宝	24	贾庆林	11	汪 洋	10	赵紫阳	95
吴佩孚	11	王洪文	21	赵紫阳	26	陈 云	18	曾庆红	11	张德江	10	林 彪	89

Table 3 lists the organizations that are frequently referred in the Party Congresses at different stages. The frequency reflects the role and impact of the organizations at that stage. For example, the organizations such as "Communist Party of China (中国共产党)" and "Central Committee(中央委员会)" are referred at all stages due to their importance. During Stage 2 and Stage 3, "People's Liberation Army of China (中国人民解放军)" and "Central Advisory Commission (中央顾问委员会)" appear for the first time in the documents of respective stage and are mentioned multiple times. After Stage 4 the names of the organizations are stable.

3   Topic Analysis of Party Congress Documents

Precious section presents the basic statistics of words and entities used in the reports. This section is to mine topics and their evolution.

3.1   Topic Modeling and Topic Analysis

LDA is used to generate topics for each stage and the overall documents. The vocabulary of political terms constructed during data processing and FoolNltk are used for text segmentation, and the gensim library[5] in Python is used to generate the topics of each stage with the premise of setting the number of topics to 4 based on the coherence test.

---

[5] https://pypi.org/project/gensim/.

Table 3. Statistics of organizations of Party Congresses at various stages

Stage 1		Stage 2		Stage 3	
Organizations	#	Organizations	#	Organizations	#
中国共产党	363	中国共产党	314	中央	235
中央	146	中央	258	中国共产党	147
国民党	145	中央委员会	141	中央委员会	134
中央执行委员会	144	中央政治局	54	中央政治局	55
苏维埃	93	中国人民解放军	42	中央纪律检查委员会	35
中央委员会	64	国民党	33	中央顾问委员会	29
中国共产主义青年团	51	国务院	21	中央书记处	25
省委员会	47	中华人民共和国	19	中华人民共和国	16
县委员会	36	国民党政府	19	国务院	15
Stage 4		Stage 5		Stage 6	
Organizations	#	Organizations	#	Organizations	#
中央	551	中央	262	中央	188
中央委员会	402	中国共产党	244	中国共产党	183
中国共产党	360	中央委员会	236	中央委员会	177
中央纪律检查委员会	212	中央政治局	169	中央政治局	76
中央政治局	158	中央纪律检查委员会	36	中央纪律检查委员会	54
中央顾问委员会	112	中共中央军事委员会	32	全国政协	20
全国政协	52	全国政协	25	中央政治局常务委员会	13
中央军事委员会	48	国务院	20	中央军事委员会	9
国务院	42	中央军事委员会	10	全国人大常委会	6

Table 4 lists the topics and main words for all documents from the 1st to 19th CPC National Congress. Based on these words, we label these four topics as "Revolutionary movement(革命运动)", "Party & Political system(党政建设)", "National development (国家发展)", and "Societal development (社会发展)".

1. **Revolutionary movement**: This topic includes the words related to the early revolutionary movements, such as the women's movement, the youth movement, the peasant movement, etc.
2. **Party & Political system**: This topic concentrates mainly on building the fundamental system of the CPC, including the Party's political, discipline, ideological, and organizational constructions.
3. **National development**: This topic is relevant to national development, including some related words such as modernization, reform and opening-up, and Socialism with Chinese Characteristics.
4. **Societal development**: This topic which includes the words like "well-off society" and "harmonious society", suggests that the CPC's contribution to the societal development, such as the economic and societal progress.

Similarly, we generate the topic by LDA of those documents at each stage. The number of topics is 4. The results are listed in Table 5. "youth movement",

Table 4. The topics generated from the 489 documents of the whole Party Congress

Topics label	Topic words
Revolutionary movement (革命运动)	Women(妇女), youth(青年), peasants(农民), revolution(革命), agriculture(农业), movement(运动), line(路线), masses(群众), imperialism(帝国主义), proletariat(无产阶级)
Party & Political system(党政建设)	CPC Central Committee(中共中央), military(军事), Political Bureau of the Central Committee(中央政治局), governance(执政), resolutions(决议), Party Constitution(党章)
National development (国家发展)	Reform and opening up(改革开放), modernization(现代化建设), Socialism with Chinese characteristics(中国特色社会主义), important ideas(重要思想), system(体制), promotion(推进)
Societal development (社会发展)	Promotion(推进),enterprise(企业),小康社会(well-off society), 创新(innovation),开放(opening up),执行(implementation), 和谐社会(harmonious society),检查(inspection),群众(masses)

"women's movement", "worker-peasant alliance", and "institution construction" represent the four topics of the documents in Stage 1. Consistent with the result of word frequency, three among 4 topics refer to the new democratic revolutionary. In the documents of Stage 2, the four topics are related to national institution construction and economic development before and after the establishment of the new China, such as the development of industry and agriculture. In Stage 3, in addition to the topics related to the construction of party and national institution, the report documents also contain the "Modernization Construction", related to promoting societal development. Likewise, besides institutional construction, the subsequent stages cover topics related to the reform and opening-up, harmonious society and common prosperity. These topics are closely associated with the primary goals of societal development and modernization.

3.2   Topic Evolution Analysis

With generated topics, we investigate the potential correlations among those topics. In order to explore the relationship among the topics of report documents, we utilize the Jaccard similarity to determine the correlation between each topic. Niwattanakul et al. [17] discovered that the Jaccard similarity can effectively convey the correlation between two-word sets. The calculation formula is as follows:

$$J(A, B) = \frac{|A \bigcap B|}{|A \bigcup B|},$$

where A and B denote the set of topic words generated by LDA. The greater the Jaccard similarity is, the more words are shared between the two topics and the higher the correlation is. Hence, the algorithm of topic evolution graph is generated through the following steps [11]:

Table 5. The topics of each stage's documents

St	Topic label	Main words
1	Youth movement	Students(学生), struggle(斗争), imperialism(帝国主义)
	Women's movement	Women(妇女), opposition(反对), labour(劳动)
	Worker-peasant Alliance	Trade unions(工会), class struggle(阶级斗争), land(土地)
	Institution construction	Congresses(大会), programs(程序), supervision(监督)
2	Economic development	Five-Year Plan(五年计划),economy(经济),industry(工业)
	Institution construction	PBCC(中央政治局), secretary(书记), line(路线)
	Agricultural Cooperation	Agriculture(农业), cooperatives(合作社),output(产量)
	Ideological Construction	Thought(思想),New Democracy(新民主主义)
3	Rectification	Rectification(整改), criticism(批评), discipline(纪律)
	Election	Confirmation(确认), election(选举), promotion(推选)
	Institution construction	Deliberation(审议), resolution(决议), modification(修改)
	Modernization	Modernization(现代化), economy(经济), society(社会)
4	Ideological construction	Chinese characteristics(中国特色), Socialism(社会主义)
	Organization construction	Presidium(主席团),Political Consultative(政协)
	Reform and opening-up	Reform and opening up(改革开放), market(市场)
	Resolution and decision	Amendment(修正案), discussion(讨论), promotion(推举)
5	Institution construction	Committee(委员会), secretariat(书记), resolution(决议)
	Institutional reform	Governing(执政), management(管理), perfection(完善)
	Societal development	Harmonious society(和谐社会),moral(道德),service(服务)
	Primary-level organization	Rural(农村), agriculture(农业), government(政府)
6	Political construction	Rule of law(法治),constitution(宪法), implement(执行)
	Common prosperity	Market(市场), wellbeing(惠及), embodiment(体现)
	Work style construction	Governance(治理),anti-corruption(反贪),legislation(法规)
	Institution construction	Judiciary(司法), mechanism(机制), improvement(完善)

1. Generate topics. We generate overall topics set Z for 489 documents and local topics set T_i for the documents of stage i, where $i = 1, 2..., n$. The top 50 words with the highest probability are chosen to represent each topic. This task has been conducted as described in Sect. 3.1.
2. Determine topic lines. For local topic $t_{ij} \in T_i$, where $j = 1, 2..., |T_i|$, and overall topic $z_k \in Z$, where $k = 1, 2..., |Z|$, the Jaccard similarity $J(t_{ij}, z_k)$ is used to assess the similarity between the local topics and the overall topic. We set the threshold α to determine their similarity. If $J(t_{ij}, z_k) > \alpha$, then topic t_{ij} and topic z_k are considered similar.
3. Determine merging and splitting of topics. The similarity of topic $z_{i,j} \in T_i$ and $z_{i+1,l} \in T_{i+1}$ is computed by Jaccard similarity $J(z_{i,j}, z_{i+1,l})$. A threshold β is used to determine the connection between topic $z_{i,j}$ and topic $z_{i+1,l}$. If $J(t_{i,j}, t_{i+1,l}) > \beta$, part of topic $t_{i,j}$ are continuous at topic $t_{i+1,l}$.
4. Generate the topic evolution graph. Using the topic and relationship computed above, we generate a Sankey diagram to illustrate the topic evolution,

where the nodes in the graph represent the topics, and the edges represent the connection of topics.

Here we set the threshold α to 0.25 and β to 0.15. The topic evolution graph of the Party Congress reports is as shown in Fig. 4, where the nodes represent topics at each stage, and the width of edge represents the similarity between the connected topics. The legends in Fig. 4 correspond to the four overall topics of the documents, with blue standing in for the "Revolutionary movement", green for the "Party & Political institution", yellow for the "National development", and orange for the "Societal development". The local topics with the same color means that they are similar to the same overall topic.

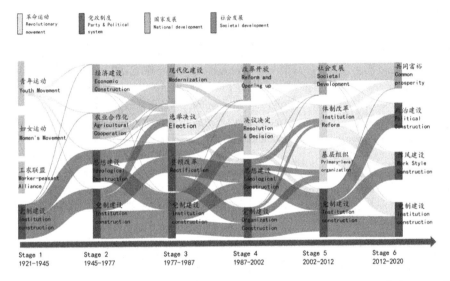

Fig. 4. Topic evolution at each stage. The four legends are the topic of the overall document. The topic color in each stage corresponds to the overall topic with the highest correlation, which is calculated by Jaccard similarity.

As illustrated in Fig. 4, only the topics of Stage 1 are related to revolutionary movements. Those topics at the following stages focus on societal development, national development, and Party system construction. Obviously no flows from the three topics related to the revolutionary movement in Stage 1 to any topics in Stage 2. Actually, the obvious transition in the topics clearly reflects the notable differences in priorities of tasks and statuses of the CPC. From Stage 2 to Stage 6, there are considerable connections between the topics of adjacent stages, indicating the presence of the same words among these topics. In other words, a clear evolution exists among these topics. It is noteworthy that while the topics are undergoing changes, their primary content is essentially consistent, relating to development and institutional construction. Thus, through the topic evolution diagram, we observe the evolution of the topic from the CPC Party

Congress documents over the past century. There are significant changes from Stage 1 to Stage 6, with some topics disappearing, while the topic of Party system construction is continuous. The topics become stable after the reform and opening-up, national development, and economic development.

4 Word Co-occurrence Network Analysis

The analysis of LDA topic evolution in the previous section shows a trend among topics in each stage. In this section, we take a closer look at how words have changed over time. Based on prior research and Sogou political dictionaries, we create a political vocabulary with 235 words in Sect. 2.1. These words hold significant meaning in the history of the CPC. We construct a co-occurrence network for the words, establishing a connection between two words when they appear within the same paragraph. To explore the occurrence time of words, we classify them according to the stage at which they first occur. The word co-occurrence network is as shown in Fig. 5.

The top 30 words with highest degree are listed in Table 6. The word "Marxism-Leninism(马克思列宁主义B)" has the highest degree in the word co-occurrence network, indicating its significance in the report documents. Words like "socialism(社会主义C)", "reform and opening up(改革开放O)", "economic construction(经济建设L)", and "modernization (现代化建设P)", highlight the importance of economic development. Moreover, some words associated with party system construction exhibit a high degree, such as "Mao's Thought(毛泽东思想J)", "Deng's theory(邓小平理论S)", and "Three Represents(三个代表T)".

Table 6. Words and degrees of word co-occurrence network

Word	Degree	Word	Degree	Word	Degree
马克思列宁主义B	263	毛泽东思想J	143	实事求是	127
社会主义C	197	中央委员	143	重要思想	126
中国共产党	176	国民经济	141	民主政治	125
代表大会	155	社会发展	141	基本路线	125
改革开放O	152	党的建设	139	统一战线	125
经济建设L	151	科学发展观	136	对外开放	124
党的领导	149	贯彻落实	132	小康社会	124
中国特色社会主义	148	邓小平理论S	132	民主集中制	122
中央委员会	144	市场经济	129	初级阶段	122
现代化建设P	144	三个代表T	127	中共中央	122

As shown in Fig. 5, the documents of Stage 1 incorporate a large number of words about Party and national construction, such as "Communist Part

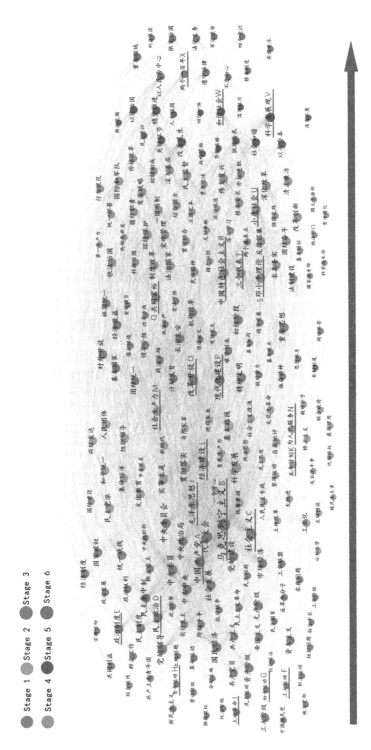

Fig. 5. The Co-occurrence Network of words. The node color corresponds to the stage in which the word occurs for the first time.

of China(中国共产党A)", "Marxism-Leninism(马克思列宁主义B)", "social-ism(社会主义C)", "democratic politics(民主政治D)","political institution(政治制度E)", etc. These words have higher degrees, illustrating that they also regularly appear in other stages and are of central roles since the Party's founding. In addition, the words like "labour movement(工人运动F)", "women's movement(妇女运动G)", "youth movement(青年运动H)", and "land revolution(土地革命I)" are found in Stage 1, whereas their degrees are smaller, indicating that they almost exclusively appear in Stage 1 and are infrequently mentioned in later stages. In Stage 2, words like "Mao's Thought(毛泽东思想J)", "Five-year plan(五年计划K)", "economic construction(经济建设L)","social productivity(社会生产力M)" and "serve the people(为人民服务N)" reflect new ideologies and objectives about the societal and economic development. In Stage 3 and Stage 4, during the initial dedication to modernization and reform and opening-up, several corresponding new words emerged, such as "reform and opening up(改革开放O)", "modernization (现代化建设P)", "common prosperity(共同富裕Q)", "Socialism with Chinese characteristics(中国特色社会主义R)", "Deng's theory(邓小平理论S)", "Three Represents(三个代表T)","well-off society(小康社会U), etc. In the subsequent stages, the words like "Scientific Outlook on Development (科学发展观V)", "harmonious society(和谐社会W)", "two centenary goals(两个一百年X)", etc., suggest that the CPC aims to elevate the national development to a higher level. The emergence of these new words implies that the CPC's governing philosophy is adaptable to the times. Furthermore, the words that appear at each stage are found to co-occur with preceding words, demonstrating coherence in the CPC's governing philosophy.

5   Discussion and Conclusions

The Party Congress report documents outline the Party's main focuses and the forthcoming phase's objectives, tasks, and implementation principles. A detailed analysis of these documents helps us understand the CPC's construction and governing philosophy.

We collect the official documents from the database of the previous National Congress of the CPC and divide the Party Congress into six stages based on history and leaderships. For each stage, descriptive statistical analysis including the word cloud and named entity, are used to analyze the content of the Party Congress documents and the characteristics in the leadership collective and important organizations. Furthermore, to explore the evolution within Party Congress documents at each stage, we construct the topic evolution graph using LDA and Jaccard similarity, and generate a word co-occurrence network to analyze the changes in the ideology and action of the CPC over time. The ultimate results indicate that the governing philosophy of the CPC exhibits a progressive evolution over time and a holistic coherence.

More generally, by providing the topic evolution chart and word co-occurrence network, our study affords a variety of insights into the working philosophy of the CPC. For example, in terms of national economic development, a discussion that initially revolves around industry and agriculture, subsequently transitions into a deliberation on modernization, and ultimately shifts towards reform and opening-up. We recognize that although the words may change, these topics consistently revolve around the discussion of the national economy. In our study, this relationship is established through LDA and topic associations. Although continuity is observable in the CPC's history, strategical transformation is also present. Our study highlights a distinct transition from the 20th century to modernity in different periods, driven by China's prevailing national conditions.

Different from the traditional approach to political texts, we apply an analytical framework to study major official Party Congress documents. Not only the working philosophy of the CPC at different phases is illustrated clearly, its evolution is also highlighted by the visualization of topic evolution and word co-occurrence network. We hope to find more effective ways for better understanding toward the CPC by interdisciplinary viewpoints.

References

1. Yang, G.: The communist party of China and the Chinese road to modernisation. Econ. Polit. Stud. **10**(1), 1–8 (2022)
2. Liu, S., Xiong, X.: The Chinese path to modernisation: its universality and uniqueness. Econ. Polit. Stud. **11**(1), 1–16 (2023)
3. Li, S.N.: Text analysis of the report from the first congress to the 18th congress of the CPC - on the transformation of the communist Party of China from a revolutionary party to a ruling party. Inherit. Innov. **03**, 16–17 (2016). (In Chinese)
4. Wang, Z.: The change from the revolutionary party to the ruling party -the analysis based on the party constitution of the eighth to the eighteenth national congress of the CPCl. Anhui Rural Revitalization Stud. **8**(04), 32–37 (2017). (In Chinese)
5. Guo, D.J., Wu, H.B.: How the communist party of china fulfill its original mission in the past 100 years: based on the text analysis of all previous congress reports. J. Soc. Sci. Hunan Normal Univ. **50**(03), 22–33 (2021). (In Chinese)
6. Wu, J.P.: The modern evolution of the idea of party building since the reform and opening up - based on the text analysis of the report of the party congress and the plenary session of the party central committee. Journal of the Party School of C.P.C. Qingdao Municipal Committee and Qingdao Administrative Institute, (2019). (In Chinese)
7. Dong, S.T., Yang, L.J.: The change of the democratic discourse system of the communist party of china since the reform and opening-up based on the text measurement and discourse analysis of the report of the 12th to 19th party congresses. Theory and Reform (2019). (In Chinese)
8. Li, J.B., Qi, J.: Research on the theory of common prosperity of Chinese socialism in the new era - text analysis on the reports of the previous party congress since the reform and opening up. J. Chongqing Univ. Technol. Soc. Sci. 1–22 (2022). (In Chinese)
9. Blei, D.M., Ng, A.Y., Jordan, M.I.: Latent dirichlet allocation. J. Mach. Learn. Res. 3(Jan), 993–1022 (2003)
10. Blei, D.M.: Probabilistic topic models. Commun. ACM **55**(4), 77–84 (2012)

11. Yan, Z., Tang, X.: Understanding shifts of public opinions on emergencies through social media. In: Chen, J., Huynh, V.N., Nguyen, G.-N., Tang, X. (eds.) KSS 2019. CCIS, vol. 1103, pp. 175–185. Springer, Singapore (2019). https://doi.org/10.1007/978-981-15-1209-4_13
12. Huang, X.H., Lu, Y., Tang, X.J.: Multi-perspective analysis of public opinion related to COVID-19 based on online media. J. Syst. Sci. Math. Sci. **41**(08), 2182–2198 (2021)
13. Rule, A., Cointet, J.P., Bearman, P.S.: Lexical shifts, substantive changes, and continuity in State of the Union discourse, 1790–2014. Proc. Natl. Acad. Sci. **112**(35), 10837–10844 (2015)
14. Wei, W., et al.: Textual topic evolution analysis based on term co-occurrence: a case study on the government work report of the State Council (1954–2017). In: 2017 12th International Conference on Intelligent Systems and Knowledge Engineering (ISKE), 1–6. IEEE (2017)
15. Barna, I., Knap, Á.: Analysis of the thematic structure and discursive framing in articles about Trianon and the holocaust in the online hungarian press using LDA topic modelling. Nationalities Pap. **51**(3), 603–621 (2023)
16. de Campos, L.M., Fernandez-Luna, J.M., Huete, J.F., et al.: LDA-based term profiles for expert finding in a political setting. J. Intell. Inf. Syst. **56**, 529–559 (2021)
17. Niwattanakul, S., Singthongchai, J., Naenudorn, E., et al.: Using of Jaccard coefficient for keywords similarity. Proc. Int. Multiconf. Eng. Comput. Scientists **1**(6), 380–384 (2013)

Analysis of Factors Influencing the Formation of Agricultural Science and Technology Collaborative Innovation Network: Empirical Evidence from ERGM

Shanshan Hu[1,2] and Zhaogang Fu[1,3(✉)]

[1] Lingnan Normal University, Zhanjiang 524048, Guangdong, China
fuzg@lingnan.edu.cn
[2] Guangdong Coastal Economic Belt Development Research Center, Zhanjiang 524048, Guangdong, China
[3] South China University of Technology, Guangzhou 510006, Guangdong, China

Abstract. Drawing upon data regarding cooperative patents applied for and authorized in the agricultural industry between 2019 and 2022, this paper conducts an analysis of the collaborative innovation network in agricultural technology involving agricultural enterprises, universities, and research institutes. From a social network perspective, this study empirically examines the impact relationships within agricultural science and technology collaborative innovation networks, employing the Exponential Random Graph Model (ERGM) across two dimensions: endogenous network configuration and exogenous node attributes. The findings reveal that both internal and external factors contribute to the dynamics of the agricultural technology collaborative innovation network, resulting in a significant sparsity effect. This effect gives rise to triangular and star-shaped structures that facilitate resource sharing. Furthermore, the agricultural science and technology collaboration network exhibits a high degree of openness and regional assortativity, enhancing the likelihood of connections between organizations. However, the influence of innovation capability and organization type on network promotion is less pronounced.

Keywords: Agricultural technology collaboration · Innovation network · Social networks · ERGM

1 Introduction

Diverse resources and knowledge-sharing platforms have opened up new avenues for scientific and technological innovation in the agricultural sector. They enable various innovative entities to progress towards networked collaboration, fostering an innovation network that integrates government, industry, academia, research, and application. The establishment of cooperative platforms, such as agricultural science and technology parks and incubators, has accelerated cross-disciplinary and cross-industry collaborative innovation, emerging as powerful drivers of innovation. Simultaneously, this innovation

J. Chen et al. (Eds.): KSS 2023, CCIS 1927, pp. 230–245, 2023.
https://doi.org/10.1007/978-981-99-8318-6_16

network not only establishes a standardized and formal cooperative structure among agricultural enterprises, universities, and research institutes but also achieves synergy through resource, knowledge, and technology sharing. This collaborative innovation network effectively mitigates the risks and reduces the costs associated with research and development activities while enhancing resource utilization efficiency. Consequently, collaborative innovation networks have become a prominent focus area. In recent years, the integration of the agricultural and knowledge industries has grown increasingly interconnected, presenting a complex landscape of networked cooperation. Within the context of open innovation, the agricultural technology collaborative innovation network, as a diverse, intricate, and uncertain system, encounters challenges in realizing collaborative effects. This system involves multiple actors with distinct characteristics, and various innovative elements interact and collaborate, making it challenging to attain the desired synergistic outcomes. Throughout the collaborative process, each entity is influenced by various factors, including the individual attributes of network nodes, environmental variations, and the structural characteristics of network connections. These factors, in turn, impact the collaborative outcomes. Therefore, identifying the key determinants affecting the formation of collaborative innovation networks and elucidating their degree of influence, as well as their internal mechanisms, is a fundamental step in promoting collaborative effects. This understanding plays a pivotal role in enhancing collaborative innovation networks, achieving organizational innovation objectives, and boosting innovation efficiency.

In light of this, this study utilizes the agricultural technology industry as a case study to explore the factors that drive the formation of collaborative innovation networks among diverse organizations, including agricultural enterprises, universities, and research institutes, and assesses the extent of influence exerted by each factor.

2   Literature Review

2.1   Collaborative Functions of Agricultural Science and Technology

A large number of scholars have conducted research on the functions of technology systems, and collaborative innovation networks are mostly based on interdisciplinary cooperation, including cooperation between different disciplines and cooperation between similar subject entities with different advantages (Hermans et al., 2015). A discipline is a collection of knowledge that humans divide into categories based on the object of knowledge in their understanding and research. It is a relatively independent scientific knowledge classification system (Zastempoeski and Cyfert, 2021). Some scholars have explained the collaborative function of agricultural technology, and Wang (2021) pointed out that agricultural technology collaborative innovation networks have been widely applied in different regions and have shown great potential. The collaborative function of agricultural technology is to achieve the integration of interdisciplinary knowledge through this management model.

2.2 Research Methods for Collaborative Innovation Networks

The relationships between innovation subject nodes such as research institutions exhibit significant heterogeneity, forming a complex network system (Meshesha, 2022). Fieldsend et al. (2020) utilized fuzzy evaluation and found that the structure and relationships of innovation networks are one of the key factors affecting their formation. Scott (2015) used expert scoring to study the impact relationship and mechanism of innovation factors. Ingram et al. (2020) found that structural equation models have strong consistency constraints, making it difficult for data obtained through questionnaire surveys to pass model validation. Feder (1998) proposed a concise expression for the system backbone matrix. Frank and Strauss (1986) proposed an ERGM, which can comprehensively consider endogenous structural factors and exogenous factors such as actor relationships in the network.

2.3 Cooperative Models of Collaborative Innovation Network Entities

Cooperation is the foundation for forming collaborative innovation networks. Yongabo and Goktepe (2021) applied game theory to construct an infinite repeated game model for knowledge collaboration between enterprises, and conducted in-depth analysis of the conditions for knowledge collaboration. A systematic model of industry university research cooperation has been established, which can serve as a basis for policy formulation by the participating parties in industry university research cooperation (Smith et al., 2021; Kiminami et al., 2022). Alirah et al. (2018) divide the cooperation models of innovation network entities into technology transfer through industry university research cooperation, short-term project-based cooperation, and joint construction of business entities.

2.4 Research Review

Previous research on collaborative innovation within and between enterprises, as well as among industry, academia, and research institutes, has produced substantial findings, offering both theoretical and methodological support for the study of collaborative innovation networks in agricultural science and technology. Agricultural collaborative innovation networks possess unique characteristics in terms of their composition, objectives, functions, and modes of cooperation. While existing studies have empirically explored the impact mechanisms of collaborative innovation networks on innovation activities, they have generally overlooked the antecedent variables influencing the formation of these networks. Some scholars have delved into research on the motivations behind the formation of collaborative innovation networks, albeit primarily employing traditional statistical methods to analyze how network nodes affect the single factor of network formation. Hence, this article integrates endogenous and exogenous variables within a unified research framework. It treats the probability of node connections within the collaborative innovation network as the dependent variable and employs network configuration and node attribute indicators as explanatory variables. The aim is to analyze the influencing factors of the agricultural technology collaborative innovation network comprehensively.

3   Research Design

3.1   Data Sources

Patent has become an important tool for enterprises to protect their own rights and interests, and joint patent application has also become an important process to match the knowledge output needs and application needs of enterprises, universities and research institutions. This study selected agricultural enterprises, universities and research institutions participating in agricultural industry patent cooperation as research samples to conduct network structure analysis and empirical research.

Data source: We searched the invention patents applied for and authorized by 31 provinces, autonomous regions and municipalities in China (excluding Hong Kong, Macao and Taiwan) during 2019–2022 from the Patent Search website of the State Intellectual Property Office of China (CNIPA) as research objects.

Data screening: We screened the retrieved patent data and selected patents related to collaborative innovation in agricultural technology as the sample data set. In the selection process, we refer to the OECD Agricultural Industry IPC classification number, patent abstract input "agriculture", patent claims input "technology" and other patent information to ensure that the selected patents are related to collaborative innovation in agricultural technology.

Partner identification: Based on the information of the patent applicant and patent holder, we identify patents involving cooperation between multiple units, and identify the name and nature of each partner (such as agricultural enterprises, universities, research institutes, etc.).

Data organization: We organized the selected patent data of cooperative inventions into a data table, and added some basic information of each cooperative unit (such as the location and size of the unit) to facilitate subsequent data analysis and statistics.

Among them, the total number of cooperative authorized invention patents was 6,245, excluding patent data that did not meet the requirements (that is, excluding duplicate projects, individuals, Hong Kong, Macao, Taiwan and foreign applicants), a total of 4,782 cooperative authorized patents, and 1973 patent applicants (network nodes).

3.2   Variable Selection and Description

When constructing the ERGM model, it is necessary to select appropriate network variables to characterize the structural characteristics of the agricultural technology collaborative innovation network. Commonly used variables include the degree of the node (i.e. the number of connections), the attributes of the node (such as organizational type, geographical location, etc.), and the type of cooperative relationship (two-way cooperation, one-way cooperation, etc.). In addition, it is also necessary to manipulate the original data and convert the nodes and edges in the network into binary variables or continuous variables to facilitate the establishment of ERGM model.

Endogenous Network Configuration Variables. Based on the characteristics of the collaborative innovation network in this study, and drawing on previous research, we selected typical network configuration indicators from existing studies as endogenous variables. Among them, the number of edges (E_d) is a control variable in network modeling and plays a reference role in each model; Interactive Triangle Structure (T_i), Geometric Weighting Degree (G_w), Geometric Weighted Edge Sharing Partner (G_e), and Geometric Weighted Binary Partner (G_D) are dependency terms in network configuration, and their existence can prevent network degradation. The interactive triangle structure depicts the tendency of three nodes to form a connectivity triangle structure, that is, the structure in which three nodes are partners in a collaborative innovation network, which can affect the probability of network edge formation; The expansion effect and star structure of the geometric weighting observation network, i.e. the degree of redundancy formed in the network, represent the tendency of each agricultural enterprise (university or research institute) as a central node to directly connect with multiple other nodes in the collaborative innovation network; Geometric weighting characterizes the transitive features corresponding to clustering in a network from different perspectives, referring to the situation where a node forms a triangular structure and shares edges with multiple other nodes; Geometrically weighted binary partners measure the number of binary groups with shared partnerships, referring to the tendency of agricultural enterprises, universities, and research institutes to have at least one path 2 network connection, or the tendency of nodes to act as intermediaries in the network.

Exogenous Node Attribute Variables. Among the node attribute variables, there are both variables that measure the characteristics of the node itself and interaction terms that can reflect the interaction characteristics of two interconnected members in the network. These characteristics of the node play different roles in the formation of collaborative innovation networks. Therefore, exogenous factors are considered from two aspects: the node's own characteristics and feature interaction terms.

Firstly, the attribute variables of the node itself are selected as agricultural technology collaborative innovation capability (C_i), cooperation breadth (C_b), and cooperation depth (C_d). The reason is that the collaborative innovation network focuses on the mutual cooperation between participating entities, and the collaborative innovation capability reflects the organization's ability to integrate and absorb resources, while cooperation breadth and cooperation depth are important indicators to measure its cooperation openness; Only by building a cooperative openness that matches the integration and absorption capabilities of innovative entities can they fully leverage their resource and location advantages in the collaborative network. Among them, collaborative innovation capability is represented by the number of cooperative patents owned by nodes, and the larger the indicator, the easier it is to attract more nodes to connect with it; The breadth of cooperation among nodes in a network refers to the number of nodes they are connected to, while the depth of cooperation refers to the number of times that node is connected to all other nodes.

Secondly, the variables of organization type matching (O_m) and matching of subordinate region (M_R) are selected from the interaction items of node attributes. The reason is that the entities participating in the collaboration have the same or different organization types and affiliation regions. The same organization type makes them have similar organizational goals, while the affiliation regions make them have the same opportunities to obtain different resources. These innovative entities' willingness to seek cooperation in the network will be more similar. Among them, the same organizational type refers

Table 1. ERGM variable description table.

	Variable	Measuring method	Effect	Abbreviation
Endogenous network Configuration variables	Edge	Number of network edges	/	E_d
	Interactive triangle	Number of interactive triangles	Closed	T_i
	Geometric weighting	Number of star structures	Extensibility	G_w
	Geometrically weighted binary shared partner	The number of structures in the network where nodes undertake intermediary functions	Intermediation	G_b
	Geometrically weighted edge sharing partner	Number of structures where two organizations have common partners	Transitivity	G_e
Exogenous node attribute variables	Collaborative innovation capability	Number of cooperative authorized patents owned by nodes	Matthew's effect	C_i
	Cooperation breadth	Degrees of nodes for removing duplicate edges	Matthew's effect	C_b
	Cooperation depth	The degree of nodes with non overlapping edges divided by the breadth of cooperation	Matthew's effect	C_d

(*continued*)

Table 1. (*continued*)

	Variable	Measuring method	Effect	Abbreviation
	Organization type matching	The number of edges at both ends belonging to the same organization	Assortativity	O_m
	Matching of subordinate regions	The number of edges where both ends belong to the same province	Assortativity	M_s

to two nodes belonging to agricultural enterprises (universities or research institutes); If the regions belong to the same configuration, it means that two nodes are in the same province. The specific representation and measurement method of each variable, namely the variable description of ERGM, are shown in Table 1.

Some endogenous network configuration variables can be shown (see Fig. 1).

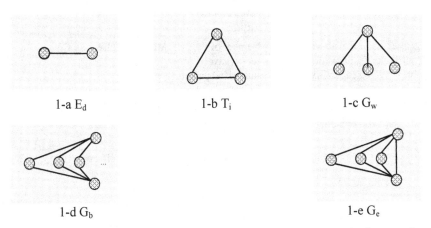

1-a E_d 1-b T_i 1-c G_w

1-d G_b 1-e G_e

Fig. 1. The macro button chooses the correct format automatically. Schematic diagram of some endogenous network configuration variables

3.3 Research Methods and Model Design

This study used ERGM (Exponential Random Graph Models) method to empirically study the influencing factors of the formation of agricultural technology collaborative innovation network. ERGM is a network model-based analysis method that can better simulate the structure and variation of non-random networks. Compared with other network analysis methods, ERGM can better adapt to the diversity of network nodes, and is not limited by network size, density, core nodes, etc. It can also accurately simulate the probability distribution of network nodes and edges, and fit the network structure based on maximum likelihood method. However, ERGM model has high requirements on data quality, and data missing or noise will affect the analysis results of the model (Jiang et al., 2023).

The agricultural technology Collaborative innovation network is a non-random network, for example, there are different types of nodes, such as experts, government officials, and farmers, and the connections between the nodes are not random. ERGM method can adapt to this non-random network well, reflect the structure and change patterns of agricultural technology collaborative innovation network more accurately, and speculate and discuss these patterns based on data. At the same time, ERGM method can accurately test the hypothesis of the network and infer the evolution and expansion law of the network. Therefore, it is scientific and feasible to use ERGM method to analyze agricultural science and technology collaborative innovation network.

The general form of ERGM is:

$$P(Y|y) = \frac{exp\{\vartheta_t g(y, X)\}}{k} \tag{1}$$

In formula (1), Y represents a random network set of collaborative innovation networks. When there is a connection between any two points i and j, $Y = 1$, otherwise $Y = 0$; Y represents a specific implementation in the network, that is, the observed real network structure; X represents the exogenous variable in the collaborative innovation network, and in this study, it refers to the attribute variables of each node; G (y, X) represents the vector composed of various statistics, including endogenous variables of network configuration (edge, triangular structure, star structure, etc.), node attributes (collaborative innovation ability, cooperation breadth, cooperation depth), and node attribute interaction (organization type matching, membership region matching); k represents a constant, ensuring that the probability of forming the collaborative innovation network structure is between 0 and 1; ϑ_t representing the parameter to be estimated (t represents the number of parameters to be estimated), which corresponds to the network structure statistic parameter, is the main focus of subsequent empirical research. The significance and magnitude of ϑ_t are used to determine the degree of influence of different factors on the formation of collaborative innovation networks. In addition, the goodness of fit of the model is judged by the *AIC* and *BIC* indicators. The smaller the values of the two indicators, the closer the model fits to the collaborative innovation network observed in reality. This study uses Stantnet package in R language to construct exponential random graph model and model fitting, using Markov chain Monte Carlo Maximum likelihood estimation.

Substitute formula 1 with a simple variable to obtain model 1, and construct a series of models based on model 1.

Model 1 is initial model, which is a simple random graph model that only includes the statistical terms of edges (E_d) in the network, namely the number of four types of cooperative relationships in the collaborative innovation network: agricultural enterprise agricultural enterprise, agricultural enterprise university, agricultural enterprise research institute, and agricultural enterprise university research institute. The construction of the initial model will provide a reference basis for the improvement of subsequent complex models. The corresponding ERGM model 1 is formula 2:

$$P(Y = y|X) = \frac{exp(\vartheta_1 E_d)}{k} \tag{2}$$

Model 2 adds an endogenous configuration variable dependency term to Model 1 to focus on the complex relationships that exist in the network. In Model 2, we introduced four indicators: T_i, G_w, G_b, and G_e to test the impact of star shaped and triangular structures formed by the connections between universities, enterprises, and research institutes in the network on the formation of collaborative innovation relationships. Specifically, indicator T_i is used to capture the triangular structure that exists in the network, that is, the situation where agricultural enterprises cooperate with universities and research institutes simultaneously in cooperation. Indicator G_w measures the intensity of cooperation between nodes, that is, the weight of the cooperative relationship. G_b and G_e are used to examine the impact of binary shared partnerships and edge shared partnerships of nodes, respectively. The corresponding ERGM model 2 is formula 3:

$$P(Y = y|X) = \frac{exp(\vartheta_1 E_d + \vartheta_2 T_i + \vartheta_3 G_w + \vartheta_4 G_b + \vartheta_5 G_e)}{k} \tag{3}$$

Model 3 is a model that adds exogenous node attribute variables to Model 1. On the one hand, starting from the characteristics of nodes themselves, Model 3 characterizes whether the probability of forming cooperative relationships is influenced by C_i, C_b, C_d, and to what extent. On the other hand, starting from the interaction terms of node features O_m and M_s, Model 3 estimates the number of edges with the same attribute endpoints and the impact of these same attributes on the probability of forming new cooperative relationships in the network. The corresponding ERGM model 3 is formula 4:

$$P(Y = y|X) = \frac{exp(\vartheta_1 E_d + \vartheta_6 C_i + \vartheta_7 C_b + \vartheta_8 C_d + \vartheta_9 O_m + \vartheta_{10} M_s)}{k} \tag{4}$$

Model 4 is a comprehensive model that incorporates network configuration variables and node attribute variables into the same statistical model, comprehensively analyzing the impact of different types of statistical items on the formation of collaborative innovation relationships. By combining network configuration variables and node attribute variables, Model 4 can simultaneously consider the impact effects of different factors on the formation of collaborative innovation networks, thereby more comprehensively explaining the structure and characteristics of the network. The corresponding ERGM model 4 is formula 5:

$$P(Y = y|X) =$$
$$\frac{exp(\vartheta_1 E_d + \vartheta_2 T_i + \vartheta_3 G_w + \vartheta_4 G_b + \vartheta_5 G_e + \vartheta_6 C_i + \vartheta_7 C_b + \vartheta_8 C_d + \vartheta_9 O_m + \vartheta_{10} M_s)}{k}$$
$$\tag{5}$$

4 Empirical Analysis

4.1 Network Description

After constructing the model, we conducted a basic feature analysis of the agricultural technology collaborative innovation network, laying the foundation for subsequent empirical research on the formation motivations of the network. When constructing a collaborative innovation network, we take agricultural enterprises, universities, and research institutes as nodes, and the patent cooperation relationship between the two as edges, forming a complex network structure, as shown in Fig. 2.

Figure 2 shows that there are a large number of "star structures" and "interaction triangles" in the collaborative innovation network, and there are also a large number of nodes acting as "intermediaries" between different innovation entities. This indicates that the different configuration variables in the network play an important role in the collaborative innovation network, and also indicates that the collaborative innovation network has geometric binary shared partners and geometric weighted edge shared partners. The network exhibits significant "core edge" characteristics, where a few centrally located organizations have closer connections with other organizations, occupying most of the network's resources, while the connections between edge nodes are very sparse. This indicates that in the collaborative innovation network, the positions of different nodes are different, and their collaborative innovation capabilities, cooperation breadth, and cooperation depth are not balanced, resulting in significant differences. Compared to edge nodes, innovation entities at the core position perform significantly in indicators such as the number of cooperative projects, partners, and patents.

The analysis of these basic characteristics provides important clues for us to deeply study the formation motivations of agricultural science and technology collaborative

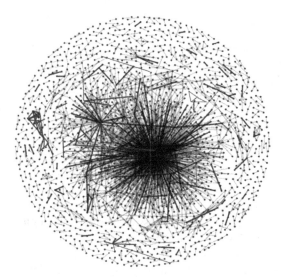

Fig. 2. Collaborative Innovation Network Relationship of Agricultural Science and Technology

innovation networks. In subsequent empirical research, we will further explore the influencing factors and mechanisms behind these characteristics, in order to reveal the evolution laws and optimization paths of agricultural technology collaborative innovation networks. Through in-depth analysis of these network characteristics, we will provide scientific basis for decision-makers such as government departments, enterprises, and research institutions, promote the development of agricultural technology innovation, and promote sustainable progress of rural economy and society.

4.2 Empirical Result

From network data analysis, the collaborative innovation network covers 1973 nodes, including 1568 agricultural enterprises, 207 universities, and 198 research institutes. This indicates that enterprises occupy a dominant position in collaborative innovation, far exceeding the number of universities and research institutes. Overall, the network density and average degree are relatively low, at 0.002 and 2.358, respectively. This means that the connection relationships in the network are relatively sparse, with an average of less than 3 partners per organization. In addition, the average path length of the network is 4.005, and the longest path is 10, indicating that the average distance between nodes in the network is shorter, but there are also some farther connections. The average clustering coefficient of the network is 0.469, indicating that the connectivity of the network is not high enough and the collaborative relationships between nodes are relatively scarce. Through comprehensive analysis, it was found that there were differences in collaborative innovation capabilities, cooperation breadth, and cooperation depth among universities, agricultural enterprises, and research institutes from 2019 to 2022. These three types of entities form different network structures in the agricultural technology collaborative innovation network.

Through the analysis of the above network structure indicators, we can observe that the overall structure of the collaborative innovation network presents the characteristics of sparse and low connectivity. This means that in the collaborative innovation network of agricultural science and technology, the cooperative relationships are not intensive or frequent, and the connections between nodes are relatively few. In order to deeply understand the influencing factors leading to this network structure, we need to analyze the degree of influence of internal and external factors on the formation of the network, so as to provide theoretical basis for improving the quality of collaborative innovation network and adjusting the cooperation mode among enterprises, universities and research institutions. However, empirical analysis is needed to more fully understand the impact of network configuration and node attribute variables on network formation, and the extent to which they affect it. In this study, we plan to conduct an empirical study using four constructed models (Model 1, Model 2, Model 3, and Model 4) to evaluate their utility in explaining the structure of collaborative innovation networks. This will help us to better understand the dynamic process of network formation and provide strong data support for improving the management and development of agricultural science and technology collaborative innovation network.

We intend to employ Stantnet within the R software to estimate parameters for four distinct models. Model 1 will serve as the foundational model, while the other three models will incorporate various statistical terms for simulation purposes. We will utilize the ERGM for parameter estimation across these models. The outcomes of ERGM parameter estimation are documented in Table 2. This analysis will facilitate a deeper understanding of network structure formation and evolution, as well as the influence of diverse statistical factors on collaborative innovation networks. These research findings will furnish valuable insights for enhancing the management and development strategies of agricultural technology collaborative innovation networks.

Table 2. ERGM parameter estimation results (N = 1973).

Abbreviation	Model 1	Model 2	Model 3	Model 4
E_d	−6.459*** (0.019)	−9.493*** (0.131)	−9.238*** (0.076)	−12.448*** (0.324)
T_i		1.683*** (0.036)		0.956*** (0.337)
G_w		6.192*** (0.512)		5.971*** (0.463)
G_b		0.067*** (0.001)		0.007*** (0.002)
G_e		1.364*** (0.089)		2.243*** (0.129)
C_i			0.004*** (0.000)	0.003 (0.002)
C_b			0.077*** (0.000)	0.053*** (0.258)
C_d			0.004*** (0.000)	0.005** (0.001)
O_m			-0.205*** (0.005)	-0.035 (0.231)
M_s			2.318*** (0.053)	1.942*** (0.142)
AIC	37492	35783	31942	30489
BIC	37901	32880	32063	30841

Remarks: *, * *, and * * indicate significant differences at 10%, 5%, and 1%, respectively.

In the ERGM, positive coefficients indicate that the observed agricultural technology collaborative relationships are more likely to form than in the predicted network, implying that certain structures or attributes promote the formation of the network. Conversely, negative coefficients suggest that the probability of collaborative relationships forming in the observed network is lower, indicating an inhibitory effect on network formation. Based on the results in Table 2, it is evident that Models 2, 3, and 4 exhibit a decrease in fit indices AIC and BIC compared to Model 1, and these decreases are statistically significant. Notably, Models 3 and 4 demonstrate a more substantial decrease, suggesting that node attribute characteristics play a more significant role in the formation of collaborative innovation networks. Model 4, with the highest goodness-of-fit, indicates that a comprehensive model incorporating both endogenous network structural variables and exogenous node attribute variables better captures the true collaborative innovation network, and it exhibits robustness. These findings provide valuable insights into the importance of node attributes and network structural factors in the formation of collaborative innovation networks. The specific statistical analysis results are as follows:

The negative coefficient of the edge variable indicates that the network's density is below 50%, reflecting the typical sparsity observed in the network. Specifically, the coefficient for the edge variable is -12.448 (p < 0.001), implying that with each additional edge introduced into the network, the likelihood of forming a new collaborative relationship diminishes, with a probability of less than 1. This suggests that the establishment of cooperative relationships among agricultural enterprises, universities, and research institutes actually reduces the likelihood of forming new connections within the network.

In the parameter estimation related to network configuration variables, a notable coefficient of 0.956 (p < 0.001) emerges for the dependency term. This finding signifies a discernible tendency within the network to adopt interactive triangular structures. Specifically, when two organizations, denoted as A and B, share a common node denoted as C, the probability of A and B establishing a collaborative relationship increases by a substantial factor of 1.238 (exp(0.956) = 1.238) when compared to the likelihood of forming two random connections. Furthermore, the coefficient associated with GW stands at 5.971 (p < 0.001), suggesting a significant trend towards expansion within the agricultural science and technology collaborative innovation network. In practical terms, this implies that a single enterprise can simultaneously engage in collaborations with multiple other enterprises, universities, or research institutions, thereby facilitating the establishment of a greater number of collaborative relationships. Moreover, the positive coefficients for G_b and G_e carry notable implications. They underscore the network's inherent inclination towards both open and closed triangular structures. These structural characteristics play a substantial role in facilitating the formation of new cooperative relationships within the collaborative innovation network. These findings provide valuable insights into the network's underlying dynamics and collaborative tendencies.

In our analysis, the collaborative innovation capability, quantified by the number of existing cooperative patents within an organization, exhibits a positive coefficient of 0.004, although it lacks statistical significance. This suggests that within the agricultural technology domain, an increase in collaborative innovation capability does not substantially heighten the likelihood of initiating new cooperative relationships between two organizations. Essentially, when the number of cooperative patents held by agricultural enterprises, universities, or research institutions increases, its impact on establishing cooperative relationships with other organizations does not reach statistical significance. These findings imply that other factors or mechanisms may play a more pivotal role in driving the formation of such cooperative ties within the agricultural technology sector.

The coefficient associated with cooperation breadth is 0.053 (p < 0.001), translating to an exponent of exp(0.053) = 1.062. This finding holds significant implications, indicating that expanding the scope of cooperation plays a pivotal role in fostering the development of collaborative innovation networks. To elaborate, for every additional partner that enterprises, universities, or research institutes engage with within the network, there is a 6.2% increase in the likelihood of establishing new cooperative relationships. The broadening of cooperation breadth underscores that organizations are more inclined to forge fresh collaborative ties when they engage with a greater number of other organizations. This, in turn, promotes the formation and growth of collaborative innovation within the network.

The coefficient pertaining to cooperation depth stands at 0.005 ($p < 0.05$), corresponding to an exponent of $\exp(0.005) = 1.023$. This finding holds substantial significance, indicating that augmenting the depth of node cooperation serves as a catalyst for the establishment of fresh cooperative relationships. In precise terms, with each increment in cooperative relationships between a node and other nodes within the network, there is a 2.3% upsurge in the likelihood of initiating new cooperative relationships. This underscores the notion that within collaborative innovation networks, organizations engaged in more extensive and intimate cooperation with other nodes are predisposed to the formation of novel cooperative ties.

In terms of node assortativity, the coefficient associated with organization type assortativity is negative, albeit statistically non-significant. This suggests that organizations of similar nature do not exhibit a pronounced inclination to establish cooperative relationships. In essence, a company is more predisposed to collaborate with diverse entities such as universities or research institutes, rather than forming partnerships with organizations of the same type. Conversely, the regional assortativity coefficient is 1.942 ($p < 0.001$), equating to an exponent of $\exp(1.942) = 5.628$. This finding signifies that the likelihood of collaborative relationships emerging between agricultural enterprises, universities, and research institutes within the same province is approximately 5.628 times greater than that between organizations located in different provinces. In other words, organizations are notably inclined to forge collaborative ties within their geographical region, underscoring the influential role of geographical proximity in fostering collaborative innovation, as demonstrated once more.

In summary, our analysis highlights the positive impact of cooperation breadth and depth on promoting collaborative innovation network formation. Additionally, organization type assortativity and regional assortativity exert some influence on network formation. These findings hold significant implications for gaining deeper insights into the mechanisms driving collaborative innovation networks and guiding their optimization and enhancement. Moreover, our research reaffirms the superiority of Model 4 as a comprehensive model, better suited for accurately representing real collaborative innovation networks.

5   Conclusion

This study employs ERGM to empirically examine factors influencing collaborative innovation networks in agricultural science and technology. It focuses on selecting dependent variables that shape these networks. The key findings and insights are as follows:

Collaborative relationships among agricultural enterprises, universities, and research institutions result from a dynamic interplay of endogenous and exogenous factors. This underscores the importance of considering node attributes and existing collaborations in network formation.

Collaborative innovation networks in agriculture exhibit sparsity, where increasing network edges doesn't necessarily lead to more cooperative relationships.

Organizations should prioritize forming star and triangle-shaped cooperative relationships to optimize resource utilization within the collaboration process.

In agricultural technology collaborative innovation networks, enhancing node collaborative innovation capability may not increase new cooperative connections significantly. To maintain a balanced approach, agricultural entities, universities, and research institutions should foster their independent innovation capabilities, engage in autonomous research and development, and manage potential resource redundancy caused by excessive cooperation.

These findings offer valuable insights into the dynamics of collaborative innovation networks in agricultural technology and provide practical guidance for organizations aiming to enhance collaboration and innovation in this sector.

Acknowledgements. The authors are grateful for financial support from the MOE (Ministry of Education in China) Project of Humanities and Social Sciences (20YJCZH030), the Guangdong Basic and Applied Basic Research Foundation (2023A1515011616), the Zhanjiang Philosophy and Social Sciences Project (ZJ23YB09, ZJ23ZZ12) for this study.

References

Hermans, F., Klerkx, L., Roep, D.: Structural conditions for collaboration and learning in innovation networks: using an innovation system performance lens to analyse agricultural knowledge systems. J. Agric. Educ. Ext. **21**(1), 35–54 (2015)

Zastempoeski, M., Cyfert, S.: The role of strategic innovation activities in creating Spanish agriculture companies' innovativeness. Agricultural Economics-Zemedelska Ekonomika **68**(6), 230–238 (2021)

Wang, D., Zhao, X., Du, X., et al.: Research on ecological evolution of national agricultural science, technology and innovation system. China Soft Sci. **372**(12), 41–49+83 (2021)

Meshesha, A.T., Birhanu, B.S., Ayele, M.B.: Effects of perceptions on adoption of climate-smart agriculture innovations: empirical evidence from the upper Blue Nile Highlands of Ethiopia. Int. J. Climate Change Strateg. Manag. **14**(3), 293–311 (2022)

Fieldsend, A.F., Cronin, E., Varga, E., et al.: Organisational innovation systems for multi-actor co-innovation in European agriculture, forestry and related sectors: diversity and common attributes. NJAS-Wageningen J. Life Sci. **92**, 1–11 (2020)

Ingram, J., Gaskell, P., Mills, J., et al.: How do we enact co-innovation with stakeholders in agricultural research projects? Managing the complex interplay between contextual and facilitation processes. J. Rural Stud. **78**, 65–77 (2020)

Feder, G.: Land policies and farm productivity in Thailand. Johns Hopkins University Press, no. 6, pp. 76–82 (1998)

Frank, O., Strauss, D.: Markov graphs. J. Am. Stat. Assoc. **81**(395), 832–842 (1986)

Yongabo, P., Goktepe, H.D.: Emergence of an agriculture innovation system in Rwanda: Stakeholders and policies as points of departure. Ind. High. Educ. **35**(5), 581–597 (2021)

Smith, H.E., Sallu, S.M., Whitfield, S., et al.: Innovation systems and affordances in climate smart agriculture. J. Rural Stud. **87**, 199–212 (2021)

Kiminami, L., Furuzawa, S., Kiminami, A.: Exploring the possibilities of creating shared value in Japan's urban agriculture: using a mixed methods approach. Asia-Pac. J. Reg. Sci. **6**, 541–569 (2022)

Alirah, E.W., Mulubrhan, A., Hildegard, G., et al.: Agricultural innovation systems and farm technology adoption: findings from a study of the Ghanaian plantain sector. J. Agric. Educ. Ext. **24**(1), 65–87 (2018)

Jiang, B., He, Q., Zhai, Z.: Anomaly detection and access control for cloud-edge collaboration networks. Intell. Autom. Soft Comput. **37**(2), 2335–2353 (2023)

Trust Building in Cross-border E-Commerce: A Blockchain-Based Approach for B2B Platform Interoperability

Zhihong Li and Guihong Qiao(✉)

South China University of Technology, 381 Wushan Road, Tianhe District, Guangzhou 510641, Guangdong, China
scutqgh@outlook.com

Abstract. Cross-border E-commerce (CBEC) currently faces a significant trust problem among merchants. Existing research in this field mainly focuses on addressing intra-platform trust issues, while little attention has been given to resolving inter-platform trust challenges. The advent of blockchain technology, with its features of decentralization, transparency, autonomy, and security, offers promising potential for the advancement of cross-border e-commerce while also presenting its own set of challenges. To address this research gap, this paper proposes a trustworthy blockchain-based cross-border B2B e-commerce platform (referred to as BCBEP) based on blockchain technology. Firstly, various sources of distrust are analyzed to gain a comprehensive understanding of the problem. Building on this analysis, blockchain-enabled smart contracts are employed to address the decentralized benefit distribution issue. Secondly, a credit-based incentive mechanism is introduced to attract merchant interest and facilitate timely connections between merchants and their target customers. Thirdly, several key mechanisms are developed to showcase the effectiveness of the proposed platform. To demonstrate the practicality and feasibility of BCBEP, we conduct a case study evaluating an existing cross-border B2B e-commerce platform, "TradeIndia." The findings from this study provide valuable insights into the potential benefits and applicability of the proposed BCBEP platform in addressing inter-platform trust issues and enhancing cross-border e-commerce operations.

Keywords: Blockchain · Incentive mechanism · Distrust · Cross-border B2B e-commerce

1 Introduction

With the rapid advancement of information technology, there has been a continuous promotion of the global information process. Recent research reports on cross-border e-commerce [1] have revealed new trends in the industry. Notably, the traditional patterns of large and long-term orders are gradually giving way to fragmented small and medium-sized orders, characterized by shorter durations. This shift has led to a substantial increase in market volume and the emergence of what is now known as the "New Normal" in international trade orders. As a result, an increasing number of international trades are turning to e-commerce, making cross-border e-commerce the new trend in international trading. This shift is driving adjustments in industrial structures and fostering steady economic growth [2].

Despite its widespread popularity, cross-border e-commerce faces various challenges, including political culture, organizational economy, technical law, and others, with one of the most significant hurdles being the issue of distrust between merchants [3, 4]. In response to the evolving trade landscape, mainstream cross-border e-commerce platforms have expanded their trade channels in different regions by collaborating with partner platforms. However, this cooperation process is not without challenges, as information asymmetry arising from varying levels of infrastructure development in different regions often leads to mutual distrust between the involved parties. This trust problem is further exacerbated when merchants come from culturally diverse areas [5]. Moreover, many cross-border e-commerce platforms rely on merchant commissions and value-added services as their primary profit model [2]. However, under a climate of distrust, this profit model can exacerbate benefit distribution problems, potentially harming the cooperation between platforms.

The increasing growth of cross-border e-commerce has brought significant attention to the issue of trust in recent years. While existing studies have primarily focused on analyzing the trust relationships within individual platforms, particularly in online transactions between strangers [6–9]. However, nowadays there has been very little research which addresses the inter-platform trust problem between merchants in cross-border B2B e-commerce.

The rapid advancement of blockchain technology, characterized by decentralization, transparency, autonomy, and security, presents significant opportunities for the growth of cross-border e-commerce. This paper aims to address the trust issues between B2B participants in cross-border e-commerce by proposing a blockchain-based cross-border B2B e-commerce platform (referred to as BCBEP). This solution serves as a design guide for the development of decentralized cross-border e-commerce platforms. We present a comprehensive system architecture and discuss key mechanisms of the BCBEP platform, highlighting the interactions between participants. Additionally, we introduce a credit-based incentive mechanism to encourage user participation, ensuring the security and stability of the system. Through this work, we hope to pave the way for more trustworthy and efficient cross-border B2B e-commerce transactions.

In this study, we present significant contributions in the field of cross-border e-commerce. Firstly, we are pioneers in utilizing advanced blockchain technology to tackle the benefit distribution and inter-platform trust issues in cross-border e-commerce, thereby filling a research gap in this area. Secondly, our proposed decentralized system architecture and incentive mechanism demonstrate technical innovation. Lastly, our solution offers valuable guidance for the development of decentralized e-commerce platforms, with implications that are relevant to both researchers and practitioners.

This paper provides a comprehensive investigation into the application of blockchain technology in cross-border B2B e-commerce. The study begins by reviewing relevant literature in cross-border e-commerce and blockchain technology. Subsequently, a five-tier blockchain system architecture for cross-border B2B e-commerce is proposed, address-ing key research issues. The paper further presents various key mechanisms and solutions of the BCBEP system based on blockchain technology. A case study on BCBEP is also included to demonstrate its practicality. Finally, the conclusions and future research directions are discussed in detail in the concluding section.

2   Related Work

2.1   Cross-border E-commerce

With the advancement of information technology, online trading has successfully over-come language and geographical barriers, resulting in significant reductions in trans-action costs [10]. Empirical studies have also demonstrated that physical distance's impact on cross-border e-commerce development in the European Union market, with its complex language environment, has been significantly reduced [2]. Presently, small to medium-sized enterprises (SMEs) play a major and vital role in cross-border e-commerce, offering them substantial access to overseas market opportunities [11]. How-ever, due to the uneven levels of infrastructure development in different countries and regions, e-commerce participants must swiftly adapt to the evolving global trade land-scape, surmount trade barriers [3], and actively seek trade channels. Consequently, cross-border e-commerce B2B merchants engage in foreign trading through collaboration with partner platforms. In this context, the importance of trust between merchants has been widely acknowledged as a critical factor influencing the success of e-commerce [12].

Researchers have made various attempts to address trust issues in e-commerce. Early studies have highlighted the significance of trust between organizations for maintain-ing long-term relationships in electronic data interchange [13]. From a technological and security perspective, Ratnasingam's research indicates that technical trust origi-nates from relationship trust, which, in turn, aids in the development of trust in B2B e-commerce relationships [6, 7]. Koh utilized information signal theory to examine how information indices and signals impact the buyer-supplier trust relationship in global B2B [8]. Kumar proposed a cognitive-based trust model to assess the credibility of companies in the B2B electronic market [14]. Chang analyzed the impact of third-party certification, reputation, and return policy on the establishment of an e-commerce trust mechanism from an empirical perspective [12], finding that third-party certification has a notice-able effect on trust establishment when the company's reputation is unknown. Overall, existing research has underscored the crucial role of trust in e-commerce and identi-fied significant factors related to trust. However, most studies have focused solely on the intra-platform situation, overlooking trust issues arising from merchant cooperation, which this work aims to address with the introduction of blockchain technology.

2.2   Blockchain Technology

Blockchain is a distributed ledger technology that offers a naturally trusted platform with-out the need for additional third-party intermediaries [15]. Its inception can be traced back to the article "Bitcoin: A Peer-to-Peer Electronic Cash System" [16]. Broadly, blockchain refers to distributed ledger technology based on its structure, encompassing distributed consensus, privacy and security protection, peer-to-peer communication technology, network protocols, and smart contracts [16, 17]. In a distributed system like blockchain, nodes do not need to trust each other to transmit information. This technology facilitates peer-to-peer transactions and resolves the issues of high transaction costs and data security prevalent in centralized systems [18]. Currently, blockchain deployment can be categorized into three types based on different means of participation: public blockchain, private blockchain, and consortium blockchain [19]. Public chains, also known as non-permissioned chains, allow all users to join and read the data information in the block and participate in the consensus process. They represent a completely decentralized blockchain, with examples including the Bitcoin and Ethereum blockchains. In contrast, permissioned chains consist of private and consortium blockchains. Private chains are entirely controlled by organizations to support peers joining the network [20]. Consortium blockchains are partly private and polycentric, maintained by multiple organizations [21]. For instance, Fabric, an open-source project of Hyperledger, deploys permissioned blockchains. Compared to the Bitcoin blockchain, Fabric is optimized for scalability and throughput [22]. EduCTX is a consortium blockchain-based higher education credit platform, providing globally consistent information for nodes on the chain [19]. Our proposed BCBEP architecture utilizes a consortium blockchain approach employing the practical Byzantine fault-tolerant system (PBFT), a distributed consensus protocol commonly used in consortium blockchains like the Hyperledger project led by IBM [22].

Smart contracts, also known as chain codes, find a secure space for their contract codes and states within the blockchain. Once a contract is written into the blockchain, its execution and verification process become self-managed without any need for intervention from the contract initiator. Moreover, changes in the contract status are not determined by a single node, but rather by multiple nodes [23]. The advantages of smart contracts are depicted in Fig. 1.

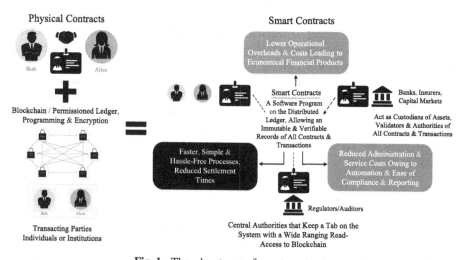

Fig. 1. The advantages of smart contracts

2.3 Application Based on Blockchain

Blockchain technology has found applications in various fields, including transportation, medical health, education, and e-commerce. In the transportation sector, Hawlitsche [24] conducted a literature review on the use of blockchain in the sharing economy, highlighting its potential to address trust issues and reduce reliance on platform providers. Yong Yuan [25] developed an intelligent traffic management system based on blockchain technology. In the medical health area, blockchain is used for secure storage of patients' electronic medical records [26] and for managing medical data access [27]. In the traditional education industry, blockchain can optimize the credit accumulation system for higher education [19].

In the realm of e-commerce, blockchain technology has also made significant strides. Frey et al. [28] proposed a security recommendation system for e-commerce, employing blockchain to support secure multi-party computation and safeguard user privacy. Sidhu et al. [29] designed an enterprise blockchain e-commerce system to tackle centralized payment issues. Dennis et al. [30] established a generic reputation system based on blockchain technology for use in traditional e-commerce platforms. Schaub et al. [31] advocated for the integration of blockchain technology in e-commerce reputation systems to enhance trust between buyers and sellers by preventing fake reviews. It is important to note that existing applications of blockchain in e-commerce primarily focus on individual platforms. Our work stands out by exploring the application of blockchain technology across multiple platforms.

3   The Architecture of the Proposed BCBEP Platform

In this section, we present a comprehensive five-tier system architecture based on blockchain for cross-border B2B e-commerce, as illustrated in Fig. 2. We provide a detailed overview of the functions associated with each layer of the system architecture and briefly discuss the key technical aspects involved in each layer. The primary aim of BCBEP is to establish a more trustworthy platform environment for cross-border e-commerce participants. Users are required to undergo authentication before being added to the existing system. The goods involved in transactions interact with the blockchain network through the Internet of Things (IoT), which enables the seamless transmission of information to the new block. Furthermore, users who can provide product traceability information can submit their information transaction to the new smart contract [23]. The contract then assesses the user's credit based on the blockchain protocol and establishes their credit rating on the platform. Once the transaction garners consensus from all nodes, the system augments the user's credit accordingly, in accordance with the contract's provisions.

3.1 Infrastructure Layer

This layer primarily serves as a repository for various essential equipment utilized in cross-border B2B e-commerce services, such as vehicle logistics and commodity assets. By incorporating IoT technology into the BCBEP system, the majority of commodities

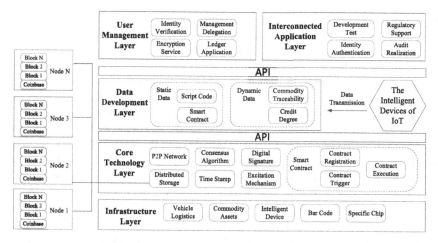

Fig. 2. The architecture of BCBEP platform.

can undergo digital asset services through the blockchain, ensuring that all digital information is securely linked. Additionally, specific chips store all data related to production, processing, and factory delivery, which are then connected to the blockchain network via smart hardware devices integrated into the IoT [18]. For instance, when manufacturing an electronic product, the merchant can utilize the IoT smart device to upload pertinent information about all the machine's parts, quality parameters, and logistics details to the blockchain network. This data is then recorded in a secure, effective, and tamper-proof blockchain ledger, allowing for complete traceability. However, the merchant's identity determines whether all this information should be registered on our established blockchain network.

3.2 Core Technology Layer

This layer encompasses all the key technologies utilized in the BCBEP system, including P2P networks, timestamps, consensus algorithms, digital signatures, and smart contracts. In BCBEP, nodes can be classified into client nodes and lightweight nodes (e.g., mobile and smart devices), both of which possess equal rights within the distributed network. The client node acts as the primary form of existence in the network, storing the complete ledger of the blockchain distributed data. On the other hand, lightweight nodes typically hold only a small portion of the complete ledger, but they can access the required ledger information from other nodes at any time through the internal protocol of the system. All ledger information is generated periodically in relation to commodity trading data using a hash algorithm (e.g., SHA256 algorithm) and is stored in the form of a Merkle structure tree within the corresponding block, with the hash pointer pointing to the timestamp of the write block's initial time [24]. The inclusion of timestamps enables the traceability of the blockchain, making the traceability of commodities possible.

Facilitating consensus among all nodes without requiring trust is a critical consideration in the design of distributed systems [32]. To date, various consensus algorithms have been developed for different blockchain systems. Among these, the consortium blockchain is considered the most suitable deployment method for B2B scenarios, with PBFT being a well-regarded consensus algorithm for the consortium blockchain. Additionally, improved algorithms like Raft and the Byzantine fault-tolerant system are also incorporated in this layer. To encourage more nodes to join the network, we have devised a unique incentive mechanism. Initially, the concept of credit is introduced, which involves nodes changing the public's cognitive psychology, gaining user approval for their business behavior, and consequently, obtaining support and opportunities. The user's credit level is reflected in their priority level during searches within the BCBEP system. Each node in the BCBEP system can build its credit through three methods. First, the initial credit of the user is established upon registering to join the blockchain network, evaluated based on the user's official authentication information in the real world. Second, the data contribution made by users to the traceability information of their products, encompassing all relevant information from production to sale, is considered. Third, users can score each other's credit after a commodity transaction occurs, but only authenticated nodes within the blockchain network have the right to score. These three situations effectively involve the process of credit tokenization. Nodes that receive more scores attain a higher credit rating. When other users search for goods, merchants with higher credit ratings are displayed at the top position, indirectly resolving the trade channel docking issue. The credit incentive mechanism serves as the primary driving force behind the BCBEP system.

3.3 Data Development Layer

This layer serves as the repository for all data information in the BCBEP system, encompassing various static data such as script codes, smart contracts, as well as dynamically updated data like product traceability information and user credit. The central function of this layer is the smart contract service, which integrates data management logic, application logic, business rules, and contract terms into our proposed blockchain system, significantly enhancing its scalability. Moreover, the process of credit tokenization is executed through a smart contract based on predefined code. When a node exercise voting rights, it triggers the conditions required for the execution of the smart contract, automatically updating the credit degree of other nodes accordingly. Additionally, a dedicated application programming interface (API) is established within BCBEP to enable external systems to access smart contract services, platforms, and data [33].

3.4 User Management Layer

This layer is primarily divided into three parts concerning user management. Firstly, authentication: The digital identity of each node is maintained within the blockchain network. When a node joins BCBEP, it undergoes registration and authentication of its identity information. Secondly, management authority: This involves access control, such as authority management based on smart contracts, user information, and blockchain. Additionally, hierarchical authority control aligns with higher governance requirements, better meeting the supervision and audit demands of various countries. Nodes have the ability to set permissions on the traceability information of their own commodities, and the entire blockchain network can also set permissions on external system access; permissions are restricted to accessing data in the federation. Lastly, optional encryption service: Users have the freedom to select and utilize various encryption algorithms independently to address potential security concerns arising from future technological advancements in the blockchain system.

3.5 Interconnected Application Layer

This layer primarily serves as the cross-layer (cross-chain) service provider. As part of the blockchain service, regulatory and audit departments can join the network as blockchain nodes, offering regulatory support, implementing audits, and ensuring network stability and security. In the present scenario, an increasing amount of commodity transaction data is stored in the cloud. With the introduction of sub-chain service functionality, privacy data can be uploaded to the blockchain, granting nodes secure access to the data on the chain. The cross-chain service mainly involves the data exchange capability of smart contracts between different blockchains [15]. Through this cross-chain service, the BCBEP system can offer alternative digital currency trading methods, making payment for goods more convenient. Furthermore, this is expected to be the future development trend of blockchain systems.

4 Key Mechanisms of the BCBEP Platform

Blockchain technology has the potential to establish a decentralized, secure, and efficient cross-border e-commerce B2B platform. However, in the context of design research, several key issues still need to be addressed for the successful implementation of BCBEP as a blockchain technology. This section delves into four crucial scenarios and their associated research problems related to the BCBEP system, aiming to fully harness the capabilities of this innovative platform.

4.1 Node Registration to Join the BCBEP System

The stability and security of a blockchain network improve with an increasing number of network nodes. Nodes play a crucial role in maintaining the blockchain. In theory, any smart device equipped with blockchain core clients or light clients and having a complete ledger of the network can be considered a blockchain node. When a new participant seeks to join a blockchain network, they must first create a key pair and contact an existing node that is already part of the network. Upon receiving the registration request from the new participant, the existing node verifies the provided official information thoroughly [19]. Upon successful verification, the relevant identity information, including the public key of the new participant, is forwarded to the transaction pool and packaged into blocks. Additionally, an intelligent contract is signed to update the initial credit rating for the new participant, which is then broadcasted to the entire network. The initial creditworthiness, typically ranging from 0 to 100, is determined based on the official information provided by the new participant. Once the block containing the verification information is accepted by other nodes in the network, the contract terms are automatically triggered, and the new participant receives their initial credit. This completes the verification process for the identity of the new node.

4.2 Third Party Payment Transaction

In the cross-border e-commerce platform based on the BCBEP system, buyers aim to purchase goods from sellers through delivery and payment. However, due to the lack of complete trust in the merchant relationship between the two parties, a third-party payment transaction becomes a suitable solution within the blockchain network. This transaction involves multiple digital signatures added through a smart contract during commodity trading. Instead of directly transferring funds into the seller's wallet, the buyer initiates a transaction with multiple digital signatures, which is then written into the smart contract and automatically executed. The transaction requires 2 out of 3 multiple digital signatures to be triggered, with the three participants being the buyer, seller, and a third party, referred to as the "judge." The credit rating of this third party is jointly determined by the buyer and seller. In case of any disputes during the transaction process, the judge, who has a higher credit rating, acts as the mediator. When the transaction is initiated and recorded on the blockchain, regulators monitor the funds, and any two of the three participants can decide the funds' destination. Assuming both parties are honest nodes, the seller will deliver the goods according to the buyer's request. Once the buyer receives the goods and confirms the seller's signature on the transaction, the money is transferred to the seller's wallet. In this smooth process, the judge does not need to participate since there are no disputes. However, if a dispute arises, such as non-delivery or damaged goods, and both parties want access to the funds on the contract, only the signatures of the buyer and seller will not suffice for the actual payment. The judge must intervene and decide the rightful owner of the funds based on the actual situation, and both parties must collectively sign to complete the fund transfer. Thus, in case of any dispute during a transaction, the judge plays a crucial role in determining whether to pay or withhold the funds.

4.3 Credit Analysis Under Incentive Mechanism

To attract more nodes and ensure the continuous operation of the BCBEP system, we have devised a unique incentive mechanism based on user credit ratings [34]. The credit rating determines the priority level at which users can be searched, effectively addressing the technical issue of commodity channels. A user's credit rating comprises three main components, as illustrated in Fig. 3. Firstly, it includes the initial credit assigned to the user upon joining the blockchain network, evaluated based on the user's official authentication information in the real world. Secondly, the credit depends on the user's contribution of data to the traceability information. According to the blockchain protocol, users have the right to manage their personal data information, giving merchants full control over the traceability information of goods during transactions. When a node opens up the commodity traceability data entirely, the data related to commodity traceability is included in the latest block through consensus competition. Subsequently, the smart contract automatically assigns appropriate credit values to each user using all possible allocation mechanisms within the blockchain. Finally, after completing the commodity transaction, the credit of the two parties involved is taken into account. Only nodes authenticated within the blockchain have the authority to score, with each node allowed to score only once for the same transaction. The proportion of voting varies for each user, with all users starting with a voting weight of 1.

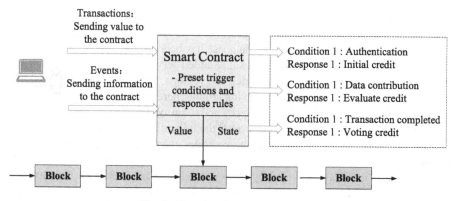

Fig. 3. The situations of users' credit

4.4 Benefit Distribution Problem

The traditional cross-border e-commerce platforms primarily rely on merchant commissions and value-added services as their main profit model. In the BCBEP system, ensuring a fair benefit distribution in cross-platform cooperation is of utmost importance. When a buyer utilizes the BCBEP system to search for sellers in cross-border regions and completes a commodity transaction, the distribution model of the involved platforms consists of the following two components. Firstly, the search commission is paid by both the buyer and the seller to their respective e-commerce platforms. Secondly, during the cross-border e-commerce transaction process, 3–20% of the retail sales of each order is reimbursed. The benefit platform and the platform each receive 50% of this reimbursement, and this entire process is governed by a smart contract, ensuring effective benefit guarantee under the supervision of all participants within the network.

5 Case Analysis

As the second-largest Internet user in the world, India boasts a trade market second only to China and the United States. TradeIndia is a cross-border B2B e-commerce website in India and currently stands as one of the largest integrated B2B e-commerce platforms in the country. Its primary goal is to provide a comprehensive range of international and Indian trade and business information, catering to numerous merchants and fostering a platform for global exchanges between sellers and buyers. Both parties can easily find suitable partners on the TradeIndia platform. Additionally, TradeIndia offers valuable product promotion services to manufacturers, importers, exporters, and service providers, both within India and worldwide. Buyers from across the globe can readily access suppliers and explore a diverse array of commodities on this platform. Despite the availability of quality products from India, the main challenge facing the TradeIndia platform lies in bridging the trust gap between buyers and sellers while also enhancing infrastructure to better facilitate cross-border B2B e-commerce.

Currently, the TradeIndia platform boasts 2.9 million registered members and offers professional platform maintenance and promotion services (Data source: http://www. cifnews.com/). This extensive member database has been established over time. Users benefit from product promotion through value-added services. However, due to buyers and sellers hailing from different countries and regions, the transaction process lacks adequate credit endorsement, often revolving around mere price inquiries. Furthermore, the platform lacks a guarantee of data privacy and security. To address these challenges, our proposed BCBEP system revolves around the utilization of blockchain technology, ensuring that transactions are executed among nodes even in an environment of mutual distrust. Every user can register through the BCBEP system and become a network node upon authentication. Data provided by users can only be accessed with their explicit authorization (user management layer), and the platform cannot use the data without permission. All nodes are interconnected, forming a decentralized peer-to-peer network with no central authority. During transactions, real-time data is verified and stored in the intelligent device within the supply chain's infrastructure layer. This allows traceability data of products to be uploaded to the blockchain network, securely encapsulated in blocks using cryptography. The entire network reaches a consensus through a consistency algorithm like PBFT and rewards users who provide traceable data with credit value (core technology layer). This process ensures real-time data availability, allowing buyers to obtain commodity data information instantly through the blockchain. Moreover, trust is fully endorsed by blockchain technology, enabling product traceability. All these processes are automated through smart contracts, requiring no human intervention. External platforms can access data through a unique API or integrate with BCBEP for cross-platform cooperation (data development layer). As part of the Belt and Road Initiative, effective regulatory measures have increased transaction trust, and relevant entities join the network as nodes, participating in the consensus process (interconnected application layer). The application of blockchain technology in cross-border e-commerce will significantly impact the future development trends of the industry, reshaping the cross-border e-commerce trading landscape.

6 Conclusion

Traditional cross-border e-commerce platforms have primarily focused on addressing trust issues within their own platforms, mainly between merchants. In contrast, our proposed platform aims to enhance trust between different platforms and their respective merchants. In this paper, we present the BCBEP system, a decentralized e-commerce service model built upon blockchain technology, which effectively tackles the trust problem in cross-border e-commerce. We have developed a comprehensive five-layer architecture for BCBEP and explored several vital application scenarios. To foster trust among participants during the business docking process, we introduce the concept of credit and propose an innovative incentive mechanism that utilizes user credit to elevate recommended priority levels. Users can acquire credit through three distinct methods. Additionally, we conduct a case study on TradeIndia, a prominent cross-border B2B e-commerce platform in India, to demonstrate the effectiveness of our proposed BCBEP system in bridging the trust gap between platforms and safeguarding data.

As an emerging technological advancement, blockchain is gradually influencing various aspects of our lives due to its decentralized nature and other advantageous characteristics. This research has presented a compelling example of blockchain application by introducing a fully designed blockchain-based cross-border e-commerce solution, with both theoretical and practical implications. Despite the remarkable robustness of the BCBEP platform, there are some limitations that need to be acknowledged. Firstly, being in the early stages, blockchain technology still faces certain constraints, such as low efficiency, which also affect our proposed solution. It is crucial to recognize that this work is a theoretical design rather than an empirical study. Secondly, our study does not deeply explore smart contracts, as we only present a simple smart contract-based code for evaluating credit. To enhance the proposed platform, a more sophisticated smart contract implementation is necessary. Moreover, this study solely considers a simple platform scenario for illustrative purposes, which implies that scalability and performance aspects are not fully addressed. In future work, the following issues will be extended and tackled: 1) The proposed architecture will be implemented in phases on existing cross-border e-commerce platforms, and user behavior studies will be conducted to empirically test the effectiveness of the proposed solution. 2) The current architecture will analyze available consensus algorithms to ensure that the selected algorithm is suitable for addressing inter-platform trust issues. 3) While the proposed architecture is primarily designed to address inter-platform trust issues, it can be further extended and applied to other scenarios as well.

Acknowledgment. This work was supported by grants No. 72171089 and 72071083 from the National Natural Science Foundation of China, grant No. 2021A1515012003 from the Guangdong Natural Science Foundation of China, and grant No. 2021GZQN09 from the Project of Philosophy and Social Science Planning of Guangzhou in 2021.

References

1. C. B. E. Innovation.: Cross Border E-Commerce Innovation Development Report. Resource document. Billion State Power Research Institute (2018)
2. Gomez-Herrera, E., Martens, B., Turlea, G.: The drivers and impediments for cross-border e-commerce in the EU. Inf. Econ. Policy **6**(28), 83–96 (2014)
3. Ifinedo, P.: An empirical analysis of factors influencing Internet/e-business technologies adoption by SMEs in Canada. Int. J. Inf. Technol. Decis. Mak. **4**(10), 731–766 (2011)
4. Ratnasingam, P., Phan, D.D.: Trading partner trust in B2B e-commerce: a case study. Inf. Syst. Manag. **3**(20), 39–50 (2003)
5. Sinkovics, R. R., Yamin, M., and Hossinger, M.: Cultural Adaptation in Cross Border E-Commerce: A study of German Companies. J. Electron. Commer. Res. **4**(8), (2007)
6. Farhoomand, A.F., Tuunainen, V.K., Yee, L.W.: Barriers to global electronic commerce: a cross-country study of Hong Kong and Finland. J. Organ. Comput. Electron. Commer. **1**(10), 23–48 (2000)
7. Ratnasingam, P.: Trust in inter-organizational exchanges: a case study in business-to-business electronic commerce. Decis. Support. Syst. **3**(39), 525–544 (2005)
8. Koh, T.K., Fichman, M., Kraut, R.E.: Trust across borders: buyer-supplier trust in global business-to-business e-commerce. J. Assoc. Inf. Syst. **11**(13), 886–922 (2012)

9. Resnick, P., Zeckhauser, R.: Trust among strangers in Internet transactions: Empirical analysis of eBay's reputation system. In: The Economics of the Internet and E-commerce, pp. 127–157. Emerald Group Publishing Limited (2002)
10. Gomez-Herrera, E., Martens, B., Turlea, G.: The drivers and impediments for cross-border e-commerce in the EU. Inf. Econ. Policy 1(28), 83–96 (2014)
11. Fan, Q.: An exploratory study of cross border e-commerce (CBEC) in China: opportunities and challenges for small to medium size enterprises (SMEs). Int. J. E-Entrep. Innov. (IJEEI) 9(1), 23–29 (2019)
12. Chang, M.K., Cheung, W., Tang, M.: Building trust online: Interactions among trust building mechanisms. Inf. Manag. 7(50), 439–445 (2013)
13. Saunders, C.S., Clark, S.: EDI adoption and implementation: a focus on interorganizational linkages. Inf. Resour. Manag. J. (IRMJ) 1(5), 9–20 (1992)
14. Kumar, B.A.: Cognition based trust model for B2B e-market. In: International Conference on Next Generation Computing Technologies, pp. 703–707. IEEE, September 2015
15. Jiang, S., Cao, J., Wu, H., Yang, Y., Ma, M., He, J.: BlocHIE: a blockchain-based platform for healthcare information exchange. In: IEEE International Conference on Smart Computing (SMARTCOMP), pp. 49–56. IEEE Computer Society (2018)
16. Nakamoto, S.: Bitcoin: a peer-to-peer electronic cash system (2008)
17. Buterin, V.: Ethereum 2.0 mauve paper. In: Ethereum Developer Conference (2016)
18. Shafagh, H., Burkhalter, L., Hithnawi, A., Duquennoy, S.: Towards blockchain-based auditable storage and sharing of IoT data. In: Proceedings of the 2017 on Cloud Computing Security Workshop, pp. 45–50. ACM, November 2017
19. Turkanović, M., Hölbl, M., Košič, K., Heričko, M., Kamišalić, A.: EduCTX: a blockchain-based higher education credit platform. IEEE Access 6, 5112–5127 (2018)
20. Buterin, V.: Ethereum: Platform Review. Opportunities and Challenges for Private and Consortium Blockchains (2016)
21. Britchenko, I., Cherniavska, T., Cherniavskyi, B.: Development of small and medium enterprises: the EU and east-partnership countries experience (2018)
22. Androulaki, E., Barger, A., Bortnikov, V., et al.: Hyperledger fabric: a distributed operating system for permissioned blockchains. In: Proceedings of the Thirteenth EuroSys Conference, p. 30. ACM (2018)
23. Buterin, V.: A next-generation smart contract and decentralized application platform. White paper (2014)
24. Hawlitschek, F., Notheisen, B., Teubner, T.: The limits of trust-free systems: a literature review on blockchain technology and trust in the sharing economy. Electron. Commer. Res. Appl. 1(29), 50–63 (2018)
25. Yuan, Y., Wang, F.Y.: Towards blockchain-based intelligent transportation systems. In: 2016 IEEE 19th International Conference on Intelligent Transportation Systems (ITSC), pp. 2663–2668. IEEE (2016)
26. Ivan, D.: Moving toward a blockchain-based method for the secure storage of patient records. In: ONC/NIST Use of Blockchain for Healthcare and Research Workshop. Gaithersburg, Maryland, ONC/NIST, United States (2016)
27. Azaria, A., Ekblaw, A., Vieira, T., et al.: MedRec: using blockchain for medical data access and permission management. In: 2016 2nd International Conference on Open and Big Data (OBD), pp. 25–30. IEEE (2016)
28. Frey, R., Wörner, D., Ilic, A.: Collaborative filtering on the blockchain: a secure recommender system for e-commerce. In: Proceedings of the 22nd Americas Conference on Information Systems (Amcis) (2016)
29. Sidhu, J.: Syscoin: a peer-to-peer electronic cash system with blockchain-based services for e-business. In: 2017 26th International Conference on Computer Communication and Networks (ICCCN). IEEE (2017)

30. Dennis, R., Owenson, G.: Rep on the roll: a peer to peer reputation system based on a rolling blockchain. Int. J. Digit. Soc. **1**(7), 1123–1134 (2016)
31. Schaub, A., Bazin, R., Hasan, O., et al.: A trustless privacy-preserving reputation system. In: Hoepman, J.H., Katzenbeisser, S. (eds.) ICT Systems Security and Privacy Protection. SEC 2016. IFIP AICT, vol. 471, pp. 398–411. Springer, Cham (2016). https://doi.org/10.1007/978-3-319-33630-5_27
32. Tanenbaum, A.S., Van Steen, M.: Distributed systems: principles and paradigms: international edition, 2/e. J. Comput. Sci. Technol. **1**(5), 279–283 (2007)
33. Liu, Y., Lu, Q., Xu, X., et al.: Applying design patterns in smart contracts. In: Chen, S., Wang, H., Zhang, L.J. (eds.) Blockchain – ICBC 2018. ICBC 2018. LNCS, vol. 10974, pp. 92–106. Springer, Cham (2018). https://doi.org/10.1007/978-3-319-94478-4_7
34. Qin, D., Wang, C., Jiang, Y.: RPchain: a blockchain-based academic social networking service for credible reputation building. In: Chen, S., Wang, H., Zhang, L.J. (eds.) Blockchain – ICBC 2018. ICBC 2018. LNCS, vol. 10974, pp. 183–198. Springer, Cham (2018). https://doi.org/10.1007/978-3-319-94478-4_13

Exploring Group Opinion Polarization Based on an Opinion Dynamics Model Considering Positive and Negative Social Influence

Shuo Liu[✉], Xiwang Guan, and Haoxiang Xia

School of Economics and Management, Dalian University of Technology,
No. 2 Linggong Road, Dalian 116024, China
liushuo1260@mail.dlut.edu.cn

Abstract. Regarding the evolution of public opinion from plurality of opinions to polarization, as well as leading to further attacks on each other by individuals holding different opinions and generating public opinion conflicts, it attracts the focused attention of research scholars. Therefore, constructing models that capture the complexity of social influence in networks should play an important role in the design of communication systems. In this study, based on the social judgment theory in social psychology, we add the repulsion threshold mechanism to the bounded confidence opinion dynamics model, combine assimilative and repulsive social influence, and analyze the characteristics of group opinion evolution. With the agent-based model, we find that the combination of assimilative and repulsive social influence produces complex outcomes that arise in social networks. Sometimes, more assimilative social influence actually leads to more, rather than less, opinion polarization. Similarly, the propensity of users to communicate with like-minded individuals sometimes diminishes opinion polarization. While only occurring in specific parts of the parameter space, these counterintuitive dynamics are robust, as demonstrated by simulation experiments. In addition, we explore the effect of WS small-world network structure on the evolution of opinions.

Keywords: Opinion polarization · Opinion dynamics model · Negative social influence · Online social network · Complexity · Agent-based Modeling

1 Introduction

In recent years, the severity of social polarization has been climbing, and it has been identified as one of the greatest threats to the global order in The Global Risks Report [27]. The problem of social bipolarity has a greater impact on the loss of social capital and on social stability, collective well-being and declining economic productivity. Here, opinion polarization is characterized by increasing opinion differences over time between emergent subgroups in a population that is

J. Chen et al. (Eds.): KSS 2023, CCIS 1927, pp. 260–273, 2023.
https://doi.org/10.1007/978-981-99-8318-6_18

internally homogeneous and mutually distinct [3,4]. The contribution of online social networks to the opinion polarization has attracted increasing attention from research scholars [5–7].

It has been argued that the personalization of online social networks creates so-called "filter bubbles" that create a diet of information for users, limiting exposure to content with which they disagree. As a result, the lack of content that challenges users' opinions and the increased exposure to like-minded content can exacerbate users' opinions and lead to the polarization of opinions. Conversely, it has been argued that online social networks may thus contribute to the polarization of opinions and, in turn, to disruptive political events such as Brexit, the Yellow Vests movement, the 2021 Capitol riots, or the strong resistance to government measures in the recent pandemic [8,9,25]. Therefore, it is essential to consider the complexities arising from individual interactions in online social networks, potentially leading to conclusions about what generates opinion polarization and how to moderate it.

Empirical research in computational social sciences has made great strides in understanding the impact of online social networks on the micro-processes of opinion polarization. Researchers have shown that political opinions are indeed influenced by the consumption of online content [10,11]. Users do align their political opinions with the content they consume, a process we refer to here as vassimilative influence". However, there is also research that suggests the existence of a counterpart to assimilative influence, namely "repulsive influence" [12]. Assimilative social influence is the mechanism by which an individual's opinion on a certain event is inconsistent with the opinions of other members of society, and by changing the cognitive part of the incongruity so as to make it agree with the opinions of other members, and finally form the similar opinion. Repulsive social influence is when individuals strive to become more different from individuals they dislike, and this social repulsion effect is more pronounced when there are large differences in opinions between individuals [1,2]. In that study, self-identified conservative users were exposed to content from liberal sources and formed more conservative opinions [10].

The early research on group behavior by sociologists and social psychologists has provided many valuable ideas and methods for further exploring the opinion evolution [13,26]. Agent-based modeling is one of the most powerful tools available for analyzing opinion dynamics [14–16]. In an agent-based model of opinion dynamics, each agent has an opinion which might be described by either continuous or discrete variables. Relationships between agents, such as their family relations and acquaintances, are usually represented utilizing a social network. Agents can influence each other's opinions through the social relationships that exist with them, and update their opinions based on the rules. In exploring the phenomenon of opinion polarization, bounded confidence models of continuous opinions have received much attention. Reflecting the well-known aphorism "Like likes like", in the bounded confidence models, individuals only interact with others whose opinions are very similar to theirs [17], as depicted in Fig. 1(b). However, the assumption of the bounded confidence model has also been subject to significant criticism: the model assumes that individuals always ignore opinions that differ from theirs. In the real world, however, individuals are not only positively influenced by similar opinions, but also negatively influenced by divergent opinions [5,18], as depicted in Fig. 1(b).

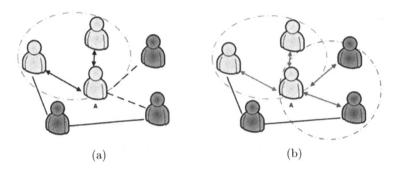

(a)   (b)

Fig. 1. Scenarios illustrating the interaction of opinions between individuals are described. Solid edges between individuals indicate neighbors that have an effect on Individual A, and dashed edges indicate neighbors that do not have an effect on Individual A. Individuals of the same color indicate similar opinions, and individuals of different colors indicate greater differences in opinions. In Fig. 1(a), individuals are only influenced by neighbors whose opinions are similar to their own. In Fig. 1(b), individual A is affected not only by neighbors with similar opinions but also by neighbors with more divergent opinions.

For this purpose, we develop an agent-based model that allows us to flexibly tune the role of both assimilative and repulsive social influences on individuals in social networks. Differing from earlier models [19,20], which had only one parameter to control the magnitude of assimilative and repulsive social influences, here we employ two thresholds to regulate both forces [5,18], and individuals can be influenced by multiple influences at the same time, which satisfies the characteristic of social media in the age of information explosion that provides easy communication for the masses. The network structure we apply is homogeneous, which is more in line with the scenario of a social network where the public, strangers to each other, discuss an event. As will be shown in this paper, the relationship between the micro-processes of assimilation and repulsion influences and the polarization of opinions is more complex than intuition suggests.

This paper demonstrates one of the most central claims of complexity science: that interactions at the micro-level can produce complex and counter-intuitive results at the macro-level. In particular, we use an agent-based model to show that increasing an individual's openness to assimilating social influences does not necessarily lead to fewer differences in opinion at the macro level: sometimes more assimilating influences at the micro level lead to more, not less, opinion bipolarization. Second, we show that increasing the likelihood that individuals are exposed to a diversity of opinions promotes opinion bipolarization. Finally, the presence of more connections between users who like each other, and thus a reduction in the relative amount of exposure leading to repulsive influences, can reduce opinion polarization under certain threshold conditions.

The following section (Sect. 2) summarizes our model incorporating assimilative and repulsive influence on individuals. Our main results are presented in Sect. 3. In Sect. 3.1, we demonstrate that sometimes increasing individuals' openness to assimilative influence can intensify opinion bipolarization. Section 3.2 and Sect. 3.3 show the effect of node degree and initial opinion distribution on opinion bipolarization and concern the robustness of these findings. We present simulation experiments testing whether the two counter-intuitive findings can be replicated when different network structures and different initial opinion distributions are set. Finally, the summary and conclusions are given in Sect. 4.

2   The Model

As mentioned above, in the classic HK model, the agent will only communicate with neighbors whose opinion difference is within the confidence threshold. But this may not be realistic. The concept used in the early bounded confidence model is similar to the social psychology concept of attitude "acceptance"; that is, an agent's opinion can change another's opinion only within the scope of its attitude acceptance. According to the social judgment theory, an individual's attitude consists of three areas: the latitude of acceptance, rejection, and non-commitment. Latitude of acceptance comprises the opinions acceptable to the individual. Latitude of rejection contains the opinions objectionable to the individual. The latitude of non-commitment consists of the opinions an individual is not committed to. Therefore, the previous model cannot effectively describe that individuals are simultaneously affected by different types of opinions.

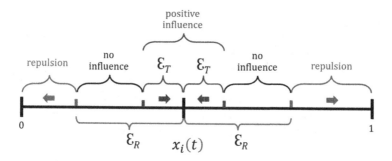

Fig. 2. The mechanism by which an individual is influenced by assimilation and repulsion.

Therefore, we propose a new opinion dynamics model that considers both positive and negative social influences on the interaction of individual opinions in the network. Consider a social network with N nodes representing the agents. The opinion in our model is continuous and uniformly randomly distributed between 0 and 1. Every agent is described by an opinion $x_i(t) \in [0, 1]$ at a discrete time t, where the subscript $i \in N$ denotes the serial number of the agent. The set $x(t) = \{x_1(t), x_2(t), \ldots, x_n(t)\}$ describes the distribution of opinions in the population. If there is a relationship between agents in the network is denoted by $w_{ij} = 1$, otherwise, it is 0. No individual is connected to him/herself, i.e., $w_{ii} = 0$. We construct two thresholds ϵ_T and ϵ_R to represent the range of opinion effects of positive influences and the range of opinion effects of negative influences, respectively, where, ϵ_R is greater than ϵ_T ($\epsilon_R > \epsilon_T$). Figure 2 shows that agent i classifies social influence types according to different types of thresholds at each time. Taking the assimilation threshold ϵ_T as the radius, neighbors within its radius have a positive influence on the agent i's opinions. Taking the repulsion threshold ϵ_R as the other radius, neighbors outside of its radius will have a negative influence on the agent i's opinions, making i strengthen his own opinions. Figure 2 illustrates the mechanism of action of an individual being positively and negatively influenced, where the individual converges towards the average of opinions that produce positive influences and away from the average of opinions that produce repulsive influences when updating their opinions. Next, we use mathematical formulas to describe the model.

To specify the social influence dynamics, we distinguish for any given point in time two subsets of the set of neighbors an individual has. More precisely, the corresponding two subsets are defined as follows.

$$
\begin{aligned}
A^+(i, x(t)) &= \{j \mid |x_i(t) - x_j(t)| \leq \epsilon_T\} \\
A^-(i, x(t)) &= \{j \mid |x_i(t) - x_j(t)| \geq \epsilon_R(\epsilon_R > \epsilon_T)\}
\end{aligned}
\tag{1}
$$

$A^+(i, x(t))$ denotes the set of neighbors who exert a positive influence on individual $i \in N$ at time t, and $A^-(i, x(t))$ denotes the set of neighbors with negative influence on individual i. We selected neighbors that positively and negatively influenced individual i by the threshold parameters ϵ_T and ϵ_R, respectively. To further specify the model dynamics, we compute separately the net pull X^+ on i's opinion exerted by the aggregation of the influences from all influential friends and the net push X^- exerted by the aggregation of the influences of all influential foes, at a given time point. Equation 2 formalize this as follows.

$$
\begin{aligned}
X^+(i, x(t)) &= \sum_{j \in A^+(i, x(t)), i \neq j} (x_j(t) - x_i(t)) * w_{ij} \\
X^-(i, x(t)) &= \sum_{j \in A^-(i, x(t)), i \neq j} (x_j(t) - x_i(t)) * w_{ij}
\end{aligned}
\tag{2}
$$

Individual opinions in the network are pulled towards the average opinion of influential friends, and the larger the proportion of influential friends among all neighbors, the stronger this attraction becomes. At the same time, individuals' opinions are pushed away from the average opinion of influential foes. The larger the proportion of influential foes among all neighbors, the stronger this push will be. Note that this push is limited by the bounds of the opinions scale.

To be able to control the extent to which the combined positive and negative changes an individual's opinion at a given time step, we introduce a parameter λ that controls the impact of these influences relative to the individual's prior opinion. The agents' opinions are updated as specified in Eq. 3, where n_i denotes the number of neighbors of agent i.

$$x_i(t+1) = x_i(t) + \frac{\lambda}{n_i} \cdot (X_i^+(t) - X_i^-(t)) \tag{3}$$

When repulsion is very strong, Eq. 3 can imply that the updated opinion adopts values beyond the bounds of the opinion scale. To ensure that opinions always remained within these bounds, we included that opinions adopt a value on the respective bound, whenever this happened.

$$x_i(t) = \begin{cases} x_i(t) = 1 & x_i(t) \geq 1 \\ x_i(t) = x_i(t) & 0 < x_i(t) < 1 \\ x_i(t) = 0 & x_i(t) \leq 0 \end{cases} \tag{4}$$

3 Results

The model we propose in this paper is nonlinear and complex, which makes it difficult to calculate and analyze. This section explains the model further by describing the results of numerical simulation experiments. Using a social network with N individuals, under different threshold characteristics of an individual, we mainly explored the evolution of group opinions over time when individual opinions in the group are both positively and negatively affected.

To answer the research questions, we studied the dynamics of the evolution of group opinion based on the proposed model from the opinion of three elements: different threshold conditions, social network structures, and the distribution of initial opinions. We construct the network with $N = 100$ individuals. The individual's opinion at time t is one-dimensional and obeys a continuous value between 0 and 1. The individual's assimilation and repulsion thresholds are also taken from 0 to 1, and each satisfies $\epsilon_T < \epsilon_R$. Each experiment runs 20 Monte Carlo simulations, and each simulation performs 10^3 time steps. $\lambda = 0.5$.

The increase of the repulsion area directly leads to the gradual bipolarization of the individuals' opinions in the group, but the expression of the number of final opinions cannot describe the distribution of the group's final opinions or the degree of group bipolarization. Therefore, this research applies the method of calculating the group bipolarization intensity Flache (2018) proposed and then explores the influence of different conditions on the evolution of group opinions from simulations [21, 22]. The degree of the group's bipolarization is obtained by calculating the variance of the opinion distances of all pairs in the group at time t by Eq. 5. In Eq. 5, $\overline{d(t)}$ represents the average opinion difference between all agents in the entire population. When the opinion difference between all agents is zero, the group has reached a consensus, which corresponds to $B_i = 0$. If the population consists of two maximally distant factions of equal size, which represents the group that has reached complete opinion bipolarization, $B_i = 1$.

$$B_i(t) = \frac{4}{N^2} \sum_{ij}^{i \in N, j \in N} (|x_i(t) - x_j(t)| - \overline{d(t)})^2 \tag{5}$$

3.1   The Effect of Different Thresholds on the Evolution of Opinions

First, we apply the structure of the WS small-world complex network, which is representative of complex networks, to explore our model. While still an abstraction of the more complex features of real online social networks, WS small-world networks combine two features observed in many real-world social networks, a high degree of local clustering of ties and a short average path length, reflecting that closely-knit local communities are typically connected via "long-range" or "weak" ties, which preserve the overall connectivity of a population [23, 24]. In the process of exploring the effect of thresholds on opinion evolution and the degree of bipolarization when opinions reach the steady state in our proposed model, we construct the WS small-world network structure containing $N = 100$ nodes, with an average node degree of $k = 6$, and a rewiring probability of $p = 0.1$. The initial opinion of the nodes is randomly uniformly distributed between 0 and 1.

Figure 3 illustrates the effect of different assimilation thresholds on the opinion evolution and the degree of bipolarization under the initial conditions described above, with a fixed repulsion threshold ($\epsilon_R = 0.7$). The value of the degree of bipolarization bi when the opinion evolution reaches the steady state in the network for the corresponding conditions is displayed in the bottom right corner of the corresponding subfigure. The assimilation thresholds in the four subgraphs of the representative opinions evolutionary process are 0.1, 0.3, 0.5, and 0.7, respectively. From Fig. 3, we find that under conditions of the presence of repulsive social influences, increasing individual openness and tolerance promotes the formation of the phenomenon of bipolarization to a certain extent. This is contrary to our intuitive conclusion that increasing an individual's trust in others decreases the phenomenon of bipolarity. Thus, the role of positive and negative social influences on opinions is complex and should be analyzed on a case-by-case basis when exploring the evolution of opinions.

To explore the robustness of this result, we conducted a simulation experiment in which we generated twenty different initial opinion distributions and network structures, and studied for each distribution the opinion dynamics for four values of the assimilation threshold ϵ_T. Figure 4 demonstrates that our finding that increasing assimilative social influence can generate more bipolarization is robust to randomized initial conditions. The figure shows box plots of the bipolarization index calculated when each of the simulation runs had reached equilibrium. When the assimilation threshold is small, at $\epsilon_T = 0.1$, the average degree of bipolarization is higher than zero but lower. In this case, the dynamics

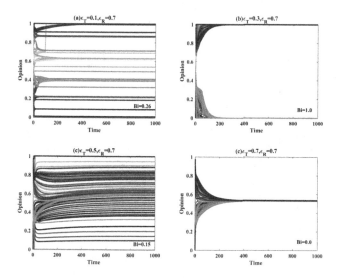

Fig. 3. Typical evolution of opinions in a WS small-world network with different assimilation thresholds. The repulsion threshold is fixed at $\epsilon_R = 0.5$. The assimilation threshold is set to $\epsilon_T = 0.1$ (panel a), $\epsilon_T = 0.3$ (panel b), $\epsilon_T = 0.5$ (panel c) and $\epsilon_T = 0.7$ (panel d), respectively.

usually generate multiple coexisting clusters of opinions, which leads to a relatively low degree of bipolarization. While, when the assimilation threshold is increased to $\epsilon_T = 0.3$, the average degree of bipolarization is higher and almost always close to 1. In this case, the dynamics influenced by extreme opinions usually produce bipolarization due to the larger assimilation threshold. As the assimilation threshold continues to increase to 0.5, individuals in the intermediate opinion state are influenced by the pull of the two end opinions to form multiple opinion clusters in the medium opinion interval, thus leading to a decrease in the degree of bipolarization. Finally, when the assimilation threshold is large enough, the network opinions reach a consensus.

Next, we analyze the effect of the repulsion threshold on the evolution of opinions and the degree of bipolarization. Figure 5 illustrates the graph of the evolution of the opinions for a fixed assimilation threshold of 0.3 ($\epsilon_T = 0.3$), and repulsion thresholds (ϵ_R) of 0.3, 0.65,0.8,0.95, respectively.

From Fig. 5 we observe a decrease in the role of negative social influence, i.e., an increase in the repulsion threshold, and a decrease followed by an increase and then a decrease in the degree of bipolarization of opinions in the network. The effect of the repulsion threshold on opinion evolution is non-monotonic. The robustness of that counterintuitive result we tested in Fig. 6 as well. The reason that the degree of bipolarization is promoted as the repulsion threshold increases is that when the repulsion threshold is large at 0.8 the positive influence is given some time to allow the individuals in the intermediate state to have time

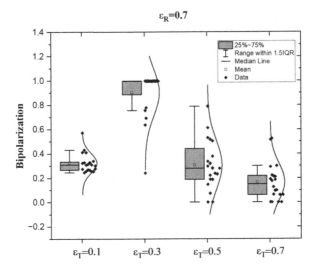

Fig. 4. Simulation experiment testing the robustness of the counter-intuitive finding that increasing the assimilation threshold ϵ_T can generate more polarization. Initial opinion values are randomly distributed between 0 and 1. In four treatments, we increased the assimilation threshold from $\epsilon_T = 0.15$ (Panel a) to $\epsilon_T = 0.3$ (Panel b), to $\epsilon_T = 0.5$ (Panel c), to $\epsilon_T = 0.7$ (Panel d). Blue areas show the interquartile range (IQR). The black dots identify the observed degree of bipolarization observed in the 20 runs per treatment. (Color figure online)

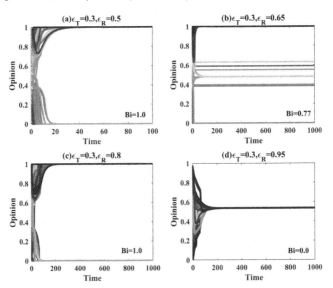

Fig. 5. Typical evolution of opinions in a WS small-world network with different repulsion thresholds. The assimilation threshold is fixed at $\epsilon_T = 0.3$. The repulsion threshold is set to $\epsilon_R = 0.5$ (panel a), $\epsilon_R = 0.65$ (panel b), $\epsilon_R = 0.8$ (panel c) and $\epsilon_R = 0.95$ (panel d), respectively.

to converge to the extreme individual opinions, and thus the phenomenon of bipolarization occurs.

3.2 The Effect of the Size of Neighbors on the Evolution of Opinions

As we know opinion updating is mainly influenced by the surrounding neighbors in the network structure. Since individuals are simultaneously influenced by a combination of their surrounding neighbors, we next explore the effect of the number of neighbors on the degree of bipolarization of opinions. Figure 7 shows for all three network structures a heat map of the average degree of bipolarization broken down by ϵ_T and ϵ_R values. Since Fig. 7 reports the average degree of bipolarization across multiple runs, bipolarization is measured when opinion

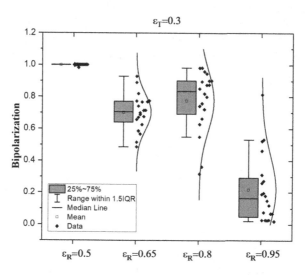

Fig. 6. Box plots of the distribution of the bipolarization measure in equilibrium, based on 20 simulation runs. In four treatments, we fixed the assimilation threshold value to 0.3, and increased the repulsion threshold from $\epsilon_R = 0.5$ (Panel a) to $\epsilon_R = 0.65$ (Panel b), to $\epsilon_R = 0.8$ (Panel c), to $\epsilon_R = 0.95$ (Panel d).

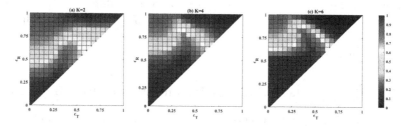

Fig. 7. Heat map of average degree of bipolarization in equilibrium for different assimilation and repulsion thresholds, the initial opinion is randomly distributed between 0 and 1. (a) $k = 2$. (b) $k = 4$. (c) $k = 6$.

evolution reaches a steady state in every condition. The initial opinion of the nodes is randomly uniformly distributed.

From Fig. 7, we find that as the average degree of the nodes in the WS small-world network structure gradually increases, the threshold range of the degree of bipolarization, when the opinions in the network reach the steady state, increases significantly. This is because the initial opinion state of the nodes in the network is uniformly randomly distributed, with the increase of node degree, the possibility of an individual encountering a neighbor with a large difference increases, and when the repulsion threshold is small, it will be easy to trigger the repulsion mechanism and thus form a bipolar state. Therefore, when the repulsion threshold is small, reducing the node degree value will weaken the degree of bipolarization to some extent. Accordingly, when the assimilation threshold is large, the area of the reached consensus state increases with the node degree. This is because when assimilation thresholds are large, increasing node averaging increases the degree to which individuals are positively socially influenced, providing the possibility of consensus. Therefore, when the node degree is large, it is important to increase the assimilation threshold to as large a value as possible.

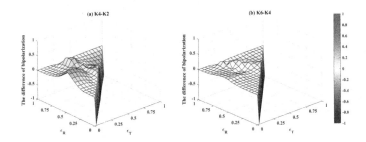

Fig. 8. Heat map of the difference in the degree of bipolarization with increasing average node degree (a) The average node degree varies from $k = 2$ to $k = 4$. (b) The average node degree varies from $k = 4$ to $k = 6$.

The specifics of whether the degree of bipolarization in Fig. 7 increases or decreases with increasing node degree in the corresponding threshold condition we show by comparison in Fig. 8. We find in all three subfigures the counterintuitive conclusion obtained above, that the effect of the assimilation threshold on the degree of bipolarization is non-monotonic. It is essential that we should analyze the situation on a case-by-case basis.

3.3   The Effect of the Initial Opinion on the Evolution of Opinions

Above the initial opinion of the network is uniformly randomly distributed. Next, we set the initial opinion distribution to a biased random distribution. The specific construction rule is that, since the WS small-world network structure is developed in the nearest-neighbor network, we divide the nodes into two groups according to the numbering order before randomly rewiring the edges; the first 50 nodes belong to one subgroup, whose initial opinion distribution is between 0 and 0.5; and the remaining 50 nodes belong to the other subgroup, whose nodes' initial opinion distribution is between 0.5 and 1. Under such conditions, the network is split into two subgroups with different opinion tendencies, and the subgroups are connected by a small number of rewiring edges. We also repeated the simulation experiment 20 times under different initial conditions and calculated the degree of bipolarization as the opinion evolved to the steady state in each of the experiments, and the averaged results are displayed in Fig. 9. The three subgraphs in Fig. 9 show the different node averaging degrees. And the results of comparing the initial opinion uniform random distribution and the biased distribution in the corresponding network structure are displayed in Fig. 10.

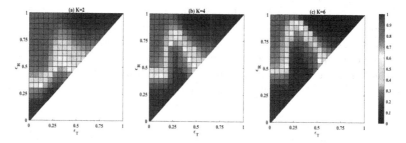

Fig. 9. Heat map of average degree of bipolarization in equilibrium for different assimilation and repulsion thresholds, initial opinions of individuals in the network are biased distributed. (a) $k = 2$. (b) $k = 4$. (c) $k = 6$.

From Figs. 9 and 10, we find that when the assimilation threshold is small (less than 0.25) or large (greater than 0.3) for different node degrees, the distribution of opinions with biased tendencies in the initial moments of the two subcommunities reduces the degree of bi-polarization due to the fact that it blocks the contact between opinions with greater differences from the perspective of the community structure; whereas when the assimilation threshold is increased to between 0.25 and 0.3, the initial opinion biased distribution would greatly contribute to the state of bipolarization, this is because the opinion differences within the sub-communities are highly likely to form the consensus within the community under that condition thus in stimulating the repulsion mechanism to form the state of opinion bipolarization.

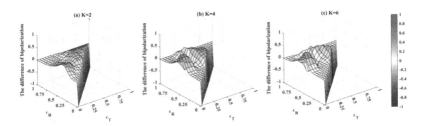

Fig. 10. Heat map illustrates the difference in opinion bipolarization between an initial opinion uniformly randomly distributed and a biased distribution under the same node average degree.

4 Conclusions

The openness of the social medium not only enables the public release of the opinions of the individuals but also makes the individuals with similar opinions aggregate, enabling the "minority" to create its own "opinion niche", thus forming an opposition opinion with the majority and providing the possibility for the "anti-silence spiral". In this paper, we add the repulsion mechanism to the bounded confidence model and then propose a new opinion dynamics model to explore the effects of individuals' simultaneous exposure to assimilative social influences and repulsion social influences on the evolution of group opinions. From the model proposed in this paper we find that increasing individuals' openness to assimilative social influences does not necessarily reduce macro-level opinion divergence: sometimes, more micro-level assimilative influences lead to more, not less, opinion bipolarization. At the same time, as repulsion social influence decreases, i.e., an increase in the repulsion threshold also produces more degrees of bipolarization. We also show that increasing personal exposure to different opinions as the degree of nodes increases promotes opinion polarization. Finally, the presence of more connections between users who like each other, thus reducing the relative amount of exposure leading to exclusionary influences, can reduce opinion polarization under certain threshold conditions. It is argued that micro-level interactions can produce complex and counterintuitive results at the macro level.

References

1. Li, Z.P., Tang, X.J.: Exogenous covariate and non-positive social influence promote group polarization. J. Manag. Sci. China **16**(3), 73–81 (2013)
2. Li, Z., Tang, X.: Polarization and non-positive social influence: a Hopfield model of emergent structure. Int. J. Knowl. Syst. Sci. (IJKSS) **3**(3), 15–25 (2012)
3. Mäs, M., Flache, A.: Differentiation without distancing. Explaining bi-polarization of opinions without negative influence. PloS one **8**(11), e74516 (2013)
4. Weron, T., Szwabiński, J.: Opinion evolution in divided community. Entropy **24**(2), 185 (2022)

5. Liu, S., Maes, M., Xia, H., Flache, A.: When intuition fails: the complex effects of assimilative and repulsive influence on opinion polarization. Adv. Complex Syst. **25**(08), 2250011 (2022)
6. Keijzer, M.A., Mäs, M.: The complex link between filter bubbles and opinion polarization. Data Sci. **5**(2), 139–166 (2022)
7. Lee, J.K., Choi, J., Kim, C., Kim, Y.: Social media, network heterogeneity, and opinion polarization. J. Commun. **64**(4), 702–722 (2014)
8. Finkel, E.J., et al.: Political sectarianism in America. Science **370**(6516), 533–536 (2020)
9. Dandekar, P., Goel, A., Lee, D.T.: Biased assimilation, homophily, and the dynamics of polarization. Proc. Natl. Acad. Sci. **110**(15), 5791–5796 (2013)
10. Bail, C.A., et al.: Exposure to opposing views on social media can increase political polarization. Proc. Natl. Acad. Sci. **115**(37), 9216–9221 (2018)
11. Del Vicario, M., Zollo, F., Caldarelli, G., Scala, A., Quattrociocchi, W.: Mapping social dynamics on Facebook: the Brexit debate. Soc. Netw. **50**, 6–16 (2017)
12. Flache, A., et al.: Models of social influence: towards the next frontiers. J. Artif. Soc. Soc. Simul. **20**(4), 1–31 (2017)
13. Asch, S. E.: Studies of independence and conformity: I. A minority of one against a unanimous majority. Psychol. Monogr. Gen. Appl. **70**(9), 1 (1956)
14. Mastroeni, L., Vellucci, P., Naldi, M.: Agent-based models for opinion formation: a bibliographic survey. IEEE Access **7**, 58836–58848 (2019)
15. Zhu, H., Hu, B.: Impact of information on public opinion reversal-An agent based model. Phys. A **512**, 578–587 (2018)
16. Schweighofer, S., Garcia, D., Schweitzer, F.: An agent-based model of multidimensional opinion dynamics and opinion alignment. Chaos Interdiscip. J. Nonlinear Sci. **30**(9), Article no. 093139 (2020)
17. Zhao, Y., Zhang, L., Tang, M., Kou, G.: Bounded confidence opinion dynamics with opinion leaders and environmental noises. Comput. Oper. Res. **74**, 205–213 (2016)
18. Jager, W., Amblard, F.: Uniformity, bipolarization and pluriformity captured as generic stylized behavior with an agent-based simulation model of attitude change. Comput. Math. Organ. Theory **10**, 295–303 (2005)
19. Deffuant, G., Bertazzi, I., Huet, S.: The dark side of gossips: Hints from a simple opinion dynamics model. Adv. Complex Syst. **21**(06n07), 1850021 (2018)
20. Macy, M.W., Kitts, J.A., Flache, A., Benard, S.: Polarization in dynamic networks: a Hopfield model of emergent structure. Dyn. Soc. Netw. Model. Anal. (2003)
21. Flache, A.: About renegades and outgroup haters: modeling the link between social influence and intergroup attitudes. Adv. Complex Syst. **21**, 1850017 (2018)
22. Flache, A., Macy, M.W.: Small worlds and cultural polarization. J. Math. Sociol. **35**, 146–176 (2011)
23. Zhou, G., et al.: Outer synchronization investigation between WS and NW small-world networks with different node numbers. Phys. A **457**, 506–513 (2016)
24. Bassett, D.S., Bullmore, E.T.: Small-world brain networks revisited. Neuroscientist **23**(5), 499–516 (2017)
25. Pariser, E.: The Filter Bubble: How the New Personalized Web is Changing What We Read and How We Think. Penguin, New York (2011)
26. Sherif, M.: The psychology of social norms (1936)
27. The Global Risk Report 2023, World Economic Forum, Geneva, 2023 (2023). https://reports.weforum.org/global-risks-2023/

The Construction Method of Knowledge Graph on Photovoltaic Industry Chain

Jinshuang Zhou and Xian Yang[✉]

School of Management Science and Engineering,
Dongbei University of Finance and Economics, Dalian 116025, China
yangxian1600@126.com

Abstract. As a strategic emerging industry supported by the state, the photovoltaic (PV) industry is supported by the national industrial policy and highly valued by local governments. With the continuous development of information technology, the introduction of knowledge graph technology into the research of PV industry chain can significantly improve the efficiency of representation, retrieval, and analysis of the PV industry chain, provide a deeper understanding of the structure and development dynamics of the PV industry chain, and provide powerful support for industrial research and decision-making. Therefore, this paper builds a preliminary PV industry chain knowledge graph through a comprehensive analysis of the industries, enterprises, and products involved in the PV industry chain, and realizes data storage and visualisation using the Neo4j graph database. This paper combines the knowledge graph with the PV industry to fully explore the industry chain information, which helps to grasp the overall situation and development trend of the industry timely identify the bottlenecks and risks in the industry chain, formulate more effective risk management and countermeasures, continuously optimize the industry chain structure and promote the sustainable development of the PV industry.

Keywords: Photovoltaic industry chain · Knowledge graph · Neo4j

1 Introduction

The photovoltaic (PV) industry is one of China's strategic emerging industries, and its development is of great significance in adjusting the energy structure, promoting the energy production and consumption revolution, and promoting the construction of ecological civilization. Under the combined influence of the "double carbon" policy and corporate technological innovation, China's PV industry is gaining momentum and is being applied in many fields such as construction, agriculture, transportation and industry, and is expected to continue to penetrate into various scenarios in multiple fields.

However, the PV industry chain has both vertical and horizontal multi-dimensional extensions, vertically, the upstream and downstream relationship is more complex, the same product may have a variety of raw materials synthesis, there may be substitution between the materials, different downstream

J. Chen et al. (Eds.): KSS 2023, CCIS 1927, pp. 274–284, 2023.
https://doi.org/10.1007/978-981-99-8318-6_19

application scenarios require different product mix. Horizontally, there are many related enterprises in the upper, middle and lower reaches of the industry chain, some of which are involved in multiple industries and play a role in different areas of the industry chain. As an important part of artificial intelligence (AI), knowledge graph provides an important technical means to extract structured knowledge from massive unstructured data and use graph analysis for association relationship mining, which can effectively solve the disadvantages of the current industrial chain research such as weak visualisation ability and inconvenience in using and retrieving, and at the same time can provide a visual presentation.

2   Literature Review

The Knowledge Graph was originally conceived by Google in 2012 to improve the capabilities of the search engine and to improve the quality of search and the search experience for users [4], is a knowledge representation method that enables real-world entities, concepts and relationships to be represented graphically. It helps people to better understand and utilize knowledge by abstracting and modeling the real world and presenting the relationships between various entities and concepts in a graphical way [6].

From an academic perspective, knowledge graphs usually consist of three parts: entities, attributes, and relationships. Entities are specific things in the real world, such as people, places, organizations, etc.; attributes are characteristics or properties of entities, such as the age of a person, the latitude and longitude of a place, the size of an organization, etc.; and relationships are connections or associations between entities, such as the kinship between people and people, the distance between places and places, etc. In terms of scope of use, knowledge graphs can be divided into general knowledge graphs and domain knowledge graphs. Generic knowledge graphs mainly include encyclopedic knowledge graphs, such as DBpedia [1], Freebase [2], and Wikidata [11], and common-sense knowledge graphs, such as ConceptNet [10] and Microsoft ConceptGraph [3]. Zhai et al. [16] constructed a knowledge graph of Chinese medicine based on heterogeneous data from multiple sources, which can assist researchers in conducting innovative research in the field of Chinese medicine. Technology companies such as Tianeye Search have built business domain knowledge graphs based on industrial and commercial data. These knowledge graphs cover the basic information of enterprises and provide functions such as querying and calculating the association of enterprises. These knowledge graphs of business domains mainly focus on the relationships among enterprises, while the knowledge graphs of industry chains range from macro to micro, focusing on the scale and development of industries as well as the competition and synergy among enterprises [7]. Wang based on chemical industry chain data, used the graph database Neo4j to initially realize the construction and analysis of the knowledge graph of the chemical industry chain [13]. Xiao et al. [14] started from the urban rail transportation industry chain, analyzed the urban rail transportation industry along the upstream, middle and downstream of the industry chain, and constructed a comprehensive graph combining knowledge graphs and industry graphs.

However, there are still few studies on industry chain knowledge graphs, especially the lack of research on the systematic construction of knowledge graphs for the PV industry chain, and the existing industry chain knowledge graphs only unilaterally focus on the upstream and downstream relationships between enterprises [14] or upstream and downstream relationships between products [13]. It is difficult to form links between the enterprise level and the product level. Therefore, this paper proposes a method to build a PV industry chain knowledge graph that can connect enterprises and products, which can identify key enterprises and key links in the industry chain, so as to optimize the structure of the PV industry chain and achieve efficient development of the PV industry.

3 Construction of a Knowledge Graph of the PV Industry Chain Framework

Knowledge graphs are usually constructed in two ways, "bottom-up" or "top-down" [8]. The bottom-up approach requires that all entities be categorized and grouped. The bottom-up approach involves categorizing all entities into the most detailed sub-categories and then moving up the hierarchy to form larger categories, which is commonly used in the construction of generic knowledge graphs. A top-down approach requires defining the data schema for the schema, starting with the top-level concepts and gradually refining them downwards to form a tree-like schema, and finally corresponding the entities to the concepts, which is generally applicable to the construction of domain knowledge graphs [9]. Therefore, the PV industry chain knowledge graph, as a kind of domain graph, adopts a top-down construction mode.

The construction process is divided into three main steps, knowledge modeling, knowledge extraction and knowledge storage, and the specific knowledge graph construction process is shown in Fig. 1. In the process of building the domain knowledge graph, knowledge extraction is firstly carried out to extract valuable information from structured or unstructured data based on techniques such as natural language processing. The sub-tasks of knowledge extraction include entity extraction, relationship extraction and property extraction. Secondly, knowledge modeling, the process of building computer-interpretable knowledge models, is performed, most notably for ontology construction. An ontology is a specification that conceptualizes the real world, abstracts the characteristics of different entities and generalizes them into different classes and relationships. Finally, knowledge storage is carried out, and there are two specific methods: knowledge storage based on table structure and knowledge storage based on graph structure. Since a knowledge graph emphasizes the entities and the relationships existing between entities, i.e., it is a knowledge graph with a directed graph structure, it is more suitable to choose knowledge storage based on graph structure. The graph database is flexible, has great scalability and is easier to update and maintain [12]. Therefore, in this paper, the Neo4j graph database is used as a tool to store the knowledge graph of the PV industry chain.

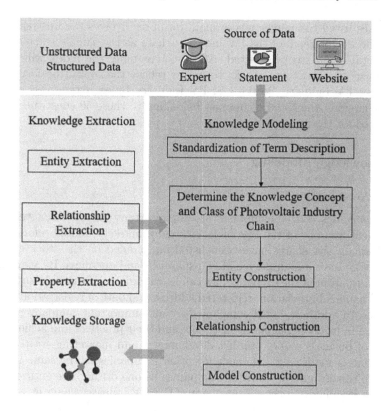

Fig. 1. Framework for building a knowledge graph of the PV industry chain

4   A Method for Building a Knowledge Graph of the PV Industry Chain

The data used in this study mainly includes the industries involved in the PV industry, the enterprises involved in each link of the upstream, midstream and downstream of the industry chain, as well as the upstream and downstream relationships between the main products of the enterprises. The data types studied in this paper contain both structured and unstructured data. The structured data source of the study mainly comes from the structured data crawl of the website, and the content of the crawl is mainly the company entity and its attributes, as well as the industry entity and its attributes, and the main product entity of the company. The unstructured data used in this paper is mainly the text data obtained from the PV website, and then the text data is extracted into the form of triples to build the upstream and downstream material relationship and product subcategory relationship of the product entity.

In order to provide comprehensive coverage of the upstream, midstream and downstream segments of the PV industry chain, this study uses the CSI PV Industry Index as the basis for the selection of companies. The CSI PV Industry

Index takes listed companies whose main business involves the upstream, midstream and downstream of the PV industry chain as the sample to be selected, with a total of 50 stocks selected. At the same time, the index will periodically adjust its constituent stocks to better reflect the overall performance of the current PV industry. So the CSI PV Industry Index is the most representative. Therefore, the listed companies belonging to these 50 constituent stocks are selected for the study.

4.1 Knowledge Modeling

Entity Construction

Listed Companies. Listed companies are an important part of China's economic development and one of the key achievements of market-oriented reforms. As a listed company, its shares have been listed and traded on the stock exchange and therefore have a high degree of transparency and openness. By viewing the public information of the exchange, we can learn various information such as the code, full name, abbreviation, registered address and date of registration of listed companies. In addition, listed companies are also important company representatives and industry benchmarks in China, and their practices and achievements in terms of business scope, product development and marketing strategies will also serve as a model and reference for other companies in the same industry. Therefore, this graph selects listed companies as one of the base entities and is important for analysing the development of the PV industry chain by analysing the operational status and business development of listed companies.

Industries. Another core element of the industry chain graph, is an important medium for connecting industries, companies and products. Through data such as industry indicators and hot sectors, we can better understand the structure of the industry chain and the relationships within the industry. At present, the most representative industry classification indicators in China are the Shen Wan Three-tier Industry Classification and the National Economic Classification. The classification of industries to which Chinese listed companies belong is based on financial data that has been audited by accounting firms and publicly disclosed [5]. The Shen Wan Industry Classification is used as the basis for this graph. According to the latest CSI PV Industry Index, the top five sub-sectors (Shenwan Level 3) to which listed companies belong are silicon wafers, PV cell modules, inverters, PV processing equipment and PV auxiliary materials. In addition, it also contains sub-sectors such as PV power generation and power transmission and transformer.

Products. Products mainly refer to the main scope of the company and the products that it operates. By understanding the main products that a company operates, you can position a company accordingly. In addition, there are other products that are currently required for the company to carry out its main business operations, including the raw materials required to manufacture the products.

Relationship Construction

The relationships that exist in the PV industry chain are very complex. An industry chain will involve relationships between industries and different industries, relationships between companies and companies, and relationships between various products. In this paper, through multiple extraction and induction, six relationships are finally formed, which can explain the PV industry chain more comprehensively. The specific relationships constructed and their meanings are shown in Table 1.

Table 1. PV industry chain knowledge graph construction relationships

relationships	meaning
industry_involved	Industry breakdown of the PV industry
superior_industry	Industry superordination constructed according to the Shen Wan three-level classification
industry	Industry classification of listed companies
main_products	Main products of the listed company's operations
upstream/downstream_material	The upstream and downstream relationships between products, which show the logical relationship between products, include two categories: upstream raw materials and downstream products
product_subclass	For a product, it is possible to divide it into different hierarchical categories, so that in the absence of a broad product name, the corresponding indicator can be obtained by calculating the smaller categories

Model Construction

In addition to the above-defined entity-to-entity association relationships, the knowledge graph also includes the linkage between entities and properties, i.e. entity-property-property values, and different entities have different property values. Accordingly, a schematic representation of the PV industry chain knowledge graph is constructed as shown in Fig. 2. The relationships between entities are connected by solid lines, and the properties contained in the entities are connected to the entities by dashed lines, and the specific properties contained are displayed in rectangular boxes.

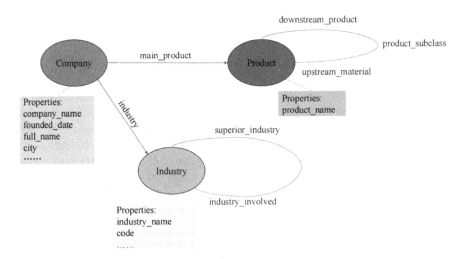

Fig. 2. Example of a PV industry chain knowledge graph model

4.2   Knowledge Extraction

The data types required for the construction of the PV industry chain knowledge graph are mainly divided into two types: structured data and unstructured data. In the process of knowledge extraction, this paper mainly adopts a rule-based pattern-matching approach and a seed-based heuristic approach. The data we use is mainly derived from the text data on the photovoltaic website. We treat it beforehand so that it meets the requirement of the rule-based pattern-matching approach. Figure 3 shows an example of a regular expression template for upstream raw material relationship extraction.

Regex1 : "(?P<productA>.*?) is the raw material of (?P<productB>.*?)"

Regex2 : "The raw material of (?P<productB>.*?) is (?P<productA>.*?)"

Regex3 : "The upstream raw material of (?P<productB>.*?) is (?P<productA>.*?)"

Fig. 3. Regular expression template

For the relationships of product subclasses, this paper uses a seed heuristic for extraction. Specifically, some high-quality entity-relationship pairs are prepared as initial seeds, such as < "inverter", "product subclass", "grid-connected inverter" >. Using these seeds as a basis, we match relevant sentences in a large-scale corpus and extract some reliable templates, such as "A is a kind of B". These templates are then used to discover further examples.

In this paper, the objects of knowledge graph quality assessment mainly involve three types of individual objects, namely concepts, entities, and attributes, as well as the relationships between concepts, the relationship between concepts and entities, and the relationship between entities, which can evaluate all aspects of the knowledge graph [15]. Due to the small amount of data used in this paper, all the used data have been manually detected, so the high quality of the knowledge graph can be guaranteed.

4.3   Knowledge Storage

By processing and extracting the data from the PV industry, a PV industry chain knowledge graph containing 1053 nodes and 998 relationships was finally formed. The details of the data are shown in Table 2.

Table 2. PV industry chain knowledge graph data information

type	name ·	quantity
node	company	50
	industry	39
	product	964
relationship	industry	50
	industry_involved	8
	superior_industry	28
	main_product	516
	upstream/downstream_material	191
	product_subclass	250

The language used to represent, query and update the Neo4j graph database is Cypher, a Neo4j-specific query language. An example of a node creation and relationship statement using the Cypher language is shown in Fig. 4.

```
CREATE (n:label {property:'property value'}) RETURN n
    MATCH (p{property:'property value'})
    MATCH (q{property:'property value'})
    CREATE (p)-[r:relationship name] -> (q)
```

Fig. 4. Cypher language creation statements

This paper uses Py2neo to connect to the Neo4j database and import the data in bulk into the Neo4j graph database for storage and visual presentation.

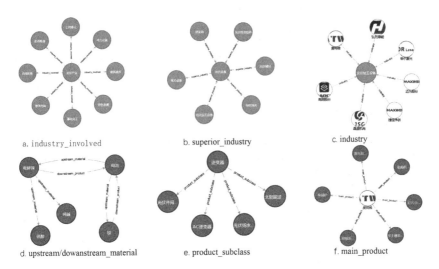

Fig. 5. Partial extraction of knowledge graph relations in the PV industry chain (Color figure online)

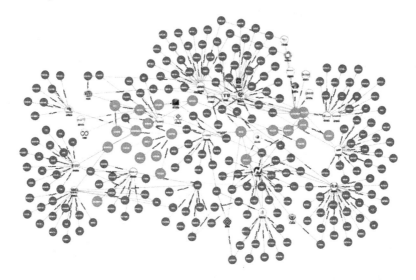

Fig. 6. PV industry chain knowledge graph (partial) (Color figure online)

Finally, the relationships of the PV industry chain were partially extracted, and the six extracted relationships are partially illustrated in Fig. 5, where orange nodes represent industries, blue represent listed companies and red represent products. The final constructed PV industry chain knowledge graph (partly) is shown in Fig. 6. The diagram provides information on the details of the industries involved in the PV industry, the upstream and downstream relationships in the industry, the overlap status of each company's main products, and the correlation relationships that exist between upstream and downstream materials between products.

5   Conclusion

This paper proposes a method to construct PV industry chain knowledge graph, by collecting and processing multi-source heterogeneous PV industry chain data, extracting entity, relationship and property knowledge, modeling the knowledge and finally inputting it into Neo4j graph database for data storage and visualisation. This paper creates a knowledge graph connected between the enterprise level and the product level that is missing from existing studies. The results show that it is feasible to apply industry chain knowledge graph to the PV industry. From the perspective of the industry as a whole, the construction of PV industry chain knowledge graph can assist in improving the layout of the PV industry, adjusting the industrial structure and providing a scientific basis for various decisions. From an individual perspective, the graph can provide a basis for government policies and financial subsidies; for enterprises, the graph can support their development direction and business expansion; for investors, the graph can also provide reference for their investment decisions. Therefore, the construction of an industry chain knowledge graph has an important role and significance for the development of China's PV industry.

There are some limitations in this paper, such as the data of PV industry is changing in real time, how to improve the knowledge graph and build a dynamically evolving knowledge graph of PV industry chain is a problem to be considered in the next research. Moreover, the amount of data used in this paper is relatively small, and the quality of the constructed knowledge graph has mainly been manually detected. The main purpose of knowledge graph is to apply, so the application effect should be the best test method. So in future studies, we will expand our dataset and the subsequent in-depth research will apply the knowledge graph, and then prove its quality and reliability.

Acknowledgements. This work is supported by the National Natural Science Foundation of China (72101051, 72293565, 72293563), and School Project (PT202139).

References

1. Bizer, C., et al.: Dbpedia-a crystallization point for the web of data. J. Web Semant. **7**(3), 154–165 (2009)
2. Bollacker, K., Cook, R., Tufts, P.: Freebase: a shared database of structured general human knowledge. In: AAAI, vol. 7, pp. 1962–1963 (2007)
3. Ji, L., Wang, Y., Shi, B., Zhang, D., Wang, Z., Yan, J.: Microsoft concept graph: mining semantic concepts for short text understanding. Data Intell. **1**(3), 238–270 (2019)
4. Lin, L., et al.: Research on visualization system based on enterprises knowledge graph construction. J. Qingdao Univ. (Natl. Sci. Ed.) **32**(1), 55–60 (2019)
5. Liu, H.: Blockbuster open source: the construction ideas and data of the 100,000-level industrial chain map for listed companies are open (2021). https://zhuanlan.zhihu.com/p/429110045. Accessed 23 June 2023
6. Liu, Y., et al.: A survey on the visual analytics of knowledge graph. J. Comput.-Aided Des. Comput. Graph. **35**(1), 23–26 (2023)

7. Mao, R., et al.: Construction of knowledge graph of industry chain based on natural language processing. J. China Soc. Sci. Tech. Inf. **41**(3), 287–299 (2022)
8. Ruan, T., et al.: Research on the construction and application of vertical knowledge graphs. Knowl. Manag. Forum **1**(03), 226–234 (2016)
9. Shao, H., et al.: Knowledge Graph from Scratch Techniques, Methods and Cases. China Machine Press (2021)
10. Speer, R., Havasi, C.: ConceptNet 5: a large semantic network for relational knowledge. In: Gurevych, I., Kim, J. (eds.) The People's Web Meets NLP. TANLP, pp. 161–176. Springer, Heidelberg (2013). https://doi.org/10.1007/978-3-642-35085-6_6
11. Vrandečić, D., Krötzsch, M.: Wikidata: a free collaborative knowledgebase. Commun. ACM **57**(10), 78–85 (2014)
12. Wang, D., et al.: Study on the construction of knowledge map of sweet cherry industry in China. China Fruits (01), 104–108 (2023)
13. Wang, S.: Knowledge graph design and implementation for chemical industry chain. Chem. Eng. Manag. **23**, 47–48 (2021)
14. Xiao, E., Xue, F., Luo, J.: Industrial agglomeration spatial analysis and knowledge synthesis map construction of urban rail transit. Mod. Urban Transit **5**, 102–107 (2023)
15. Xiao, Y., et al.: Knowledge Graph Concepts and Techniques. Publishing House of Electronics Industry (2020)
16. Zhai, D., et al.: Constructing TCM knowledge graph with multi-source heterogeneous data. Data Anal. Knowl. Discov. **7**(09), 146–158 (2023)

How to Improve the Collaborative Governance of Public Opinion Risk in Emergencies? The View of Tripartite Evolutionary Game Analysis

Ning Ma[1] , Yijun Liu[1,2] (✉) , and Mingzhu Wang[1,2]

[1] Institute of Science and Development, Chinese Academy of Sciences, Beijing 100190, China
yijunliu@casisd.cn
[2] School of Public Policy and Management, University of Chinese Academy of Sciences, Beijing 100049, China

Abstract. Negative public opinions caused by rumors, fear and hostility during public emergencies can often exacerbate conflicts and hinder communication, thus creating new challenges for the management of emergencies. Taking the creation of a public opinion risk-control system as its entry point, this article builds a multi-subject, tripartite public opinion risk-control system to handle public opinion risks in situations of public emergencies. Based on this, a tripartite evolutionary game model is created to systematically analyze the strategy-making process undertaken by the tripartite for the purpose of achieving collaborative control. Through quantitative calculations and analysis, this study finds that when the decision makers have undertaken successful strategies and the policy enforcers have taken sensible actions, the general public are unlikely to be affected by rumors, as they will choose to not believe these or to refute these. Under such circumstances, the control of public opinion risks can yield optimal results.

Keywords: Public emergencies · Public opinion risk · Collaborative governance · Tripartite evolutionary game · Simulation modeling

1 Introduction

Every time a major public emergency occurs, it is essential for the government and related state departments to quickly respond to the situation by creating appropriate strategies and effectively controlling and diverting negative public opinions. Without a timely response, public opinion risks of all types are likely to arise, subsequently endangering public safety. For example, global extreme weather, global pollution incidents, global public health events, as the major emergency, has had a significant impact on the global economy and society.

Taking the creation of a public opinion risk-control system as its entry point and by clarifying the roles of multiple risk-control bodies and the relations among these bodies, this article builds a multi-subject, tripartite collaborative control system for handling public opinion risks arising from public emergencies. Such a system involves the joint effort of decision makers, policy enforcers, and general public. Then, based on the tripartite evolutionary game theory, this work creates a tripartite evolutionary game model involving the three risk-control bodies, with a particular focus on studying the different roles played by each of the three bodies in their attempt to control public opinion risks arising from public emergencies. Proceeding from the interactions between the three risk-control bodies, this study analyzes the correlation between the actions and strategies undertaken by the bodies and proposes optimal combinations of strategies.

2 Related Work

Evolutionary games refer to players' participation in constant feedback activities in dynamic systems, in which the participants are mutually constrained, mutually dependent, and mutually beneficial, with the aim of reaching a dynamic balance through learning and imitation [1]. With the development of evolutionary game theory, it has been gradually seeped to human behaviors and combined with other theories to apply in different fields [2]. Many scholars have conducted in-depth investigations on the applicability of the evolutionary game model, extending two-party evolutionary games [3] to tripartite evolutionary games [4]. Additionally, scholars have also carried out simulation studies by combining evolutionary game analysis with analytical methodologies such as the system dynamics model [5] and network theory [6]. Some studies have combined evolutionary game analysis with the classical susceptible-infected-removed (SIR) model to explore the trends of information diffusion [7], social hotspot propagation [8], rumor propagation [9], etc.

Social governance covers many areas. In environmental protection and management, research has investigated the interactions between evolutionary game players such as enterprise and consumers [10], UR and Enterprise [11], government and green startups [12], tripartite evolutionary gaming involving the government, general public and companies, etc. [13] For investigating the management of online communities, some studies have adopted the perspectives of the stakeholders through a tripartite evolutionary game model involving web regulatory bodies, webcasting platforms, and webcasting users [14]. The evolutionary game model was developed from the perspective of cooperative supervision between the government and consumers [15]. From the perspective of controlling Internet rumors, other studies have investigated the evolutionary gaming process among the government, Internet service providers and those spreading rumors [16].

3 Tripartite Evolutionary Game Model

3.1 Model Framework

Based on the components of the traditional pyramid-shaped social governance structure, namely, the decision makers, the policy enforcers and the general public, the current study modifies the two-way organizational model of "receiving orders from above and collecting information from below" and introduces a tripartite governance system characterized by a goals-implementation-feedback-optimization process (Fig. 1).

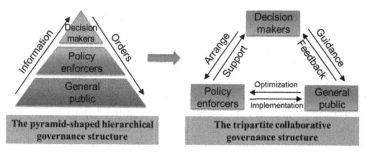

Fig. 1. The governance system for managing public opinion risks during emergencies

3.2 Basic Hypotheses

Behavioral Strategies of the Game Players. *Strategies of the decision makers.* In relation to the strategies undertaken by the decision makers, there are two possible outcomes: strategy success and strategy failure. In other words, the decision makers' strategy set S_M = {strategy success S_{M1}, strategy failure S_{M2}}. In this hypothesis, x is the success rate of the decision makers' strategies, and $(1 - x)$ is the failure rate of the strategies, where x is a function of time t. *Strategies of the policy enforcers.* As the concrete bodies for the execution of orders, policy enforcers often have two options: action and inaction. Thus, the policy enforcers' strategy set S_E = {action S_{E1}, inaction S_{E2}}. *Strategies of the general public.* As the general public, the hearers will face challenges in making correct decisions when they are exposed to public emergency-related rumors concerning their own safety and the safety of their property. Normally, they can either choose to identify and refute rumors or to believe and spread rumors. Subsequently, the general publics' strategy set S_H = {identify and refute rumors S_{H1}, believe and spread rumors S_{H2}} (Fig. 2).

Fig. 2. Strategic choices of the collaborative risk-control bodies in the tripartite evolutionary game model

Model Variables and Parameters. *Variables and parameters for the decision makers.* R_M: The decision makers have achieved strategy success, managed to effectively contain public opinion risks, and created potential benefits by raising the government's reputation among the general public. C_M: This variable represents the time, labor and other costs needed for decision makers to make prudent decisions and achieve strategy success. L_M: The decision makers have suffered strategy failures, which have in turn damaged the government's reputation among the general public, causing a negative impact on its image. αC_M: Despite strategy failures resulting from limited effort and resources, the decision makers may still generate hidden benefits because they can use conserved effort and resources to attend to other tasks. L_{MZ}: In addition to strategy failures by decision makers, inactive policy enforcers may cause further damage to the government's reputation and can thus increase the hidden costs for future social governance.

Variables and parameters for the policy enforcers. R_E: The policy enforcers effectively responded to the situation by taking responsible and sensible actions and created potential benefits by winning the support of the public as well as gaining recognition from the decision makers. γC_E: This variable represents the policy enforcers' labor and time costs for finding sensible ways to handle risks, by constantly organizing meetings to discuss issues and sending related personnel to carry out tasks, etc. γ indicates the reasonableness of the control measures undertaken by the policy enforcers to manage the risks. The more sensible the measures, the higher are the policy enforcers' costs for managing the risks, with the value of γ being $0 < \gamma < 1$. C_{EL}: Should strategy failures occur, the policy enforcers' labor, time and other costs for finding sensible ways to handle the risks will further increase. βP_H: The policy enforcers have acted responsibly by imposing fines on those general public who have engaged in behaviors, such as spreading rumors recklessly. After the collected fines enter the state revenue, a β percentage of the total amount of the fines collected will be returned to the policy enforcers to fund future work. The value of β is $0 < \beta < 1$. I_E: This variable represents the benefits created through policy enforcers' inaction, such as leisure time. L_E: This variable represents the potential losses resulting from the policy enforcers' inaction, such as complaints from the general public, and the policy enforcers' damaged reputation among the public.

Variables and parameters for the general public. S_H: When hearing rumors, some rational general public are able to refute the rumors and, as a result, will be rewarded accordingly by the policy enforcers. $(1-\gamma) C_H$: To be able to identify and refute rumors, the rational general public will have to analyze the information, make corresponding judgments, spend time searching for information on the Internet, etc. The more sensible the policy enforcers' risk-control measures, the less are the general publics' costs for identifying rumors. R_H: With the policy enforcers' inaction, those believing and spreading rumors may obtain extra illicit gains. P_H: This variable represents the monetary costs of those who blindly believe in rumors and help to spread rumors as they are fined for their behaviors. L_H: This variable represents the monetary losses as a result of the reckless purchases made by those who blindly believe in rumors and help to spread them.

3.3 Payoff Matrix

Based on the above hypotheses, a tripartite evolutionary game model involving the participation of the three risk-control bodies, namely, the decision makers, the policy enforcers, and the general public, was created. The possible combinations of the strategies undertaken by the three risk-control bodies are provided in Table 1 below.

Table 1. The payoff matrix of the combination of strategies undertaken by the tripartite

Combination of strategies undertaken by the decision makers, the policy enforcers, and the general public	Benefits of the decision makers, the policy enforcers, and the general public
(strategy success, action, identify and refute rumors)	$(R_M - C_M, R_E - \gamma C_E, S_H - (1 - \gamma)C_H)$
(strategy success, action, believe and spread rumors)	$(R_M - C_M, R_E - \gamma C_E + \beta P_H, P_H - L_H)$
(strategy success, inaction, identify and refute rumors)	$(R_M - C_M, I_E - L_E, S_H - C_H)$
(strategy success, inaction, believe and spread rumors)	$(R_M - C_M, I_E - L_E + \beta P_H, R_H - P_H - L_H)$
(strategy failure, action, identify and refute rumors)	$(\alpha C_M - L_M, R_E - \gamma C_{EL}, S_H - (1 - \gamma)C_H)$
(strategy failure, action, believe and spread rumors)	$(\alpha C_M - L_M, R_E - \gamma C_{EL} + \beta P_H, P_H - L_H)$
(strategy failure, inaction, identify and refute rumors)	$(\alpha C_M - L_{MZ}, I_E - L_E, S_H - C_H)$
(strategy failure, inaction, believe and spread rumors)	$(\alpha C_M - L_{MZ}, I_E - L_E + \beta P_H, R_H - P_H - L_H)$

4   Evolutionary Stable Strategies

4.1   The Replication Dynamic Equation for the Tripartite

Taking $U_{M(1)}$ as the expected gains obtained following the decision makers' strategy success, $U_{M(0)}$ as the expected gains achieved based on the decision makers' strategy failures, and U_M as the average gains, then:

$$U_{M(1)} = R_M - C_M \tag{1}$$

$$U_{M(0)} = y(\alpha C_M - L_M) + (1 - y)(\alpha C_M - L_{MZ}) = -yL_M + \alpha C_M - (1 - y)L_{MZ} \tag{2}$$

$$U_M = xU_{M(1)} + (1 - x)U_{M(0)} \tag{3}$$

Based on this, the replication dynamic equation for the decision makers' strategy success is thus as follows:

$$F(x) = \frac{d(x)}{d(t)} = x(1 - x)\left(U_{M(1)} - U_{M(0)}\right) = x(1 - x)\left[R_M - C_M + yL_M - \alpha C_M + (1 - y)L_{MZ}\right] \tag{4}$$

When $y = \frac{R_M - C_M - \alpha C_M + L_{MZ}}{L_{MZ} - L_M}$, $F(x) \equiv 0$, all states are stable; When $y \neq \frac{R_M - C_M - \alpha C_M + L_{MZ}}{L_{MZ} - L_M}$, taking $F(x) = 0$, one thus obtains two stable states: $x = 0$ and x

= 1. The derivative of F(x) is as follows:

$$\frac{dF(x)}{dx} = (1-2x)[R_M - C_M + yL_M - \alpha C_M + (1-y)L_{MZ}] \quad (5)$$

Then, two situations will arise:

When $y > \frac{R_M - C_M - \alpha C_M + L_{MZ}}{L_{MZ} - L_M}$, $\frac{dF(x)}{dx}|_{x=0} > 0$, $\frac{dF(x)}{dx}|_{x=1} < 0$, x = 1 is the evolution-ary stable point; When $y < \frac{R_M - C_M - \alpha C_M + L_{MZ}}{L_{MZ} - L_M}$, $\frac{dF(x)}{dx}|_{x=0} < 0$, $\frac{dF(x)}{dx}|_{x=1} > 0$, x = 0 is the evolutionary stable point.

The same reasoning leads to the fact that the replication dynamic equation for the policy enforcers' strategies for actions is as follows:

$$F(y) = d(y)/d(t) = y(1-y)\left(U_{E^{(1)}} - U_{E^{(0)}}\right) = y(1-y)[R_E - x\gamma C_E - (1-x)\gamma C_{EL} - I_E + L_E] \quad (6)$$

When $x > \frac{R_E - \gamma C_{EL} - I_E + L_E}{\gamma C_E - \gamma C_{EL}}$, $x\frac{dF(y)}{dy}|_{y=0} > 0$, $\frac{dF(y)}{dy}|_{y=1} < 0$, y = 1 is the evolutionary stable point; When $x < \frac{R_E - \gamma C_{EL} - I_E + L_E}{\gamma C_E - \gamma C_{EL}}$, $\frac{dF(y)}{dy}|_{y=0} < 0$, $\frac{dF(y)}{dy}|_{y=1} > 0$, y = 0 is the evolutionary stable point.

The same reasoning leads to the fact that the replication dynamic equation for the general publics' strategies for actions is as follows:

$$F(z) = d(z)/d(t) = z(1-z)\left(U_{H^{(1)}} - U_{H^{(0)}}\right) = z(1-z)\big[S_H - (1-y\gamma)C_H + P_H + L_H - (1-y)R_H\big] \quad (7)$$

When $y > \frac{C_H - S_H - P_H - L_H + R_H}{\gamma C_H + R_H}$, $\frac{dF(z)}{dz}|_{z=0} > 0$, $\frac{dF(z)}{dz}|_{z=1} < 0$, z = 1 is the evolu-tionary stable point; When $y < \frac{C_H - S_H - P_H - L_H + R_H}{\gamma C_H + R_H}$, $\frac{dF(z)}{dz}|_{z=0} < 0$, $\frac{dF(z)}{dz}|_{z=1} > 0$, z = 0 is the evolutionary stable point.

Based on the above process, one can obtain a dynamic trend diagram showing the decision makers' strategies, as shown in Fig. 3.

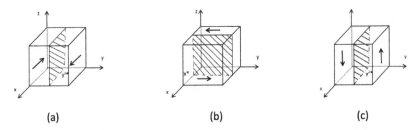

(a)   (b)   (c)

Fig. 3. The evolutionary gaming process of strategies.

4.2 The Equilibrium Points and the Stability of the Model

Equilibrium Points. Combining Eqs. (4), (6) and (7), one can obtain a three-dimensional evolutionary dynamical system:

$$
\begin{cases}
F(x) = \frac{d(x)}{d(t)} = x(1-x)\left(U_{M^{(1)}} - U_{M^{(0)}}\right) \\
\quad = x(1-x)\left[R_M - C_M + yL_M - \alpha C_M + (1-y)L_{MZ}\right] = 0 \\
F(y) = \frac{d(y)}{d(t)} = y(1-y)\left(U_{E^{(1)}} - U_{E^{(0)}}\right) \\
\quad = y(1-y)[R_E - x\gamma C_E - (1-x)\gamma C_{EL} - I_E + L_E] = 0 \\
F(z) = \frac{d(z)}{d(t)} = z(1-z)\left(U_{H^{(1)}} - U_{H^{(0)}}\right) \\
\quad = z(1-z)\left[S_H - (1-y\gamma)C_H + P_H + L_H - (1-y))R_H\right] = 0
\end{cases}
\tag{8}
$$

$$
\begin{cases}
R_M - C_M + yL_M - \alpha C_M + (1-y)L_{MZ} = 0 \\
R_E - x\gamma C_E - (1-x)\gamma C_{EL} - I_E + L_E = 0 \\
S_H - (1-y\gamma)C_H + P_H + L_H - (1-y)R_H = 0
\end{cases}
\tag{9}
$$

Solving equation system (8), one can obtain nine equilibrium points: $E_1(0, 0, 0)$, $E_2(0, 0, 1)$, $E_3(0, 1, 0)$, $E_4(0, 1, 1)$, $E_5(1, 0, 0)$, $E_6(1, 0, 1)$, $E_7(1, 1, 0)$, $E_8(1, 1, 1)$, and $E_9(x^*, y^*, z^*)$. Among these, $E_9(x^*, y^*, z^*)$ is the solution for equation system (9). In reference to Selten's research on dynamic replicator systems, both evolutionary stable strategy combinations and asymptotically stable states are pure strategy Nash equilibria. Therefore, it is only necessary to discuss the asymptotic stability of E1-E8.

5   Simulation Results and Cases

5.1   The Strategy Scenarios

Taking the outcomes of the tripartite's management of public opinion risks during public emergencies, one can divide the strategy scenarios into two categories: positive strategy scenarios (1, 1, 1) and negative strategy scenarios (0, 0, 0). Through a comparative approach, the analysis examines the important roles played by the decision makers and the policy enforcers in the management of public opinion risks. Specifically, this examination is achieved by comparing (0, 0, 1) with (0, 1, 1) and comparing (1, 0, 1) with (1, 0, 0).

The Positive Strategy Scenario. When $\begin{cases} R_M - C_M + L_{MZ} - \alpha C_M > 0 \\ R_M - C_M + L_M - \alpha C_M > 0 \\ R_E - \gamma C_{EL} - I_E + L_E > 0 \end{cases}$, the equilibrium point is asymptotically stable, with $E_4(1, 1, 1)$ being the evolutionary stability strategies of the system, representing the decision makers' strategy success, the policy enforcers' effective actions and the general publics' ability to identify and refute rumors. In this scenario, the decision makers' strategy success and the policy enforcers' effective actions can both yield higher gains, and the general public will choose to identify and refute rumors unconditionally. This is the most ideal and positive scenario for the tripartite's collaborative control of public opinion risks during public emergencies (Fig. 4).

The Negative Strategy Scenario. When $\begin{cases} R_M - C_M + L_{MZ} - \alpha C_M < 0 \\ R_M - C_M + L_M - \alpha C_M < 0 \\ R_E - \gamma C_{EL} - I_E + L_E < 0 \\ S_H - C_H + P_H + L_H - R_H < 0 \end{cases}$, the equi-

librium point is asymptotically stable, with $E_1(0, 0, 0)$ being the evolutionary stability strategies of the system, representing the decision makers' strategy failures, the policy enforcers' failure to act responsibly, and the general publics' decisions to believe and spread rumors. In this scenario, the decision makers' strategy failures imply lower action costs. With the decision makers' strategy failures, the policy enforcers' strategies for acting responsibly can impose higher action costs (Fig. 5).

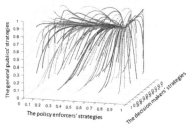

Fig. 4. Simulation modeling of the positive strategy scenarios

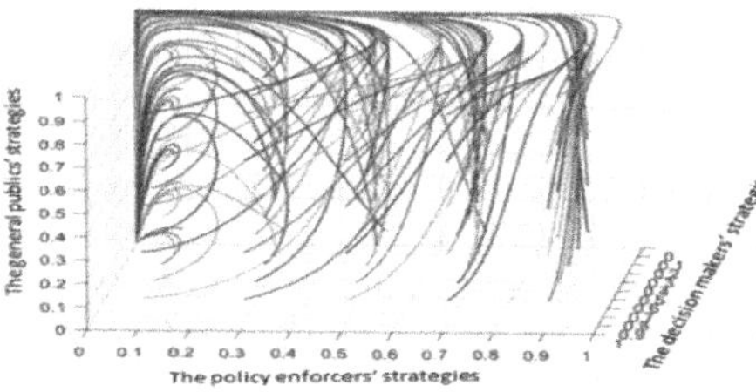

Fig. 5. The simulation modeling of the negative strategy scenarios

The Importance of the Decision Makers. When $\begin{cases} R_M - C_M + L_{MZ} - \alpha C_M < 0 \\ R_M - C_M + L_M - \alpha C_M < 0 \\ R_E - \gamma C_{EL} - I_E + L_E < 0 \\ S_H - C_H + P_H + L_H - R_H > 0 \end{cases}$,

the equilibrium point is asymptotically stable, with $E_2(0, 0, 1)$ being the evolutionary stability strategies of the system, representing the decision makers' strategy failures, the policy enforcers' failure to act responsibly, the general publics' ability to identify and refute rumors. In this scenario, strategy failures imply lower action costs for decision makers, while, following the decision makers' strategy failures, the policy enforcers' strategies for acting responsibly can induce higher action costs. However, for the general public, their decision to identify and refute rumors can instead yield higher gains (Fig. 6-a).

When $\begin{cases} R_M - C_M + L_{MZ} - \alpha C_M > 0 \\ R_M - C_M + L_M - \alpha C_M < 0 \\ R_E - \gamma C_{EL} - I_E + L_E > 0 \end{cases}$, the equilibrium point is asymptotically sta-

ble, with $E_3(0, 1, 1)$ being the evolutionary stability strategies of the system, representing the decision makers' strategy failures, the policy enforcers' effective actions and the general publics' ability to identify and refute rumors. This scenario implies higher hidden costs for the decision makers and the policy enforcers when they fail to create successful strategies and to act responsibly. Additionally, the policy enforcers' effective actions can instead yield higher gains. Under such circumstances, the general public choose to identify and refute rumors unconditionally (Fig. 6-b).

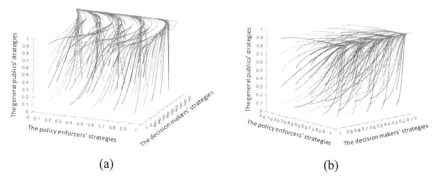

(a)   (b)

Fig. 6. Simulation modeling of the importance of the decision makers

The Importance of the Policy Enforcers. When $\begin{cases} R_M - C_M + L_{MZ} - \alpha C_M > 0 \\ R_E - \gamma C_{EL} - I_E + L_E < 0 \\ S_H - C_H + P_H + L_H - R_H > 0 \end{cases}$,

the equilibrium point is asymptotically stable, with $E_4(1, 0, 1)$ being the evolutionary stability strategies of the system, representing the decision makers' strategy success, the policy enforcers' failure to act responsibly, and the general publics' ability to identify and refute rumors. This scenario implies higher gains for the decision makers and the general public when they are able to undertake successful strategies and to identify and refute rumors. However, for the policy enforcers, acting responsibly can only generate lower gains (Fig. 7-a).

When $\begin{cases} R_M - C_M + L_{MZ} - \alpha C_M > 0 \\ R_E - \gamma C_{EL} - I_E + L_E < 0 \\ S_H - C_H + P_H + L_H - R_H < 0 \end{cases}$, the equilibrium point is asymptotically

stable, with $E_4(1, 0, 0)$ being the evolutionary stability strategies of the system, representing **the decision makers' strategy success**, the policy enforcers' failure to act responsibly, and the general publics' decisions to believe and spread rumors. This scenario implies higher hidden costs for the decision makers when they fail to create successful strategies and to take effective actions. For the policy enforcers, acting effectively and responsibly can only yield lower gains. Consequently, with the policy enforcers being inactive, the general publics' decisions to believe and spread rumors can instead generate higher gains (Fig. 7-b).

Comparing the above two equilibrium points, without the policy enforcers' effective and sensible actions, it is not guaranteed that the general public will be capable of identifying and refuting rumors, even when the decision makers have created successful strategies because the general public may still decide to believe and spread rumors.

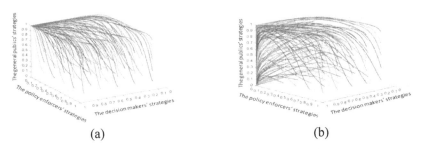

Fig. 7. Simulation modeling of the importance of the policy enforcers

5.2 Typical Case Studies

Taking previous global public health emergencies of international concern (PHEICs) into consideration, this section analyzes the strategies undertaken by some countries and regions during their management of PHEIC events.

The Decision Makers' Strategies. With the increased benefits generated through their strategy success, the decision makers create increasingly more successful strategies and produce less and less strategy failures (Fig. 8). This suggests that the higher gains yielded by strategy success can largely motivate the decision makers to invest more effort in creating good strategies.

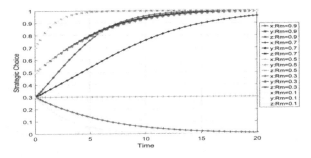

Fig. 8. Simulation modeling of decision makers' strategies

From 2013 to 2014, the global number of wild-type poliovirus (WPV) infection increased by 86%. Take Afghanistan as an example. Due to the ongoing armed conflict, the decision makers within the government are unable to dedicate sufficient effort to manage the pandemic. Neither can the policy enforcers guarantee vaccinations for the population because of economic pressure. As a result of the influences of religion and the lack of knowledge on pandemic risk management, the general public may also refuse to receive the vaccination.

The Policy Enforcers' Strategies. With the increased benefits generated through their sensible actions, the inactive policy enforcers become increasingly more active (Fig. 9). Thus, by increasing the policy enforcers' gains when they act responsibly, the policy enforcers can take more effective and more sensible measures to handle public opinion risks.

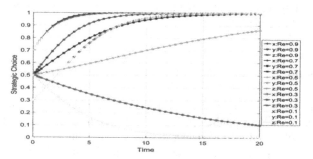

Fig. 9. Simulation modeling of the policy enforcers' strategies

In August 2014, Zika infections occurred in Brazil. As there were no fatal cases at the time, the then Brazilian Minister of Health declared that Zika was only a mild disease. Moreover, due to the lack of knowledge within the medical community on the relationship between Zika virus and neurological disorders, such as microcephaly among newborns, many Latin American countries had a very limited understanding of the Zika virus and only took limited risk management measures.

The General Publics' Strategies. With the increased benefits generated through general public identifying and refuting rumors, the general publics' strategies shift from believing and spreading rumors to identifying and refuting rumors (Fig. 10). Thus, by rewarding the general public for identifying and refuting rumors and by recruiting Internet celebrities who are able to recognize and dispel rumors and guide other web users (general public) to participate more actively in refuting rumors, the Internet community can facilitate the self-management of public opinion risks.

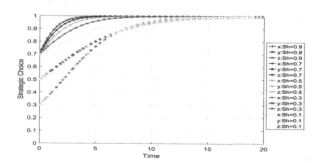

Fig. 10. Simulation modeling of the general publics' strategies

In August 2014, the WHO declared this outbreak of Ebola in West Africa as a PHEIC. The weaknesses of the local government and the lack of knowledge regarding the Ebola virus disease among the community, religious and political leaders who were tightly connected with the general public caused the spread of the rumor that "Ebola is God's punishment for 'penetrating homosexualism'" among the local population. The subsequent religious and cultural conflicts caused by the rumor also largely hindered the management of the pandemic.

6 Conclusion

In terms of its theoretical significance, with the aim of addressing the major problems during the management of public opinion risks during major public emergencies, this study created a collaborative governance system for handling public opinion risks arising from public emergencies. This system was created using a practical point of view, as it explores the fundamental causes of public opinion risks by adhering to the laws underlying the organization of society. In relation to its practical significance, the current work has built a tripartite evolutionary game model, created replication dynamic equations for each of the risk-control bodies, and analyzed the stability of the overall tripartite evolutionary game model. This study also used simulation modeling to analyze the scenarios of the different combinations of strategies.

With regard to further investigations, in relation to methodology, future studies should consider incorporating other simulation modeling technologies and methods, such as system dynamics models and network theory, to further analyze the key parameters of the tripartite evolutionary game model in the collaborative governance system through simulation modeling. System dynamics models can further refine the interconnections between the risk-control bodies, and complex network theory can be used to demonstrate the dissemination process of public opinion risk information. At the data level, future investigations should further explore the practical applicability of the above theoretical model by combining the data retrieved through case studies, questionnaires, social media, etc.

Acknowledgements. The authors acknowledge financial support from the National Natural Science Foundation of China (No. 72074206, 72074205, T2293772).

References

1. Sethi, R.: Strategy-specific barriers to learning and nonmonotonic selection dynamics. Games Econ. Behav. **23**(2), 284–304 (1998)
2. Wang, Z., Jusup, M., Wang, R.W., et al.: Onymity promotes cooperation in social dilemma experiments. Sci. Adv. **3**(3), e1601444 (2017)
3. Wang, L., Schuetz, C.G., Cai, D.: Choosing response strategies in social media crisis communication: an evolutionary game theory perspective. Inf. Manag. **58**(6), 103371 (2021)
4. Sheng, J., Zhou, W., Zhu, B.: The coordination of stakeholder interests in environmental regulation: lessons from China's environmental regulation policies from the perspective of the evolutionary game theory. J. Clean. Prod. **249**, 119385 (2020)

5. Shan, S., Duan, X., Ji, W., et al.: Evolutionary game analysis of stakeholder behavior strategies in 'Not in My Backyard' conflicts: effect of the intervention by environmental Non-Governmental Organizations. Sustain. Prod. Consum. **28**, 829–847 (2021)
6. Xu, X., Hou, Y., Zhao, C., et al.: Research on cooperation mechanism of marine plastic waste management based on complex network evolutionary game. Mar. Policy **134**, 104774 (2021)
7. Xiao, Y., Wang, Z., Li, Q., et al.: Dynamic model of information diffusion based on multidimensional complex network space and social game. Phys. A **521**, 578–590 (2019)
8. Xiao, Y., Song, C., Liu, Y.: Social hotspot propagation dynamics model based on multidimensional attributes and evolutionary games. Commun. Nonlinear Sci. Numer. Simul. **67**, 13–25 (2019)
9. Askarizadeh, M., Ladani, B.T., Manshaei, M.H.: An evolutionary game model for analysis of rumor propagation and control in social networks. Phys. A **523**, 21–39 (2019)
10. Zhao, R., Zhou, X., Han, J., et al.: For the sustainable performance of the carbon reduction labeling policies under an evolutionary game simulation. Technol. Forecast. Soc. Change **112**, 262–274 (2016)
11. Hao, X., Liu, G., Zhang, X., et al.: The coevolution mechanism of stakeholder strategies in the recycled resources industry innovation ecosystem: the view of evolutionary game theory. Technol. Forecast. Soc. Change **179**, 121627 (2022)
12. Eghbali, M.A., Rasti-Barzoki, M., Safarzadeh, S.: A hybrid evolutionary game-theoretic and system dynamics approach for analysis of implementation strategies of green technological innovation under government intervention. Technol. Soc. **70**, 102039 (2022)
13. Cui, M.: Tripartite evolutionary game analysis for environmental credit supervision under the background of collaborative governance. Syst. Eng.-Theory Pract. **41**(3), 713–726 (2021)
14. Li, Y.B., Zhang, J.R.: Evolutionary game of webcast governance strategy: stakeholder perspective. Econ. Manag. **34**(2), 25–31 (2020)
15. Wu, B., Cheng, J., Qi, Y.: Tripartite evolutionary game analysis for "Deceive acquaintances" behavior of e-commerce platforms in cooperative supervision. Phys. A **550**, 123892 (2020)
16. Hu, H., Guo, X.J., Liang, Y.R.: Game analysis of the three-way evolution of network rumor control under the major epidemic based on prospect theory. Inf. Sci. **39**(7), 45–53 (2021)

Author Index